Symmetrical Analysis Techniques for Genetic Systems and Bioinformatics:
Advanced Patterns and Applications

Sergey Petoukhov
Russian Academy of Sciences, Russia

Matthew He
Nova Southeastern University, USA

Medical Information Science REFERENCE

MEDICAL INFORMATION SCIENCE REFERENCE
Hershey · New York

Director of Editorial Content: Kristin Klinger
Senior Managing Editor: Jamie Snavely
Assistant Managing Editor: Michael Brehm
Publishing Assistant: Sean Woznicki
Typesetter: Mike Killian, Sean Woznicki
Cover Design: Lisa Tosheff
Printed at: Yurchak Printing Inc.

Published in the United States of America by
Medical Information Science Reference (an imprint of IGI Global)
701 E. Chocolate Avenue
Hershey PA 17033
Tel: 717-533-8845
Fax: 717-533-8661
E-mail: cust@igi-global.com
Web site: http://www.igi-global.com/reference

Library of Congress Cataloging-in-Publication Data

Symmetrical analysis techniques for genetic systems and bioinformatics : advanced patterns and applications / Sergey Petoukhov and Matthew He, authors.

p. ; cm.
Includes bibliographical references and index.
Summary: "This book compiles studies that demonstrate effective approaches to the structural analysis of genetic systems and bioinformatics"--Provided by publisher.

ISBN 978-1-60566-124-7 (hardcover)
1. Genetics--Mathematical models. 2. Bioinformatics. I. Petoukhov, S. V. (Sergei Valentinovich) II. He, Matthew.
[DNLM: 1. Genetic Code--genetics. 2. Computational Biology--methods. 3. Genetic Techniques. 4. Models, Theoretical. QU 470 S987 2010]
QH438.4.M3S96 2010
572.80285--dc22

2009010497

British Cataloguing in Publication Data
A Cataloguing in Publication record for this book is available from the British Library.

Table of Contents

Section 3
Algebras of Genetic Codes

Preface

Modern science connects many basic secrets of living matter with the genetic codes. Biological organisms belong to a category of very complex natural systems, which correspond to a huge number of biological species with inherited properties. But surprisingly, molecular genetics has discovered that all organisms are identical to each other by their basic molecular-genetic structures. Due to this revolutionary discovery, a great unification of all biological organisms has happened in science. The information-genetic line of investigations has become one of the most prospective lines not only in biology, but also in science as a whole. A basic system of genetic coding has become strikingly simple. Its simplicities and orderliness presented challenges to specialists from many scientific fields. Bioinformatics considers each biological organism as an ensemble of information systems which are interrelated to each other. The genetic coding system is the basic one. All other biological systems must be correlated to this system to be transmitted to next generations of organisms.

The natural technology of genetic coding is a major and most effective technology of life on our planet. Using this natural technology, huge biomass of living matter with unique and valuable properties is produced around the world. Bioinformatics and biotechnology have been applied to many areas such as biology, medicine, and life sciences. Bioinformatics knowledge is used to manufacture biological organisms with new properties, to extend human life, to diagnose and treat disease, to clone organisms, to develop new computer technologies, to create new materials with unique characteristics, and so on. It seems that all fields of human life will be influenced in the future by progress in bioinformatics.

Modern science recognizes a key meaning of information principles for inherited self-organization of living matter. In view of this, the following statements have appeared in the recent literature.

Notions of "information" or "valuable information" are not utilized in physics of non-biological nature because they are not needed there. On the contrary, in biology notions "information" and especially "valuable information" are main ones; understanding and description of phenomena in biological nature are impossible without these notions. A specificity of "living substance" lies in these notions. (Chernavskiy, 2000)

If you want to understand life, don't think about vibrant, throbbing gels and oozes, think about information technology. (Dawkins, 1991).

Here one should add that modern informatics is an independent branch of science, which possesses its own language and mathematical formalisms and exists together with physics, chemistry, and other scientific branches. A problem of information evolution of living matter has been investigated intensively in the last decades in addition to studies of the classical problem of biochemical evolution.

One of the effective methods of cognition of complex natural system, including the genetic coding system, is the investigation of symmetries. Modern science knows that deep knowledge about phenomenological relations of symmetry among separate parts of a complex natural system can tell many important things about the evolution and mechanisms of these systems. Physics and other natural sciences have great numbers of successful applications of a symmetry method. Principles of symmetry have become one of the bases of mathematical natural science. Nowadays, many physical theories, beginning from the theory of relativity to quantum mechanics, are created as theories of invariants of mathematical groups of transformations, in other words as theories of special kinds of symmetry. The study of symmetries and asymmetries in molecular structures is one of the important branches of chemistry. For example, functional differences between the right forms of molecules and the left forms of molecules in living organisms have become known to mankind due to investigations of symmetry in biological molecules. Principles of symmetry have a new essential quality in modern science.

But not only physics and chemistry deal with principles and methods of symmetry, informatics and digital signal processing also pay great attention to them. How is theory of signal processing connected to geometry and geometrical symmetries? Signals are represented there in a form of a sequence of the numeric values of their amplitude in reference points. The theory of signal processing is based on an interpretation of discrete signals as a form of vector of multi-dimensional spaces. In each tact time a signal value is interpreted as the corresponding value of one coordinate of a multi-dimensional vector space of signals. In this way, the theory of discrete signals turns out to be the science of geometries of multi-dimensional spaces. The number of dimensions of such a space is equal to the quantity of referent points for the signal. Metric notions and all other necessary things are introduced in these multi-dimensional vector spaces for those or other problems of maintenance of reliability, speed, economy of the signal information. For example, the important notions of the energy and the power of a discrete signal appear in multi-dimensional geometry of the space of signals as forms of a square of the length of a multi-dimensional vector-signal and of a square of the length of a vector-signal divided by the number of dimensions of an appropriate space. On this geometrical basis, many methods and algorithms of recognition of signals and images, coding information, detections and corrections of information mistakes, and artificial intellect and training of robots are constructed. One can add here about the importance of symmetries in permutations of components for coding signals, in spectral analysis of signals, in orthogonal and other transformations of signals, and so on.

An investigation of symmetrical and structural analogies between computer informatics and genetic informatics is one of the important tasks of modern science in connection to the creation of DNA-computers, DNA-robotics and to a development of bioinformatics. A significant part of this book describes the study of symmetries in matrix forms of the genetic code systems ("matrix genetics"). The results of this study are new examples of the usefulness of symmetry investigations in natural systems. In this book, we first present matrix methods of presentation and the analysis of molecular ensembles of the genetic code systems. Secondly, we present special multi-dimensional matrix algebras related to the genetic code and describe the importance of phenomenological symmetries in matrix forms of presentation of the genetic code. Furthermore, we present advanced patterns and applications.

THE CHALLENGES

A biological meaning of genetic informatics is reflected in the brief statement: *"life is a partnership between genes and mathematics"* (Stewart, 1999). But what kind of mathematics has partner relations with the genetic code and what kind of mathematics is behind genetic phenomenology which includes a great noise-immunity of the genetic code? This question is one of the main challenges in mathematical natural sciences today. A significant part of the challenge is the question of an adequate mathematics for the phenomenon of degeneracy of the genetic code. A character of this degeneracy is reflected in symmetrical patterns of black-and-white mosaics of genetic matrices of 64 triplets (for example, see a genetic matrix with a black-and-white symmetrical mosaic on Figure 2.2. in Chapter 2).

Why do genetic matrices of 64 triplets posess such symmetrical mosaics? Is degeneracy of the genetic code an accidental choice of nature? Is it provided by substantial mathematics of the genetic code? Is the construction of the genetic code non-accidental at all? The last question is essential because the famous hypothesis by F. Crick (1968) about "the frozen accident" in the origin of the genetic code has supposed that the first accidental system of coding, which possessed satisfactory features, was reproduced in biological evolution with its further evolutionary improvements.

We are searching for scientific answers to facilitate an analysis of the genetic code phenomenology from the viewpoint of mathematics of discrete signal processing, of computer informatics, and of noise-immunity coding in digital communication. This book describes substantial answers to these questions by means of discovering deep connections of the genetic code with hypercomplex numeric systems and their matrix algebras (which can be multi-dimensional algebras of operators simultaneously). These multi-dimensional algebras and their relevant geometries are interpreted in relation to multi-dimensional vector spaces of bioinformatics (or bioinformation vector spaces). An example of such an algebra is the 8-dimensional Yin-Yang-algebra (or the bipolar algebra), which is the algebra of degeneracy of the genetic code and which is described in Chapter 7. Recent progress in the determination of genomic sequences yields many millions of gene sequences now. But what do these sequences tell us and what generalities and rules govern them? The modern situation in the theoretic field of genetic informatics can be characterized by the following citation:

What will we have when these genomic sequences are determined? What do we have now in the 10 million nucleotide of sequence data determined to date? We are in the position of Johann Kepler when he first began looking for patterns in the volumes of data that Tycho Brahe had spent his life accumulating. We have the program that runs the cellular machinery, but we know very little about how to read it. Bench biologists, by experiment and by close association with the data, have found meaningful patterns. Theoreticians, by careful reasoning and use of collections of data, have found others, but we still understand frustratingly little. (Fickett & Burks, 1989)

Kepler is mentioned here not without reason. The history of science shows the importance of cognitive forms of presentation of phenomenological data to find regularities or laws in this phenomenology. The work by Kepler is the classical example of an important meaning of a cognitive form of presentation of phenomenological data. He did not make his own astronomic observations, but he found the cognitive form of presentation in the huge astronomic data from the collection of Tycho Brahe. This discovered form, which was connected to the general idea of movements along ellipses, allowed him to formulate the famous Kepler's laws of planetary movements relative to the Sun. Owing to this cognitive form, Kepler and Newton have led us to the law of Newtonian attraction.

A discovery of such a cognitive form of presentation in the case of the phenomenology of genetic code systems is one more challenge, which arises from the very beginning in the course of attempts to find regularities among a huge number of genetic data and to create a relevant theory. Matrix genetics proposes a new cognitive form of presentation of phenomenological data in the field of genetic informatics. This cognitive matrix form gives new tools to analyze and to model ensembles of the genetic code as well. It paves the way for a worthy attempt at answering the mentioned challenges.

SEARCHING FOR A SOLUTION

This book presents a matrix form of presentation of the genetic code as an effective cognitive form of presentation of relevant phenomenological data. An initial choice of such a form of presentation of molecular ensembles of the genetic code is explained by the following main reasons:

- Information is usually stored in computers in the form of matrices.
- Noise-immunity codes are constructed on the basis of matrices.
- Quantum mechanics utilizes matrix operators, connections with which can be detected in matrix forms of presentation of the genetic code. The significance of matrix approach is emphasized by the fact that quantum mechanics has arisen in a form of matrix mechanics by W. Heisenberg.
- Complex and hypercomplex numbers, which are utilized in physics and mathematics, possess matrix forms of their presentation. The notion of number is the main notion of mathematics and mathematical natural sciences. In view of this, investigation of a possible connection of the genetic code to multi-dimensional numbers in their matrix presentations can lead to very significant results.
- Matrix analysis is one of the main investigation tools in mathematical natural sciences. The study of possible analogies between matrices, which are specific for the genetic code, and famous matrices from other branches of sciences can be heuristic and useful.
- Matrices, which are a kind of union of many components in a single whole, are subordinated to certain mathematical operations, which determine substantial connections between collectives of many components. Such connections can be essential for collectives of genetic elements of different levels, as well.

The authors utilize a presentation of molecular ensembles of genetic multiplets in the form of a Kronecker family of genetic matrices $[C\ \ A;\ U\ \ G]^{(n)}$, where C, A, U, G are nitrogenous bases cytosine, adenine, uracil, guanine, and (n) is a Kronecker power. The genetic matrix $[C\ \ A;\ U\ \ G]^{(3)}$ contains all 64 triplets in an ordering arrangement, which is comfortable and effective to study degeneracy of the genetic code. Kronecker families of square matrices are utilized in the theory of noise-immunity coding and of discrete signal processing. Applying these matrix families to genetic informatics is justified by a discrete character of the genetic code. This matrix form has allowed us to derive the following main results:

- new phenomenological rules of evolution of the genetic code;
- the connections of the genetic code structures with multi-dimensional numeric systems;
- multi-dimensional algebras for modelling and for analysing the genetic code systems;
- Hadamard matrices and matrices of a hyperbolic turn in the Kronecker family of genetic matrices;
- parallels with quantum computers;
- hidden interrelations between the golden section and parameters of genetic multiplets;

- relations between the Pythagorean musical scale and an important class of quint genetic matrices which show a molecular genetic basis with a sense of musical harmony and of aesthetics of proportions;
- cyclic algebraic principles in the structure of matrices of the genetic code;
- generalized hypercomplex numeric systems, which are new for mathematical natural sciences and which allow one to model a binary opposition of male and female beginnings on the level of genetic-molecular ensembles;
- materials for a chronocyclic conception, which connects structures of the genetic system with chrono-medicine and a problem of the internal clock of organisms;
- parallels with famous symbolic tables of the Ancient Chinese book "I Ching" which declares a cyclic principle in nature and which is very important for all Oriental medicinc (acupuncture, pulse diagnostics of Tibetan medicine, and so on);
- a new answer to the fundamental questions–"why are there 4 letters in the genetic alphabet?" and "why 20 amino acids?"

One of the most important results is that degeneracy of the genetic code agrees with the 8-dimensional algebra, which is unknown in modern mathematical natural science. This algebra and the elements of its multi-dimensional geometry are presented in Chapters 7 and 11. After the discovery of non-Euclidean geometries and of Hamilton quaternions, it is known that different natural systems can possess their own geometry and their own algebra. The genetic code is connected with its own multi-dimensional numerical system or the multi-dimensional algebra. This genetic algebra can be considered as the pre-code or the mathematical model of the genetic code. This algebra allows one to reveal hidden peculiarities of the structure and evolution of the genetic code. The genetic code has its own forms of ordering. It seems that many difficulties of modern bioinformatics are connected with utilizing inadequate algebras, which were developed for completely different natural systems. Hamilton had similar difficulties in his attempts to describe 3D-space transformations by means of 3-dimensional numbers while this description needs 4-dimensional quaternions. We proposed a new algebraic system for bioinformatics and for mathematical biology. The described results are interesting from the viewpoint of many modern tasks: creating computers from DNA molecules; understanding the genetic system as a quantum computer; creating new kinds of neurocomputers and cellular automata on the basis of principles of genetic code systems.

A set of these results and proposed matrix methods in the field of genetic forms a new scientific discipline–"matrix genetics," which is related to symmetrical analyses and visual patterns of bioinformatics closely. This book can be considered as an introduction to matrix genetics. The main intended audiences are students and scientists in the fields of genetics, bioinformatics, theoretical biology, mathematical biology, computer informatics, neurocomputing, theory of symmetries, biotechnology, mathematics, theoretical physics, medicine, physiology, psychophysics, art design, music, cellular automata. Our mathematical approaches and results about structural peculiarities of genetic code systems increase knowledge and further investigations for many scientists and students. The presented genetic matrices and their ensembles are interesting not only by beautiful mathematical properties but, first of all, by their reflection of fundamental phenomenology of the genetic code. Therefore, science will return to them in future at different levels of knowledge again and again.

ORGANIZATION OF THE BOOK

The book is organized into twelve chapters. A brief description of each chapter follows.

Chapter 1 is devoted to symmetrical analysis for genetic code systems. The genetic coding possesses noise-immunity. Mathematical theories of noise-immunity coding and discrete signals processing are based on matrix methods of representation and analysis of information. These matrix methods, which are connected closely with relations of symmetry, are borrowed for a matrix analysis of ensembles of molecular elements of the genetic code. This chapter describes a uniform representation of ensembles of genetic multiplets in the form of matrices of a cumulative Kronecker family. The analysis of molecular peculiarities of the system of nitrogenous bases reveals the first significant relations of symmetry in these genetic matrices. It permits one to introduce a natural numbering of the multiplets in each of the genetic matrices and to give the basis for further analysis of genetic structures. The connection of the numerated genetic matrices with famous matrices of dyadic shifts is demonstrated.

Chapter 2 describes symmetries of the degeneracy of the vertebrate mitochondrial genetic code in the mosaic matrix form of its presentation. The initial black-and-white genomatrix of this code is reformed into a new mosaic matrix when internal positions in all triplets are permuted simultaneously. It is revealed unexpectedly that for all six variants of positional permutations in triplets (1-2-3, 2-3-1, 3-1-2, 1-3-2, 2-1-3, 3-2-1) the appropriate genetic matrices possess symmetrical mosaics of the code degeneracy. Moreover the six appropriate mosaic matrices in their binary presentation have the general non-trivial property of their "tetra-reproduction," which can be utilized in particular for mathematical modeling of the phenomenon of the tetra-division of gametal cells in meiosis. Mutual interchanges of the genetic letters A, C, G, U in the genomatrices lead to new mosaic genomatrices, which possess similar symmetrical and tetra-reproduction properties as well.

Chapter 3 demonstrates results of a comparative investigation of characteristics of degeneracy of all known dialects of the genetic code. This investigation is conducted on the basis of the results of symmetrological analysis, which were described in Chapter 2, about the division of the set of the 20 amino acids into the two canonical subsets: the subset of the 8 high-degeneracy acids and the subset of the 12 low-degeneracy acids. The existence of numerical and structural invariants in the set of these dialects is shown. The derived results from the comparative investigation permit one to formulate some phenomenological rules of evolution of these dialects. These numeric invariants and parameters of code degeneracy draw attention to the formal connection of this evolution with famous facts of chrono-biology and chrono-medicine. The chronocyclic conception of the functioning of molecular-genetic systems is proposed on this basis. The biophysical basis of this conception provides connection to the genetic code structures with mechanisms of photosynthesis which produce living substance by means of utilization of solar energy. And the solar energy comes cyclically on the surface of the Earth. The revealed numeric invariants of evolution of the genetic code give new approaches to the fundamental question, why do 20 amino acids exist? We will demonstrate new patterns of the genetic code systems.

Chapter 4 is devoted to a consideration of the Kronecker family of the genetic matrices but in the new numerical form of their presentation. This numeric presentation gives opportunities to investigate ensembles of parameters of the genetic code by means of system analysis including matrix and symmetric methods. In this way new knowledge is obtained about hidden regularities of element ensembles of the genetic code and about connections of these ensembles with famous mathematical objects and theories from other branches of science. First of all, this chapter demonstrates the connection of molecular-genetic system with the golden section and principles of musical harmony.

Chapter 5 uses the Gray code representation of the genetic code C = 00, U = 10, G = 11 and A = 01 (C pairs with G, A pairs with U) to generate a sequence of genetic code-based matrices. In connection with these code-based matrices, we use the Hamming distance to generate a sequence of numerical matrices. We then further investigate the properties of the numerical matrices and show that they are doubly stochastic and symmetric. We determine the frequency distributions of the Hamming distances, building blocks of the matrices, decomposition and iterations of matrices. We present an explicit decomposition formula for the genetic code-based matrix in terms of permutation matrices. Furthermore, we establish a relation between the genetic code and a stochastic matrix based on hydrogen bonds of DNA. Using fundamental properties of the stochastic matrices, we determine explicitly the decomposition formula of genetic code-based biperiodic table. By iterating the stochastic matrix, we demonstrate the symmetrical relations between the entries of the matrix and DNA molar concentration accumulation. The evolution matrices based on genetic code were derived by using hydrogen bonds-based symmetric stochastic (2x2)-matrices as primary building blocks. The fractal structure of the genetic code and stochastic matrices were illustrated in the process of matrix decomposition, iteration and expansion corresponding to the fractal structure of the biperiodic table introduced by the authors.

Chapter 6 continues an analysis of the degeneracy of the vertebrate mitochondrial genetic code in the matrix form of its presentation, which possesses the symmetrical black-and-white mosaic. Taking into account a symmetry breakdown in molecular compositions of the four letters of the genetic alphabet, the connection of this matrix form of the genetic code with a Hadamard (8x8)-matrix is discovered. Hadamard matrices are one of the most famous and the most important kind of matrices in the theory of discrete signals processing and in spectral analysis. The special U-algorithm of transformation of the symbolic genetic matrix $[C A; U G]^{(3)}$ into the appropriate Hadamard matrix is demonstrated. This algorithm is based on the molecular parameters of the letters A, C, G, U/T of the genetic alphabet. In addition, the analogical relations is shown between Hadamard matrices and other symmetrical forms of genetic matrices, which are produced from the symmetrical genomatrix $[C A; U G]^{(3)}$ by permutations of positions inside triplets. Many new questions arise due to the described fact of the connection of the genetic matrices with Hadamard matrices. Some of them are discussed here including questions about an importance of amino-group NH_2 in molecular-genetic systems, and about possible relations with the theory of quantum computers, where Hadamard gates are utilized. A new possible answer is proposed to the fundamental question concerning reasons for the existence of four letters in the genetic alphabet. Some thoughts about cyclic codes and a principle of molecular economy in genetic informatics are presented as well.

Chapter 7 analyzes algebraic properties of the genetic code. The investigations of the genetic code on the basis of matrix approaches ("matrix genetics") are described. The degeneracy of the vertebrate mitochondrial genetic code is reflected in the black-and-white mosaic of the (8*8)-matrix of 64 triplets, 20 amino acids and stop-signals. The special algorithm, which is based on features of genetic molecules, exists to transform the mosaic genomatrix into a numeric matrix, which is the matrix form of presentation of the special 8-dimensional genetic algebra. This algebra can be named as Yin-Yang-algebra or bipolar algebra. Main mathematical properties of this genetic algebra and its relations with other algebras are analyzed together with some important consequences from the adequate algebraic models of the genetic code. Elements of a new "genovector calculation" and ideas of "genetic mechanics" are discussed. The revealed fact of the relation between the genetic code and these genetic algebras, which define new multi-dimensional numeric systems, is discussed in connection with the famous idea by Pythagoras: "All things are numbers." Simultaneously, these genetic algebras can be utilized as the algebras of genetic operators in biological organisms. The described results are related to the problem of algebraization of bioinformatics. They draw attention to the question: what is life from the viewpoint of algebra?

Chapter 8 considers the octet Yin-Yang-algebra as the model of the genetic code. From the viewpoint of this algebraic model, for example, the sets of 20 amino acids and of 64 triplets consist of subsets of "male," "female," and "androgynous" molecules, and so forth. This algebra allows one to reveal the hidden peculiarities of the structure and evolution of the genetic code and to propose the conception of "sexual" relationships among genetic molecules. The first results of the analysis of the genetic code systems from such an algebraic viewpoint speak about the close connection between evolution of the genetic code and this algebra. They include 7 phenomenological rules of evolution of the dialects of the genetic code. The evolution of the genetic code appears as the struggle between male and female beginnings. The hypothesis about new biophysical factor of "sexual" interactions among genetic molecules is proposed. The matrix forms of presentation of elements of the genetic octet Yin-Yang-algebra are connected with Hadamard matrices by means of the simple U-algorithm. Hadamard matrices play a significant role in the theory of quantum computers, in particular. It leads to new opportunities for the possible understanding of genetic code systems as quantum computer systems. Revealed algebraic properties of the genetic code allow one to put forward the problem of algebraization of bioinformatics on the basis of the algebras of the genetic code.

Chapter 9 returns to the kind of numeric genetic matrices, which were discussed in Chapters 4-6. This kind of genomatrix is not connected with the degeneracy of the genetic code directly, but it is related to some other structural features of genetic code systems. The connection of the Kronecker families of such genomatrices with special categories of hypercomplex numbers and with their algebras is demonstrated. Hypercomplex numbers of these two categories are named "matrions of a hyperbolic type" and "matrions of a circular type." These hypercomplex numbers are a generalization of complex numbers and double numbers. Mathematical properties of these additional categories of algebras are presented. A possible meaning and possible applications of these hypercomplex numbers are discussed. The investigation of these hyperbolic numbers in connection with the parameters of molecular systems of the genetic code can be considered as a continuation of the Pythagorean approach to understanding natural systems.

Chapter 10 describes data suggesting a connection between matrix genetics and one of the most famous branches of mathematical biology: phyllotaxis laws of morphogenesis. Thousands of scientific works are devoted to this morphogenetic phenomenon, which relates with Fibonacci numbers, the golden section and beautiful symmetrical patterns. These typical patterns are realized by nature in a huge number of biological bodies on various branches and levels of biological evolution. Some matrix methods are known for a long time to simulate in mathematical forms these phyllotaxis phenomena. This chapter describes connections of the famous Fibonacci (2x2)-matrices with genetic matrices. Some generalizations of the Fibonacci matrices for cases of (2^nx2^n)-matrices are proposed. Special geometrical invariants, which are connected with the golden section and Fibonacci numbers and which characterize some proportions of human and animal bodies, are described. All these data are related to matrices of the genetic code in some aspects.

Chapter 11 presents data about cyclic properties of the genetic code in its matrix forms of presentation. These cyclic properties concern cyclic changes of genetic Yin-Yang-matrices and their Yin-Yang-algebras at many kinds of circular permutations of genetic elements in genetic matrices. These circular permutations lead to such reorganizations of the matrix form of presentation of the initial genetic Yin-Yang-algebra that such matrices serve as matrix forms of presentations of new Yin-Yang-algebras. They are connected algorithmically with Hadamard matrices. New patterns and relations of symmetry are described. The discovered existence of a hierarchy of the cyclic changes of genetic Yin-Yang-algebras allows one to develop new algebraic models of cyclic processes in bioinformatics and in other related fields. These cycles of changes of the genetic 8-dimensional algebras and of their 8-dimensional numeric systems have many analogies with famous facts and doctrines of modern and ancient physiology, medicine, etc. This viewpoint proposes that the famous idea by Pythagoras (about organization of natural systems in accordance with harmony of numerical systems) should be combined with the idea of cyclic changes of Yin-Yang-numeric systems in considered cases. This second idea suggests the ancient idea of cyclic changes in nature. From such an algebraic-genetic viewpoint, the notion of biological time can be considered as a factor in coordinating these hierarchical ensembles of cyclic changes of the genetic multi-dimensional algebras.

Chapter 12 considers the topic of connections of the genetic code with various fields of culture and with inherited physiological properties which provide existence of these fields. Some examples of such physiological bases for branches of culture are described. These examples are related to linguistics, music, and physiology of color perception. Special attention is paid to connections between the genetic matrices and the system of the Ancient Chinese book "I Ching." The conception and its arguments are put forward that the famous table of 64 hexagrams of "|I Ching" reflects the notions of Ancient Chinese about music quint harmony as a universal archetype.

Sergey Petoukhov
Matthew He
March 6, 2009

Acknowledgment

The authors would like to express our gratitude to many of our colleagues who had worked with us in exploring the topics relevant to this book. Their names can be found in the references under each chapter. Only those literatures closely related to our work are listed in the references. However, due to the large extent of subject of the studies; the references cited in our book are incomplete. The authors deeply apologize for any omission.

We want to thank the Mechanical Engineering Institute of the Russian Academy of Sciences, Moscow, Russia and Farquhar College of Arts and Sciences of Nova Southeastern University, Fort Lauderdale, USA for their support. The authors would like to acknowledge the help of our colleagues from the Russian and Hungarian Academies of Sciences, from International Society of Symmetry in Bioinformatics and from the International Symmetry Association, first of all, Prof. Konstantin Frolov, Dr. Gyorgy Darvas, Prof. Yuval Ne'eman, Prof. Larry Gould, Prof. Jay Kappraff, Prof. Alexandr Koganov, Prof. Raisa Szabo, Prof. Vladimir Smolianinov, Prof. Yuriy Vladimirov, Prof. Aleksey Stakhov, Prof. Aleksandr Dubrov, and Mr. O.Tavassoly.

We are deeply indebted to our colleagues Prof. Miguel Angel Jimenez-Montano of Universidad Veracruzana, Mexico and Prof. Diego Castano of Nova Southeastern University for their review and suggestions to the final version of the manuscript.

Special thanks also go to the publishing team at IGI Global, whose contributions throughout the entire process from initial proposal to final publication have been in valuable, in particular to the IGI assistant development editor Ms. Christine Bufton who continuously provided us prompt guidance and support throughout the book editing process.

Finally, we would like to give our special thanks to our families for their love and patience that enabled us to complete this work.

Sergey Petoukhov
Russian Academy of Sciences, Russia
Matthew He
Nova Southeastern University, USA
March 6, 2009

Section 1
Symmetrical Analysis Techniques and Symbolic Matrices of Matrix Genetics

Section 1 is organized into three chapters. It presents symmetrical patterns for genetic systems, natural system of numeration of genetic multiplets, and biological evolution of degeneracy of genetic codes. A matrix representation of the genetic code is introduced to describe phenomenological symmetries of degeneracy of the Vertebrate Mitochondrial Code and to study consequences from these symmetries and other features of genetic matrices.

Chapter 1
Genetic Code:
Emergence of Symmetrical Pattern, Beginnings of Matrix Genetics

ABSTRACT

This chapter is devoted to symmetrical analysis for genetic code systems. The genetic coding possesses the noise-immunity. Mathematical theories of the noise-immunity coding and discrete signals processing are based on matrix methods of representation and analysis of information. These matrix methods, which are connected closely with relations of symmetry, are borrowed for a matrix analysis of ensembles of molecular elements of the genetic code. This chapter describes a uniform representation of ensembles of genetic multiplets in the form of matrices of a cumulative Kronecker family. The analysis of molecular peculiarities of the system of nitrogenous bases reveals the first significant relations of symmetry in these genetic matrices. It permits to introduce a natural numbering the multiplets in each of the genetic matrices and to give the basis for further analysis of genetic structures. The connection of the numerated genetic matrices with famous matrices of dyadic shifts is demonstrated.

INTRODUCTION AND BACKGROUND

Bioinformatics is defined frequently as the branch of life science that deals with the study of application of information technology to the field of molecular biology. The primary goal of bioinformatics is to increase our understanding of biological processes. The term bioinformatics was coined by Paulien Hogeweg in 1978 for the study of informatics processes in biotic systems.

The genetic code is a key to bioinformatics and to a science about biological self-organizing on the whole. The modern science faces the necessity of understanding and system explanation of mysterious features of ensemble of molecular structures of the genetic code. Why does the genetic alphabet consist

DOI: 10.4018/978-1-60566-124-7.ch001

of the four letters? Why does the genetic code encode 20 amino acids? How is the system structure of the molecular genetic code connected with known principles of quantum mechanics, which were developed to explain phenomena on atomic and molecular levels? Why has nature chosen the special code conformity between 64 genetic triplets and 20 amino acids? Can knowledge about the structural essence of the genetic code be useful for mathematical natural sciences on the whole? What kind of mathematical approach should be chosen among many possible approaches to represent and model structuralized ensembles of molecules of the genetic code?

Achievement of deep understanding the genetic code should promote an inclusion of a science about it into the field of the mathematical natural sciences. To provide it, the direction of searches should be based on fundamental mathematical methods and concepts. Methods and principles of symmetry, as well as the matrix analysis, are some of bases of modern mathematical natural sciences. While biological structures are genetically inherited, morphological structures of biological bodies are characterized by many kinds of symmetry. It is known from the history of molecular genetics that investigations of symmetry in genetic molecules have given essential results already. Revelations of new symmetric structures in molecular-genetic systems produce a set of useful heuristic associations due to analogies with known symmetric structures in other scientific fields: quantum mechanics, theory of digital communication and noise-immunity coding, geometry, etc.

Genetic coding possesses the noise-immunity, which allows descendants to be similar to their parents, despite of strong disturbances and noise in the environment of biological molecules. It reminds one of the effective noise-immunity of modern systems of digital communication and signal processing, which is reached by means of special mathematics. The mathematics is based on matrix and symmetric methods of representation and analysis of signals. It's natural to ask whether it is possible that these mathematical methods, which were developed for digital technique, can be applied in the adequate manner to studying the genetic code?

The objectives of Chapter 1 are the following:

- The explanation of the choice of symmetric and matrix methods of analysis of the genetic code as prospective and adequate methods to investigate and to model structural interrelations among various parts of the integral molecular system of the genetic code;
- The description of the main data about molecular structures of the genetic code;
- The demonstration of the possibility of representation of all sets of genetic multiplets, which differ from each other by their lengths and compositions, in the well-ordered symmetrical form by means of the Kronecker family of the genetic matrices;
- The explanation of the fact that all multiplets in this general matrix form of presentation of their sets can be numerated individually by means of taking into account the symmetrical binary sub-alphabets of the four-letter genetic alphabet;
- The revelation of the connection between the genetic matrices and the matrices of diadic shifts, which are utilized in the theory of discrete signal processing long ago as fundamentals of some special methods of analysis and synthesis of signals.

SIGNIFICANCE OF SYMMETRICAL PATTERNS FOR BIOLOGY, MOLECULAR GENETICS AND BIOINFORMATICS

Symmetry in biological systems, in particular, in forms of biological bodies caused steadfast interest of thinkers as one of the most remarkable and mysterious phenomena of nature during centuries (Thompson d'Arcy, 1942; Weyl, 1952; etc.). The works of many modern scientists are devoted to it as well. Problems of biological symmetries at a macromolecular level were considered on the special Nobel symposium (Engstrom & Strandberg, 1968), on which the important role of the concept of symmetry for biological researches was emphasized. School programs of biology already include considerations of numerous examples of rotary, transmitting and mirror symmetries, and also symmetries of scale similarity in biological bodies: flowers and sprouts of plants, support-motion systems of animals, etc.

Principles of symmetry have played the important role in the X-ray analysis of genetic molecules. It is well know the concept of the double helix of DNA has arisen in the famous works by Crick and Watson (Roller, 1974; Watson, & Crick, 1953). Besides, the living substance is traditionally compared to crystals to reveal similarities and differences between them. For example, Schrodinger (1955) considered the living substance as an aperiodic crystal. But all crystallography is based on principles of symmetry; crystallography has given a powerful impulse to development and application of methods of symmetry in mathematical natural sciences including mathematical biology. New discoveries in crystallography frequently generate new hypotheses and discussions about the role of symmetry in crystals and living substance. As an example, the discovery of quasi-crystalls (Shechtman, Blech, Gratias & Cahn, 1984), which are connected with mosaics by R. Penrose (1989, 2004), with pentagrams (the penta-symmetry) and the golden section, can serve as the example here. This discovery has drawn the attention of researchers again to 5-symmetries, which exist in biological bodies widely (colors, starfishes and so forth) and which are forbidden in classical crystallography.

The development of biological knowledge is accompanied by opening new facts of subordination of very different biological objects to principles of symmetry on very different levels of their organization. Many biological concepts, which have been affirmed in the science or which sometimes cause sharp discussions, are connected with a question about biological symmetries to some extent: the law of homologous series (Vavilov, 1922); theories of morphogenetic fields; the hypothesis by Vernadsky (1965) about non-Euclidean geometry of living matter; conceptions about morphogenetic conditionality of many psychological phenomena including the phenomenon of aesthetic preference of the golden section, which is connected with Fibonacci numbers and morphogenetic laws of phyllotaxis (see review about phyllotaxis in the books (Jean, 1994; Jean & Barabe, 2001)), etc.

Molecular biology has discovered the existence of fundamental problems of symmetry and of the left-right dissymmetry on the level of biological molecules. On the other hand the development of the theory of symmetry has put forward questions about new kinds of symmetry, for example, of non-Euclidean symmetries in biological bodies (see reviews in (Petoukhov, 1981, 1989)). Modeling the biological phenomena on the basis of modern theories of nonlinear dynamics brings into the biological models the highest symmetries, which were known before in the fields of mathematics and physics. For example, the solitonic model of the macrobiological phenomena involve symmetries of Lorentz transformations from the special theory of relativity (Petoukhov, 1999a). It is no doubt that principles of symmetry were, are and will be the major component of development of biology. We think that they will play the increasing role in theoretical biology because of their status as one of the fundamentals of modern mathematical natural sciences on the whole (Bernal, Hamilton & Ricci, 1972; Birss, 1964; Darvas, 2007; Gardner,

1991; Hahn, 1989, 1998; Hargittai, 1986, 1989; Hargittai & Hargittai, 1994; Kappraff. 2002; Leyton, 1992; Loeb, 1971, 1993; Mandelbrot, 1983; Mainzer, 1988; Miller, 1972; Moller & Swaddle, 1997; Ne'eman, 1999, 2000; Ne'eman & Kirsh, 1986; Petoukhov, 1981; Rosen, 1983, 1992, 1995; Shubnikov & Koptsik, 1974; Stewart & Golubitsky, 1992; Weyl, 1931, 1946, 1952; Wigner, 1965, 1967, 1970; etc.). Such fundamental status of principles of symmetry is connected with the famous Erlangen program by F.Klein and with the process of geometrization of physics (Lochak, 1994; Weyl, 1952). This process of geometrization has led to interpretation of many basic theories of physics as theories of symmetry: special theory of relativity, quantum mechanics, theory of conservation laws, theories of elementary particles and some other parts of modern physics are such examples.

Investigations of symmetries are the most relevant in that case, when science doesn't know how to create a theory of a concrete natural system. Biological organisms belong to a category of the very complex natural systems. The variety of organisms is very numerous. Their sorts differ each from other vastly by many aspects: by their sizes, appearances, kinds of motions, etc. But to humanity's surprise, molecular genetics has discovered that, from a molecular-genetic viewpoint, all organisms are equivalent to each other by their basic genetic structures. Due to this revolutionary discovers, a great unification of all biological organisms was happened in the science, and information-genetic line of investigations became one of the most perspective lines not only in biology, but also in the science as a whole. A basic system of genetic coding has been happened strikingly simple. Its simplicities and its orderliness throw down a challenge to specialists from many scientific fields, including specialists in a theory of symmetry and of anti-symmetry.

It should be noted that fantastic successes of molecular genetics were defined in particular by a disclosure of phenomenological facts of symmetry in molecular constructions of genetic code and by use of these facts in theoretical modeling. A bright example is a disclosure of a symmetrological fact, reflected in the famous rule by Chargaff, of an equality of quantities of nitrogenous bases in their appropriate pairs (adenine-thymine and cytosine-guanine) in molecules of DNA in different organisms. This phenomenological rule was used skillfully in a theoretic modeling of the double helix of DNA by Crick and Watson using additional symmetrological principles (Roller, 1974). Many specialists from many countries around the world work in this very attractive field of investigation of symmetries in the genetic code and bioinformatics now (Arques & Michel, 1994, 1996, 1994; Bakhtiarov, 2001; Bashfold, Tsohantjis, & Yarvis, 1997; Chernavskiy, 2000; Chi Ming Yang, 2001; Dragovich & Dragovich, 2007; Forger, Hornos, & Hornos, 1997, 1999; Frank-Kamenetskiy, 1988; Frappat, Sciarrino, & Sorba, 1998; Hargittai, 2001; He, 2001; He, Narasimhan & Petoukhov, 2005; He & Petoukhov, 2007; He, Petoukhov & Ricci, 2004; Jimenes-Montano, 2005; Karasev, 2003; Karasev, Luchinin, Stefanov, 2005; Kargupta, 2001; Khrennikov & Kozyrev, 2007; Konopelchenko & Rumer, 1975; MacDonaill, 2003, 2005; Makovskiy, 1992; Marcus, 2001, 2007; Negadi, 2005, 2006; Petoukhov, 2001-2008; Ratner, 2002; Rumer, 1968, 1975; Shcherbak, 1988; Stambuk, 1999; Stambuk, Konyevoda & Gotovac, 2005; Szabo & He, 2006; Szabo, He, Burnham & Jurani, 2005; Waterman, 1999; Yang, 2005; etc.).

From an information-theoretic viewpoint, biological organisms are informational essences. They obtain genetic information from their ancestors and transmit it to descendants. In the biological literature it is possible quite often to meet the statement that living organisms are the texts since a molecular level of their organization. Just from the information-hereditary point of view all living organisms are unified wonderfully: all of them have identical bases of system of genetic coding. A conception of informational nature of living organisms is reflected in the words: "*If you want to understand life, don't think about vibrant, throbbing gels and oozes, and think about information technology*" (Dawkins,

1991). Or another citation, which presents a similar direction of thoughts: '*Notions of 'information' or 'valuable information' are not utilized in physics of non-biological nature because they are not needed there. On the contrary, in biology notions 'information' and especially 'valuable information' are main ones; understanding and description of phenomena in biological nature are impossible without these notions. A specificity of 'living substance' lies in these notions*" (Chernavskiy, 2000).

Due to revealing the genetic code, the theoretical problem of "bio-information evolution" has arisen. This problem exists alongside with ideas about chemical evolution and is very significant for understanding biological life.

Informatics began to be used in concepts of an origin of a life and in theoretical biology in the last decades only. And now the modern science hopes to receive deeper and adequate understanding of life and its origin from positions of bioinformatics. In our opinion, modern investigations in the field of bioinformatics form the foundation of the future theoretical biology. Therefore the problem of maximal union of molecular-genetic knowledge with the mathematics of the theory of discrete signals processing is especially appropriate.

Bioinformatics can give deeper knowledge to the questions of what is life and why life exists. An investigation of symmetrical and structural analogies between computer informatics and genetic informatics is one of the important tasks of modern science in connection with a creation of DNA-computers and with development of bioinformatics. The development of bioinformatics and its applications requires appropriate mathematical models of structural ensembles of genetic elements. The methods of symmetry can be useful to create such model. This book demonstrates the usefulness of the methods of symmetry to study the genetic code and to develop effective matrix approaches in the field of genetic coding.

One should note that many attempts at construction of mathematical models or biochemical explanations of separate features of the genetic code are known. One of the most historically famous attempts of answering the question about 20 amino acids was made by G. Gamov more than 50 years ago (Gamov, 1954; Gamov & Metropolis, 1954). He supposed the explanation of the morphological character, that this quantity of amino acids is defined by the molecular configuration of the double helix of DNA, which possesses the appropriate quantity of hollows along the double helix. A few initial attempts of explanation of features of the genetic code are presented in books (Ycas, 1969; Stent, 1971).

Some mathematical and other approaches to the genetic code were proposed in the works (Chi Ming Yang, 2001; Eingorin, 2001, 2003, 2006; Dragovich & Dragovich, 2007; He, 2001; Jimenes-Montano, 2005; Karasev, 2003; Khrennikov & Kozyrev, 2007; Konopelchenko & Rumer, 1975; Laubenbacher & Sturmfels, 2008; MacDonaill, 2003, 2005; Negadi, 2005, 2006; Petoukhov, 2001-2008; Ratner, 2002; Sanchez & Grau, 2008; Shcherbak, 1988; Stambuk, 1999; Waterman, 1999; Yang, 2005; etc.). Each of these attempts was important for the general advancement of a science to cognition of a genetic code. These works were very useful because they have shown the specificity of the genetic code and its differences from many other natural systems; difficulties of modeling its features for receiving a fruitful model; a multiplicity of approaches in attempts of such modeling; an importance of the decision of this task, etc. These works have drawn the attention of many young talented researchers to this fundamental problem. In spite of many interesting publications, the general situation of understanding the genetic code is characterized by the following words, which were cited already in the preface of this book in more detail: "*What do we have now in the 10 million nucleotide of sequence data determined to date? ... We have the program that runs the cellular machinery, but we know very little about how to read it. Bench biologists, by experiment and by close association with the data, have found meaningful patterns. Theoreticians, by careful reasoning and use of collections of data, have found others, but we still under-*

stand frustratingly little" (Fickett & Burks, 1989). So, new efforts should be made to study structural organization of the genetic code from the viewpoint of informatics and mathematical natural sciences.

INFORMATION SCIENCE, THE NOISE-IMMUNITY AND THE MATRIX APPROACH TO THE GENETIC CODE

Mechanisms of genetic coding provide the high noise-immunity of transfer of the hereditary information from one generation to next generation, despite a set of disturbances and noise, which exist in biological environments. From the very beginning of discovery of the genetic code, scientists thought that structures of the genetic code are connected with the noise-immunity (noisc-proof features) of genetic systems (see review in (Ycas, 1969)). However, speaking about the noise-immunity of the genetic coding, speakers are usually limited to reference the fact of high degeneracy of the genetic code, which is capable to reduce a quantity of lethal mutations.

But modern works exist already, which suppose that an influence of the requirement of the noise-immunity on structures of the genetic code is much deeper. The given area of researches uses achievements of the mathematical theory of the noise-immunity coding, which are applied in the technique of digital communication, in attempts to understand phenomena of bioinformatics. In this area the suppositional influence of the noise-immunity can be studied by different methods and on different directions of thoughts (see, for example, (MacDonaill, 2003)). Our own researches presented in this book, which are based on the idea of deep connection between structures of the genetic code and the requirement of the noise-immunity of the genetic information, are original in research methods and revealed new facts.

Let us discuss the noise-immunity property of genetic system more attentively. It seems to be fantastic, but descendants grow similar to the ancestors due to the genetic information despite of enormous disturbances and noise in a billon of trillions of biological molecules. How is it possible to approach this problem about such fantastic noise-immunity in molecular genetics? Does modern science have any precedents of the decision of similar problems of the noise-immunity?

Yes, science has successfully decided the similar fantastic task recently: the noise-immunity transfer of photos from surfaces of other planets to the Earth. In this task electromagnetic signals, which carry data, should pass through millions kilometers of cosmic space of electromagnetic disturbances. These disturbances transform signals monstrously, but the modern mathematical technology permits to restore a transferred photo qualitatively.

The completion of this task became possible due to the theory of noise-immunity coding created by mathematicians. This theory of noise-immunity coding has appeared rather recently, initial basic work in this field was published by Hamming in 1950 (Hamming, 1980). The theory of such a coding utilizes intensively matrix mathematics including the representation of sets of signals and codes in a form of matrices and their Kronecker powers. Our book describes many interesting results in the field of molecular genetics and bioinformatics, which were obtained by authors on the basis of matrix mathematics. The investigation of the genetic code from the viewpoint of the theory of discrete signals is a natural way because of the discrete character of the genetic code.

One can note that coding in modern digital technique is usually utilized not for providing a difficulty of reading the text by the undesirable reader but for providing a technical opportunity of transfer of the discrete information with high noise-immunity, speed and reliability. The most famous example of codes is the Morse code. Of course, the modern codes are much more effective, than the Morse code.

These codes allow transferring the copious information through huge distances qualitatively. Orthogonal codes, which use Hadamard matrices, belong to the set of such codes (Ahmed & Rao, 1975; Blahut, 1985; Geadah & Corinthios, 1977; Lee & Kaveh, 1986; Peterson & Weldon, 1972; Petoukhov, 2008a, 2008b); Sklar, 2001; Trahtman, 1972; Trahtman, & Trahtman, 1975; Yarlagadda & Hershey, 1997). Any transmitted signal consists of a set of elementary signals (a component of a signal vector of an appropriate dimension). The task of the receiver in conditions of noise is the approximate definition of a concrete vector-signal, which has been sent from a known set of vector-signals (Sklar, 2001). Application of Hadamard matrices allows solving similar problems by means of a spectral decomposition of vector-signals and by means of a transfer of their spectra, on the basis of which the receiver restores an initial signal. This decomposition utilizes orthogonal functions of rows of Hadamard matrices (Ahmed & Rao, 1975).

One should emphasize the important circumstance: unlike digital technique, biological organisms solve the task not only to provide the noise-immunity simply, but to provide it in such a kind, which is suitable for transfer of this property of the noise-immunity along a chain of biological generations.

This book pays significant attention to the matrix approach to the genetic code, which has formed the special investigation field of matrix genetics. Investigations in this field reveal an important role of symmetries in structural organization of molecular ensembles of the genetic code. But why have we chosen the matrix approach to study the genetic system among many other possible approaches?

The six main reasons exist to explain this matrix choice to study the genetic code and to develop matrix genetics:

1. Information is usually stored in computers in the form of matrices;
2. Noise-immunity codes are constructed on the basis of matrices;
3. quantum mechanics utilizes matrix operators, connections with which can be detected in matrix forms of presentation of the genetic code; a significance of matrix approach is emphasized by the fact that quantum mechanics has arisen in a form of matrix mechanics by W. Heisenberg;
4. complex and hypercomplex numbers, which are utilized in physics and mathematics, possess matrix forms of their presentation. The notion of number is the main notion of mathematics and mathematical natural sciences. In view of this, investigation of a possible connection of the genetic code with multi-dimensional numbers in their matrix presentations can lead to very significant results.
5. Matrix analysis is one of the main investigation tools in mathematical natural sciences. Study of possible analogies between matrices, which are specific for the genetic code, and famous matrices from other branches of sciences can be heuristic and useful.
6. Matrices, which are a kind of union of many components in a single whole, are subordinated to certain mathematical operations, which determine substantial connections between collectives of many components; this kind of connections can be essential for collectives of genetic elements of different levels as well.

Matrix genetics studies matrix forms of presentation of the genetic code systems including genetic alphabets and sets of genetic multiplets. It studies also those phenomenological peculiarities of genetic systems which are reflected in these forms. The task of these researches consists in deeper understanding of genetic systems and inherited biological phenomena from a viewpoint of information technology and mathematical sciences.

The early work (Konopelchenko & Rumer, 1975a, 1975d) published in the most prestigious scientific journal of the USSR and in a form of the preprint in English may be considered as the pioneer work in the field of matrix genetics. This work presented the 4-letter genetic alphabet C, A, G, U/T in a form of a (2x2)-matrix and considered the second Kronecker power of this alphabetic matrix which generated a (4x4)-matrix of 16 genetic duplets for investigation of symmetrical and other properties of these genetic components.

Here we briefly note that G. Rumer was the main co-author of this pioneer article and he was a prominent Russian scientist in the field of theory of symmetry. WIth a personal recommendation by A. Einstein and P. Ehrenfest, he received a Lorentz's grant and worked as an assistant of M. Born in Gottingen in the period of 1929-1932. In the co-authorship with H.Weyl, V. Heitler and E. Teller, Rumer has created the basis of quantum chemistry. He knew 12 foreign languages. With another recommendation by A. Einstein, P. Ehrenfest, M. Born and E. Schrodinger, Rumer returned to Moscow from Gottingen in 1932 and became a professor of the Moscow State University. He is the author of a few famous books on problems of group theory and theoretical physics (Rumer, 1936, 1956; Rumer & Fet, 1970, 1977; Rumer & Ryvkin, 1972, etc.). One of his books about the relativity theory in the co-authorship with Nobel Prize winner in physics L. Landau was published in more than 20 languages around the world (Landau & Rumer, 2003). Rumer believed that properties of symmetry play an essential role in phenomenology of the genetic code. His works (Rumer, 1966, 1968; Konopelchenko & Rumer, 1975a, 1975d) on classification of codons in the genetic code, based on the principles of symmetry and linguistic reasons, have obtained a benevolent response by F. Crick. His other important works include a correlation between the structure of amino acids and the degeneracy of the genetic code (Konopelchenko & Rumer, 1975b), the wobble hypothesis by Crick in connection with the sequence of nucleotides (Konopelchenko & Rumer, 1975c), and regularities in codons (Volkenstein, & Rumer, 1966). Rumer's works have resulted in many responses all over the world. More information about Rumer and his works are presented in the article (Ginzburg, Mihailov & Pokrovskiy, 2001).

THE BASIC STRUCTURES OF THE GENETIC CODE

Due to wonderful works of many researches, the modern science knows basic phenomenological data about molecular structures of the genetic code including the four-letter genetic alphabet, 64 triplets, 20 amino acids, etc. History of molecular genetics knows attempts to understand and explain these phenomenological data from various viewpoints. For example, one can mention the famous hypothesis by G. Gamov (Ycas, 1969) about the reason for the existence of 20 amino acids. By this hypothesis, this reason is in the special configuration of DNA molecule. Some other hypothesis, which have only historical meanings also now, are considered in many text-books and historical reviews in the field of molecular genetics (Cantor & Schimmel, 1980; Chapevillle & Haenni, 1974; Karasev, 2003; Ratner, 2002; Roller,1974; Shults & Schirmer, 1979; Watson, 1968; Stent, 1971; Ycas, 1969; etc.).

All living organisms are unified wonderfully: all of them have identical molecular bases of the system of genetic coding. These bases are amazingly simple. For realization of the genetic messages, which encode sequences of amino acids in proteins, all kinds of organisms utilize in their molecules of heredity DNA (and RNA – ribonucleic acid) the "alphabet" consisting of only four "letters" or nitrogenous bases: adenine (A), cytosine (C), guanine (G), thymine (T) {or uracil (U) in RNA} (Figure 1). Linear sequences of these four letters on strings of molecules of heredity (DNA and RNA) contain the genetic

Figure 1. The complementary pairs of the four nitrogenous bases in DNA. A-T (adenine and thymine), C-G (cytosine and guanine). Hydrogen bonds in these pairs are shown by dotted lines. Black circles are atoms of carbon; small white circles are atoms of hydrogen; squares with the letter N are atoms of nitrogen; triangles with the letter O are atoms of oxygen. Amides (or amino-groups) NH_2 are marked by big circles

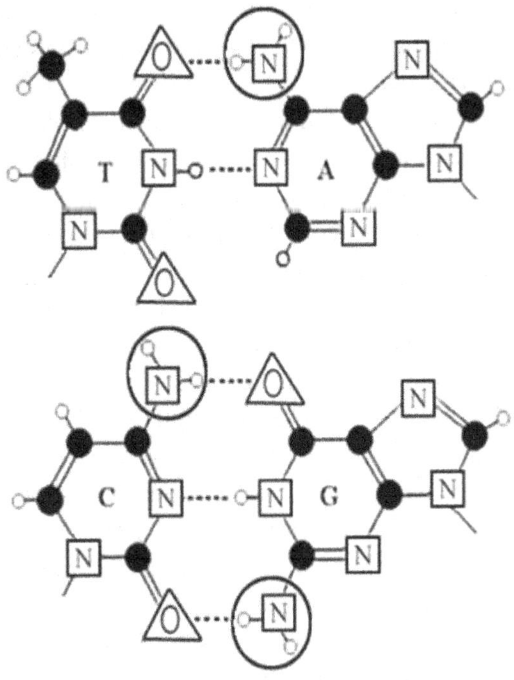

information for protein synthesis in all living bodies - from bacteria up to a whale or from a worm up to a bird and even a human. One can hear sometimes the figurative expression that the encyclopedia of life is written by four letters.

The given set of the four letters is usually considered as the elementary alphabet of a genetic code. These letters form the complementary pairs C-G and A-U (or A-T), because they stand opposite each other in molecules of heredity. The complementary letters C and G are connected by three hydrogen bonds; the complementary letters A and U (or A and T) are connected by two hydrogen bonds.

Genetic information, which is transferred by molecules of heredity, defines the primary structure of proteins of biological organisms. Each coded protein exists in the form of a chain of 20 kinds of amino acids. A sequence of amino acids in protein chain is defined by an appropriate sequence of genetic triplets. A triplet (or a codon) is a block of three neighbor nitrogenous bases, which are disposed along a filament of DNA or RNA. A sequence of amino acids in any protein is coded by an appropriate sequence of triplets (such sequence of "n" triplets is named "3n-multiplet" briefly).

The general quantity of kinds of triplets, which can be constructed from the four-letter alphabet, is equal to $4^3 = 64$. Each triplet has its code meaning: it encodes one of 20 kinds of amino acids or plays a role of a stop-signal or a start-signal for a process of a protein synthesis. Each codon has its anti-codon, which consists of the appropriate complementary letters: for example, the triplet CUG has the anti-codon GAC.

The genetic code is named "the degeneracy code" because its 64 triplets encode 20 amino acids and different amino acids are encoded by different quantities of triplets. Hypotheses about a connection between this degeneracy and the noise-immunity of the genetic information exist since time of the discovery of the genetic code. Symmetries in the structures of degeneracy of the genetic code are one of the main objects of investigation in our book. Many dialects of the genetic code exist in biological organisms and their subsystems, which differ each from other by some differences in correspondences between triplets and objects encoded by them (see details in the NCBI's site: http://www.ncbi.nlm.nih.gov/Taxonomy/Utils/wprintgc.cgi).

Proteins are the main dense component of biological organisms. Many thousands kinds of proteins exist. Each of them possesses its own individual function. In particular, all biological ferments, which provide phenomenal speeds of many biochemical reactions in organisms, are proteins. The whole harmonic system of metabolism depends on proteins. All amino acids in proteins are connected by the same type of chemical bond, which is named peptide bond.

The correspondence between triplets and objects encoded by them is usually illustrated by the table of the size (4x16), which was proposed by F. Crick half a century ago and which is reproduced in many textbooks and historical reviews in the field of molecular genetics (Cantor & Schimmel, 1980; Frank-Kamenetskiy, 1988; Roller, 1974; Stent, 1971; Watson, 1968; etc.). Each of its 64 tabular cells contains one triplet and an appropriate object (an amino acid or stop-codon) encoded by this triplet. However nobody insisted that possibilities of analytical and heuristic representation of systems of elements of the genetic code in tabular forms are exhausted by this table. Let us list the 20 amino acids, which are encoded genetically, and their traditional abbreviations, which are used in our book: Ala – alanine, Arg – arginine, Asn – asparagines, Asp - aspartic, Cys – cysteine, Gln - glutamine, Glu – glutamic, Gly – glycine, His – histidine, Ile - isoleucine, Leu – leucine, Lys – lysine, Met – methionine, Phe – phenylalanine, Pro – proline, Ser – serine, Thr – threonine, Trp – tryptophan, Tyr – tyrosine, Val – valine.

Modern science does not know why the alphabet of genetic language has four letters (it could have any other number of letters in principle)? And why just these four nitrogenous bases are chosen by nature as elements of the genetic alphabet from billions possible chemical compounds? And why the quantity of amino acids encoded by the triplets is equal to 20? In our opinion, this choice has a deep sense. Investigations of symmetries in structures of the genetic code can help to answer these and other important questions.

The problem of the heritable noise-immunity is the general one for all multi-channel systems of informatics of each organism. Many applied tasks of nanotechnology and biotechnology are connected with ensembles of genetic molecules: for example, the task of creation of DNA-computers and DNA-robotics exists (Paun, Rozenberg & Salomaa, 2006; Seeman, 2004; Shapiro & Benenson, 2006). It is necessary to study those peculiarities of ensembles of genetic molecules, which possess formal analogies with formalisms of digital informatics and its matrix mathematics.

One may ask whether these mathematical methods allow one to numerate each genetic multiplet in binary manner taking into account the natural characteristics of genetic letters A, C, G, U/T?. The main thrust of the present chapter is to consider an effective transfer of the named methods into the field of molecular genetics. Some initial constructions of matrix genetics with elements of symmetry are introduced below.

THE BINARY SUB-ALPHABETS OF THE GENETIC ALPHABET FOR NUMBERING THE MULTIPLETS IN GENETIC MATRICES

Is it possible to propose a matrix approach to represent all sets of genetic multiplets in the well-ordered general form and with an individual binary number for each multiplet on the basis of molecular features of the four letters A, C, G, U/T of the genetic alphabet? Will such general form be connected with important principles and methods of computer informatics and of the noise-immunity in digital technique?

Positive answers to these questions will be useful to analyze structural properties and symmetries of the genetic system and to reveal analogies between principles of the genetic code and computer informatics for many theoretic and applied tasks.

To get such positive answers, we will demonstrate, first of all, that symmetries in molecular characteristics of the genetic alphabet provide the existence of its binary sub-alphabets. The four letters (or the four nitrogenous bases) of the genetic alphabet represent specific poly-nuclear constructions with the special biochemical properties. The set of these four constructions is not absolutely heterogeneous, but it bears the substantial symmetric system of distinctive-uniting attributes (or, more precisely, pairs of "attribute-antiattribute"). This system of pairs of opposite attributes divides the genetic four-letter alphabet into various three pairs of letters by all three possible ways; letters of each such pair are equivalent to each other in accordance with one of these attributes or with its absence.

Really, the system of such attributes divides the genetic four-letter alphabet into various three pairs of letters, which are equivalent from a viewpoint of one of these attributes or its absence: 1) C = U & A = G (according to the binary-opposite attributes: "pyrimidine" or "non-pyrimidine", that is purine); 2) A = C & G = U (according to the attributes: amino-mutating or non-amino-mutating under action of nitrous acid HNO_2 (Wittmann, 1961; Ycas, 1969); the same division is given by the attributes "keto" or "amino" (Waterman, 1999); 3) C = G & A = U (according to the attributes: three or two hydrogen bonds are materialized in these complementary pairs). The possibility of such division of the genetic alphabet into three binary sub-alphabets is known from the book (Waterman, 1999). We will utilize these known sub-alphabets by means of a new method in the field of matrix genetics. We will attach appropriate binary symbols "0" or "1" to each of the genetic letters based on these sub-alphabets. Then we will use these binary symbols for binary numbering the columns and the rows of the genetic matrices of the Kronecker family.

Let us mark these three kinds of binary-opposite attributes by numbers $N = 1, 2, 3$ and ascribe to each of the four genetic letters the symbol "0_N" (the symbol "1_N") in case of presence (of absence correspondingly) of the attribute under number "N" to this letter. As a result we obtain the following representation of the genetic four-letter alphabet in the system of its three "binary sub-alphabets corresponding to attributes" (Figure 2).

The table on Figure 2 shows that, on the basis of each kind of the attributes, each of the letters A, C, G, U/T possesses three "faces" or meanings in the three binary sub-alphabets. On the basis of each kind of the attributes, the genetic four-letter alphabet is curtailed into the two-letter alphabet. For example, on the basis of the first kind of binary-opposite attributes we have (instead of the four-letter alphabet) the alphabet from two letters 0_1 and 1_1, which one can name "the binary sub-alphabet to the first kind of the binary attributes".

Accordingly, any genetic message as a sequence of the four letters C, A, G, U consists of three parallel and various binary texts or three different sequences of zero and unit (such binary sequences are used at storage and transfer of the information in computers). Each from these parallel binary texts, based on

Figure 2. Three binary sub-alphabets according to three kinds of binary-opposite attributes in a set of nitrogenous bases C, A, G, U. The scheme on the right side explains graphically the symmetric relations of equivalence between the pairs of letters from the viewpoint of the separate attributes 1, 2, 3

	Symbols of a genetic letter from a viewpoint of a kind of the binary-opposite attributes	C	A	G	U/T
№1	0_1 – pyrimidines (one ring in a molecule); 1_1 – purines (two rings in a molecule)	0_1	1_1	1_1	0_1
№2	0_2 – a letter with amino-mutating property (amino); 1_2 – a letter without it (keto)	0_2	0_2	1_2	1_2
№3	0_3 – a letter with three hydrogen bonds; 1_3 – a letter with two hydrogen bonds	0_3	1_3	0_3	1_3

objective biochemical attributes, can provide its own genetic function in organisms. According to our data, the genetic system uses the possibility to read triplets from the viewpoint of different binary sub-alphabets: this possibility participates in the construction of the genetic octet Yin-Yang-algebra (or the octet bipolar algebra), which serves as the algebraic model of the genetic code in Chapter 7.

NATURAL SYSTEM OF NUMBERING THE GENETIC MULTIPLETS

Genetic information is transferred by means of discrete elements: 4 letters of genetic alphabet, 64 amino acids, etc. General theory of processing of discrete signals utilizes encoding the signals by means of special mathematical matrices and spectral representation of signals with the main aim to increase the reliability and efficiency of information transfer (Ahmed & Rao, 1975; Sklar, 2001; etc). A typical example of such matrices with appropriate properties is the Kronecker family of Hadamard matrices in the equation (1).

$$H_{n+1} = [1\ 1; -1\ 1]^{(n)} \qquad (1)$$

where (n) means the integer Kronecker power. The mathematical peculiarities of Kronecker product are described below.

The Kronecker product is an operation on two matrices of arbitrary size resulting in a block matrix. The Kronecker product should not be confused with usual matrix multiplication, which is an entirely different operation. It is named after German mathematician Leopold Kronecker. If one has two square matrices $A = \| \alpha_{ij} \|$ and $B = \| \beta_{kp} \|$, where $i, j = 1,\ldots, m$ and $k, p = 1,\ldots, n$, then a square block matrix

$$C = A \otimes B = \|\alpha_{ij} * \beta_{kp}\|$$

is called the Kronecker product of the matrices A and B.

The Kronecker product of matrices arises in a natural way in a problem of searching a matrix. The eigenvalues of matrix $A \otimes B$ are equal to a product of $a_i * b_j$, where a_i and b_j are eigenvalues of the matrices A and B. It was proved that the Kronecker product of matrices A and B possesses such eigenvalues (Bellman, 1960). The Kronecker product is connected with fractal structures; these questions are described in the book (Gazale, 1999).

The simplest Hadamard matrix $H_2 = [1\ 1; -1\ 1]$ is named the kernel of this Kronecker family. Rows of Hadamard matrices form an orthogonal system of Hadamard-Walsh functions, which is used for a spectral presentation and transfer of discrete signals (Ahmed & Rao, 1975; Yarlagadda & Hershey, 1997). Quantum computers use normalized Hadamard matrixes in a role of logic gates in connection with the important role of these matrixes in the quantum mechanics (Nielsen & Chuang, 2001). Chapter 6 describes deep connections between Hadamard matrices and ensembles of elements of the genetic code.

On the basis of the idea about a possible analogy between discrete signals processing in computers and in a genetic code system, one can present the genetic 4-letter alphabet in the following matrix form P = [C A; U G]. It is obvious, that this form possesses the analogy with the kernel (equation (1)) of the Kronecker family of Hadamard matrices. Then the Kronecker family of matrices with such alphabetical kernel can be considered:

$$P^{(n)} = [C\ A;\ U\ G]^{(n)} \tag{2}$$

where (n) means the integer Kronecker power. Figure 3 shows the first matrices of such a family. One can see on this figure that each matrix contains all genetic multiplets of equal length: $[C\ A;\ U\ G]^{(1)}$ contains all 4 monoplets; $[C\ A;\ U\ G]^{(2)}$ contains all 16 duplets; $[C\ A;\ U\ G]^{(3)}$ contains all 64 triplets, etc. It should be emphasized that this book pays great attention to the genetic alphabet: we will consider the alphabetic matrices $[C\ A;\ U\ G]^{(n)}$ from different viewpoint permanently and we will construct algorithms of matrix transformations on the basis of features of the alphabetic letters A, C, G, U/T. The genetic alphabet serves as the key structure to investigate system properties of the genetic code and its dialects.

Such presentation of ensembles of elements of the genetic code in the form of Kronecker families of genetic matrices (or "genomatrices" briefly) has appeared as a useful tool to investigate structures of the genetic code from the viewpoint of their analogy with the theory of discrete signals processing and noise-immunity coding. The scientific direction, which deals with such matrix presentation of the ensembles of genetic elements and their parameters, is named "matrix genetics' briefly. The results of matrix genetics reveal hidden interconnections, symmetries and evolutionary invariants in genetic code systems (He, 2001; He & Petoukhov, 2007; He, Petoukhov & Ricci, 2004; Petoukhov, 1999b, 2001, 2003-2008). Simultaneously they show that genetic molecules are the important part of a specific maintenance of the noise-immunity and efficiency of a discrete information transfer.

Figure 3. The first genetic matrices of the Kronecker family $P^{(n)}=[C\ A;\ U\ G]^{(n)}$ with the binary numbering their columns and rows on the base of the binary sub-alphabets № 1 and № 2 fromFigure 2. The lower matrix is the genomatrix $P^{(3)}=[C\ A;\ U\ G]^{(3)}$. Each matrix cell contains a symbol of a multiplet, a binary number of this multiplet and its expression in decimal notation. Decimal numbers of columns, rows and multiplets are written in brackets

$P^{(1)}$ =

	0	1
0	C 00 (0)	A 01 (1)
1	U 10 (2)	G 11 (3)

; $P^{(2)}$ =

	00(0)	01(1)	10(2)	11(3)
00 (0)	CC 0000 (0)	CA 0001 (1)	AC 0010 (2)	AA 0011 (3)
01 (1)	CU 0100 (4)	CG 0101 (5)	AU 0110 (6)	AG 0111 (7)
10 (2)	UC 1000 (8)	UA 1001 (9)	GC 1010 (10)	GA 1011 (11)
11 (3)	UU 1100 (12)	UG 1101 (13)	GU 1110 (14)	GG 1111 (15)

	000 (0)	001 (1)	010 (2)	011 (3)	100 (4)	101 (5)	110 (6)	111 (7)
000 (0)	CCC 000000 (0)	CCA 000001 (1)	CAC 000010 (2)	CAA 000011 (3)	ACC 000100 (4)	ACA 000101 (5)	AAC 000110 (6)	AAA 000111 (7)
001 (1)	CCU 001000 (8)	CCG 001001 (9)	CAU 001010 (10)	CAG 001011 (11)	ACU 001100 (12)	ACG 001101 (13)	AAU 001110 (14)	AAG 001111 (15)
010 (2)	CUC 010000 (16)	CUA 010001 (17)	CGC 010010 (18)	CGA 010011 (19)	AUC 010100 (20)	AUA 010101 (21)	AGC 010110 (22)	AGA 010111 (23)
011 (3)	CUU 011000 (24)	CUG 011001 (25)	CGU 011010 (26)	CGG 011011 (27)	AUU 011100 (28)	AUG 011101 (29)	AGU 011110 (30)	AGG 011111 (31)
100 (4)	UCC 100000 (32)	UCA 100001 (33)	UAC 100010 (34)	UAA 100011 (35)	GCC 100100 (36)	GCA 100101 (37)	GAC 100110 (38)	GAA 100111 (39)
101 (5)	UCU 101000 (40)	UCG 101001 (41)	UAU 101010 (42)	UAG 101011 (43)	GCU 101100 (44)	GCG 101101 (45)	GAU 101110 (46)	GAG 101111 (47)
110 (6)	UUC 110000 (48)	UUA 110001 (49)	UGC 110010 (50)	UGA 110011 (51)	GUC 001100 (52)	GUA 110101 (53)	GGC 110110 (54)	GGA 110111 (55)
111 (7)	UUU 111000 (56)	UUG 111001 (57)	UGU 111010 (58)	UGG 111011 (59)	GUU 111100 (60)	GUG 111101 (61)	GGU 111110 (62)	GGG 111111 (63)

The matrix $P^{(1)}$ is the simplest representative (specimen) of a set of biperiodic matrices (or tables) of the genetic code system. It has a vertical periodicity of the matrix elements from the viewpoint of the binary sub-alphabet № 1 and it has a horizontal periodicity of the matrix elements from the viewpoint of the binary sub-alphabet № 2. It can be checked easily that all matrices $P^{(n)}$ are biperiodic matrices.

Actually any column of such a matrix consists of only the n-plets which are equivalent to each other from the viewpoint of binary sub-alphabet № 1. And any row of a matrix $P^{(n)}$ consists of those *n*-plets only, which are equivalent to each other from the viewpoint of binary sub-alphabet № 2.

The Kronecker family of genetic matrices [C A; U G]$^{(n)}$ (Figure 3) represents all genetic multiplets, if the value of "*n*" is big enough. This family includes the genomatrix of the genetic alphabet; the genomatrix of triplets, which encode the amino acids; the genomatrices of long multiplets, which encode proteins. All this natural set of genetic multiplets, which have various coding functions in the genetic system, appears coordinated with this simple Kronecker family of matrices [C A; U G]$^{(n)}$ (Figure 3).

Each genetic multiplet has its own individual binary number in the described natural system of numbering the multiplets. This multiplet also has its own disposition in the appropriate genetic matrix of the Kronecker family. It is obvious that a length of the individual binary number for a *n*-plet, which contains "*n*" letters, is equal to 2*n*: the first half of this number is the interpretation of letters of the multiplet from the viewpoint of the second binary sub-alphabet (Figure 2) and the second part is the interpretation from the viewpoint of the first binary sub-alphabet. For example, the sequence GACUU-CACGGUG, which contains 9 letters, has the individual binary number with 9x2=18 binary symbols: 100110001111/110000101101. If one should construct the catalog of genetic sequences of various lengths and composition, it can be done on the basis of the described natural system of numbering the sequences as multiplets.

All *n*-plets, which are begun with one of the four letters C, A, U, G, are disposed in one of the four quadrants of an appropriate genomatrix [C A; U G]$^{(n)}$ because of the specifics of Kronecker multiplication. If one does not pay attention to this first letter in *n*-plets of each matrix quadrant, then one can see that each quadrant reproduces a previous matrix [C A; U G]$^{(n-1)}$ of this Kronecker family. Figuratively each genomatrix of such family possesses information (or "memory") about all previous genomatrices of this family.

It should be noted that each column of the formal constructed genomatrix [C A; U G]$^{(3)}$ (Figure 3) is corresponded to one of the 8 classical octets by Wittmann (1961), which are famous in the history of molecular genetics and reflect real biochemical properties of elements of the genetic code (Ycas, 1969). This fact is the first indirect confirmation of adequacy of the given matrix approach, which reflects a natural orderliness inside of the genetic system.

Let us demonstrate now that all 64 triplets can be binary numerated in a natural manner by means of the binary sub-alphabets (Figure 2), which are based on the real structural and biochemical features of the genetic molecules. As the result of such a natural numbering, all triplets appear disposed in the genomatrix [C A; U G]$^{(3)}$ in the monotonous order on increase of their binary numbers.

Really, all columns and rows of the matrices on Figure 3 are binary numerated by the following algorithm. Their numbers are formed automatically if one interprets multiplets of each column from the viewpoint of the first binary sub-alphabet (Figure 2) and if one interprets multiplets of each row from the viewpoint of the second binary sub-alphabet. For example, from the viewpoint of the first sub-alphabet, the triplet CAU possesses the binary number 010 (all triplets of the same column possess the same binary number, which is utilized as the general number of this column correspondingly). But from the viewpoint of the second sub-alphabet, the triplet CAU possesses the binary number 001 (all triplets of the same row possess the same binary number, which is utilized as the general number of this row). One can see on Figure 3, that all columns and all rows in the genomatrix [C A; U G]$^{(3)}$ appear renumbered and disposed in an monotonic order.

In the genomatrix [C A; U G]$^{(3)}$, each of 64 triplets has its own individual number, which consists of association of binary numbers of its row and column (for example, triplet CAU has the binary number 001010, which is equal to 10 in decimal notation). This genomatrix reflects real interrelations of elements in the set of triplets: any codon and its anti-codon are disposed in inversion-symmetrical manner relative to the centre of the genomatrix (Figure 3).

And each pair "codon-anticodon" (and only such pair) has the sum of their decimal numbers, which is to equal 63 (in binary notation it is equal to 111111). For example, the triplet CAU has the decimal number 10 and the complementary triplet GUA has the decimal number 53; the sum of these numbers is equal to 63. Each sequence of triplets can be presented in the genomatrix [C A; U G]$^{(3)}$ in a form of an appropriate trajectory passing through matrix cells with these triplets in series. It is obvious that the complementary sequences on the two filaments of the double helix of DNA correspond to two appropriate trajectories in the genomatrix [C A; U G]$^{(3)}$, which are inversion-symmetrical to each other relative to its centre.

In the case of a conservation in each cell of the genomatrix [C A; U G]$^{(3)}$ (Figure 3) with binary six-digit numbers of these 64 triplets, this genomatrix coincides with the famous table of 64 hexagrams in Fu-Xi's order from the ancient Chinese "The Book of Changes" ("I Ching"), which was written a few thousand years ago (see Chapter 12). This matrix has amazed the creator of a computer G. Leibnitz (1646-1716 years). He considered himself as a creator of the system of binary notation, but in one moment he suddenly found out ancient predecessors relative to this system. Leibnitz has seen in features of the given ancient table of 64 hexagrams many features of similarity to his ideas of binary systems and universal language. "*Leibnitz has seen in this similarity ... the evidence of the preestablished harmony and unity of the divine plan for all times and people*" (Schutskiy, 1997, p. 12). Modern physics and other branches of science pay attention to "I Ching" and other ancient Oriental teachings also (see, for example, (Capra, 2000; Gell-Mann & Ne'eman, 2000). A possible connection between the genetic code and the symbolic system of "I Ching" was noted in the works (Stent, 1969; Jakob, 1974, 1977; etc.). Our results in the field of matrix genetics confirm this work. So, the described natural system of numbering the genetic triplets and their cells in the genomatrix [C A; U G]$^{(3)}$ is known for thousands years already. It can be named the ancient Chinese system from the historical viewpoint. The matrix approach to a genetic code, besides the fundamentality of object of research and matrix mathematics, unexpectedly leads to historical analogies and a problem of connection of times. We will return to this theme in more detail in Chapter 12.

It should be noted that the huge quantity $64! \approx 10^{89}$ of variants exists for dispositions of 64 triplets in the (8x8)-matrix. The modern physics estimates time of existence of the Universe in 10^{17} seconds. It means the following: if for consideration of each of these variants we spend only one second, then during all time of existence of the Universe we shall have time to consider only insignificant part from this 10^{89} variants. It is obvious that in such a situation an accidental disposition of the 20 amino acids and the corresponding triplets in a (8x8)-matrix will give almost never any symmetry in their disposition in matrix halves, quadrants and rows. One can illustrate this circumstance by the following way. Let us consider the (8x8)-matrix, the 64 cells of which are numbered one after another. Everyone can make an accidental sample of 32 natural numbers from the series of 64 values from 0 up to 63 and then mark by dark color the 32 cells with these 32 numbers. Other 32 cells with other numbers are marked by white color. The obtained black-and-white mosaic of the matrix will be asymmetric with very high probability. Figure 4 demonstrates an example of such asymmetric mosaic in the case of the accidental

Figure 4. An example of a black-and-white mosaic of the (8x8)-matrix, the cells of which are numbered one after another. The black cells correspond to the case of 32 numbers of an accidental choice (an explanation in the text)

0	1	2	3	4	5	6	7
8	9	10	11	12	13	14	15
16	17	18	19	20	21	22	23
24	25	26	27	28	29	30	31
32	33	34	35	36	37	38	39
40	41	42	43	44	45	46	47
48	49	50	51	52	53	54	55
56	57	58	59	60	61	62	63

choice of the following 32 numbers: 53, 2, 47, 62, 23, 6, 38, 11, 19, 8, 26, 12, 28, 32, 9, 36, 42, 4, 43, 33, 45, 18, 48, 24, 51, 0, 41, 55, 58, 13, 60, 3.

One may ask why nature has chosen that variant of the degeneracy of the genetic code, which fits symmetrically (regular) inside the genetic matrix $P^{(3)} = [C\ A;\ U\ G]^{(3)}$ relative to its halves, quadrants and rows (see Figure 2 in the next chapter). Chapter 2 will demonstrate that nature has divided the set of 64 triplets into two sub-sets with 32 triplets in each because of special properties of the degeneracy of the genetic code. One of these sub-sets contains the triplets with the following numbering in the described natural system of numbering the triplets: 0, 1, 2, 3, 8, 9, 10, 11, 18, 19, 22, 23, 26, 27, 30, 31, 36, 37, 38, 39, 44, 45, 46, 47, 50, 51, 54, 55, 58, 59, 62, and 63. These 32 triplets, which are shown in the next chapter on Figure 2 in the black cells, are opposed by nature to other 32 triplets in 32 white cells. The general disposition of these black and white cells in the genomatrix $[C\ A;\ U\ G]^{(3)}$ possesses the expressed symmetric characteristics considered in Chapter 2.

One can remark, that the hidden relations of symmetry between these two sub-sets of the triplets are revealed in an exclusive (alphabetical-Kronecker) variant of the disposition of 64 triplets in (8x8)-matrix, which is described above and is one of 10^{89} variants of their dispositions. The main results in the field of matrix genetics, which are described in our book, were obtained in connection with this special variant of the disposition of the triplets in the genomatrix $[C\ A;\ U\ G]^{(3)}$ from the Kronecker family of genomatrices $[C\ A;\ U\ G]^{(n)}$. Chapter 2 presents a few genomatrices additionally, which are produced from the genomatrix $[C\ A;\ U\ G]^{(n)}$ algorithmically and which possess symmetrical characteristics of the degeneracy of the genetic code as well.

THE MATRIX NUMBERING THE GENETIC MULTIPLETS AND MATRICES OF DIADIC SHIFTS

Next we describe the connection between numerated genomatrices $[C\ A;\ U\ G]^{(n)}$ (Figure 3) and those matrices of dyadic shifts, which are known in the theory of discrete signals processing long ago.

The theory of discrete signals processing utilizes widely the special mathematical operation of modulo-2 addition for binary numbers. Modulo-2 addition is one of fundamental operations for binary variables. By definition, the modulo-2 addition of two numbers, which are written in binary notation, is made in bitwise manner in accordance with the following rules:

$$0 + 0 = 0,\ 0 + 1 = 1,\ 1 + 0 = 1,\ 1 + 1 = 0 \quad (3)$$

For example, modulo-2 addition of two binary numbers 110 and 101, which are equal to 6 and 5 in decimal notation correspondingly, gives the result: $110 \oplus 101 = 011$, which is equal to 3 in decimal notation (here $\oplus$ is the symbol of modulo-2 addition).

The series of binary numbers

$$000, 001, 010, 011, 100, 101, 110, 111 \quad (4)$$

forms the so named diadic group, where the modulo-2 addition serves as the group operation (Harmut, 1989). The distance in this group of symmetry is defined as Hamming distance. Since Hamming distance satisfies the conditions of a metric group, the diadic group is the metric group. The modulo-2 addition of any two binary numbers from the expression (4) always gives a new number from the same series of the expression (4). The number 000 serves as the unit element of this group, for example $010 \oplus 000 = 010$. The reverse element for any number of this group is the number itself, for example $010 \oplus 010 = 000$.

The series of the expression (4) is transformed by the modulo-2 addition with the binary number 001 into the new series with the new sequence of the same numbers:

$$001, 000, 011, 010, 101, 100, 111, 110 \quad (5)$$

Such changes of the initial binary sequence, which are produced by modulo-2 addition of its members with any of binary numbers from the expression (4), are named "diadic shifts" (Ahmed & Rao, 1975; Harmut, 1989). If any system of elements demonstrates its connection with diadic shifts, it shows that the structural organization of his system is related to the logics of modulo-2 addition.

Let us make modulo-2 addition of binary numbers of columns and rows for all cells in the genomatrix $[C\ A;\ U\ G]^{(3)}$ on Figure 3. For example, the cell, which is disposed in the column 110 and in the row 101, obtains the binary number 011 by means of such addition. As a result, the following numeric matrix $P^{(3)}_{DIAD} = [C\ A;\ U\ G]^{(3)}_{DIAD}$ arises (Figure 5).

The (8x8)-matrix $[C\ A;\ U\ G]^{(3)}_{DIAD}$ is bisymmetrical because it is symmetrical relative to both diagonals. This matrix contains only 8 binary numbers, which is equal to 0, 1, 2, 3, 4, 5, 6, 7 in decimal notation. Each of these numbers occupies 8 matrix cells from 64 numerated cells (see Figure 3). The sum of numbers of these 8 matrix cells is equal to 252 in decimal notation for each case. For example, the number 5 occupies those 8 matrix cells on Figure 5, which are numerated individually on the Figure 3 by numbers 5, 12, 23, 30, 33, 40, 51, 58. The sum of these 8 numbers is equal to 252. The left and right halves (and the upper and lower halves) of this matrix $[C\ A;\ U\ G]^{(3)}_{DIAD}$ are inversion-symmetrical to each other in the sense of the binary inversion relative to their three-digit numbers in matrix cells (by definition, the binary inversion interchanges the binary symbols 1 and 0 to each other). For this reason, the modulo-2 addition of such binary numbers, which are disposed in any two mirror-symmetrical cells of this matrix, gives the binary number 111. For example, a cell with the number 001 in the left half of the matrix has a mirror-symmetrical cell in its right half with the number 110 always. Their sum in the sense of modulo-2 addition is equal to: $001 \oplus 110 = 111$.

Figure 5. The bisymmetrical matrix [C A; U G]$^{(3)}_{DIAD}$ of dyadic shifts; brackets contain expressions of numbers in decimal notation

	000 (0)	001 (1)	010 (2)	011 (3)	100 (4)	101 (5)	110 (6)	111 (7)
000 (0)	000 (0)	001 (1)	010 (2)	011 (3)	100 (4)	101 (5)	110 (6)	111 (7)
001 (1)	001 (1)	000 (0)	011 (3)	010 (2)	101 (5)	100 (4)	111 (7)	110 (6)
010 (2)	010 (2)	011 (3)	000 (0)	001 (1)	110 (6)	111 (7)	100 (4)	101 (5)
011 (3)	011 (3)	010 (2)	001 (1)	000 (0)	111 (7)	110 (6)	101 (5)	100 (4)
100 (4)	100 (4)	101 (5)	110 (6)	111 (7)	000 (0)	001 (1)	010 (2)	011 (3)
101 (5)	101 (5)	100 (4)	111 (7)	110 (6)	001 (1)	000 (0)	011 (3)	010 (2)
110 (6)	110 (6)	111 (7)	100 (4)	101 (5)	010 (2)	011 (3)	000 (0)	001 (1)
111 (7)	111 (7)	110 (6)	101 (5)	100 (4)	011 (3)	010 (2)	001 (1)	000 (0)

By analogical algorithm of modulo-2 addition, the whole family of matrices of dyadic shifts $P^{(n)}_{DIAD}$, where n = 2, 4, 5,…, can be constructed from the genomatrices [C A; U G]$^{(n)}$ (Figure 3). All such matrices $P^{(n)}_{DIAD}$ are bisymmetrical as well. Each of matrices $P^{(n)}_{DIAD}$ is the matrix form of presentation of a particular case of special hypercomplex numbers, which are named "hyperbolic matrions" (these hyperbolic matrions are described in Chapter 8 in more detail).

Do such matrices $P^{(n)}_{DIAD}$ have any connection with the theory of discrete signals processing? Yes, they have. The matrix [C A; U G]$^{(3)}_{DIAD}$ and other analogical matrices [C A; U G]$^{(n)}_{DIAD}$ are known in this information theory long ago under the name "matrices of dyadic shifts" (for example, see (Ahmed & Rao, 1975)). They are fundamentals of some special methods of analysis and synthesis of signals as vectors. In computer informatics, matrices of dyadic shifts are constructed by means of modulo-2 addition without utilizing Kronecker multiplication of matrices, which we have used to receive the Kronecker family of the genomatrices [C A; U G]$^{(n)}$ of all multiplets from the (2x2)-matrix of the genetic alphabet (Figure 3). One can note that the analogical (8x8)-matrix of diadic shifts is constructed from the table of 64 hexagrams of "I Ching" (Chapter 11). We will return to diadic shifts in Chapters 7 and 8 to demonstrate additionally that the logics of structures of the genetic code is connected with diadic shifts and hence with the modulo-2 addition.

It should be emphasized specially that dyadic shifts are one of the elements of interesting theory, which is described in the book about applications of methods of information theory in physics (Harmut, 1989). This theory utilizes the notions of dyadic spaces, dyadic metrics, and dyadic coordinates in a connection with special codes. Relation of the genetic code to this theory is one of the prospective topics in the field of matrix genetics for investigations in future.

Now let us pay attention to the block character of the matrices of dyadic shifts $P^{(n)}_{DIAD}$. Each $(2^n x 2^n)$-matrix $P^{(n)}_{DIAD}$ is a system of fractal kind. It contains four block matrices, each of which has the size (2x2). Two such block matrices, which are disposed along each diagonal, are identical to each other always. For this reason, the lower half of each $(2^n x 2^n)$-matrix $P^{(n)}_{DIAD}$ can be produced from its upper half algorithmically by a cyclic shift. In this sense, each block matrix $P^{(n)}_{DIAD}$ is a matrix of the cyclic shift of its (2x2)-blocks and possesses the crosswise character.

Two quadrants along the main diagonal contain identical block elements, which are $(2^{n-1} x 2^{n-1})$-matrices of a dyadic shift. Matrix cells along the second diagonal contain identical block elements in a form of $(2^{n-1} x 2^{n-1})$-matrices also, elements of which are changed only by addition of number 2^{n-1} relative to elements of the $(2^{n-1} x 2^{n-1})$-matrices along the main diagonal. In turn, these $(2^{n-1} x 2^{n-1})$-matrices are the block matrices of the cyclic shift, which possess a crosswise character, etc.

For example, the $(2^3 x 2^3)$-matrix $[C\ A;\ U\ G]^{(3)}_{DIAD}$ on Figure 5 is the block matrix of the cyclic shift relative to its (2x2)-quadrants. Identical quadrants, which are disposed along the main diagonal, are $(2^2 x 2^2)$-matrices of the dyadic shift with elements 0, 1, 2, and 3. Another kind of identical blocks in the form of the $(2^2 x 2^2)$-quadrants with elements 4, 5, 6, 7 are disposed along the second diagonal. They only differ from the first $(2^2 x 2^2)$-quadrants by addition of number 2^2 to their elements. In turn, each of these $(2^2 x 2^2)$-quadrants of the matrix $[C\ A;\ U\ G]^{(3)}_{DIAD}$ on Figure 5 is the block matrix of the cyclic shift of its (2x2)-blocks.

In connection with cyclic shifts in described genetic matrices, one can mention so named cyclic codes, which are based on cyclic shifts (Peterson & Weldon, 1972; Sklar, 2001). Cyclic codes are considered usually as one of the most interesting codes in the field of digital technique due to their mathematical properties. Some modern publications in the field of molecular genetics analyze the question about a possible important participation of cyclic codes in systems of genetic coding (Arques & Michel, 1996, 1997; Frey & Michel, 2003, 2006; Stambuk, 1999).

Returning to the crosswise character of described genetic matrices of diadic shifts $P^{(n)}_{DIAD}$ (Figure 5), which reminds one of a crosswise character of chromosomes to some extent, we note that genetic inherited constructions of physiological systems (including sensory-motion systems) demonstrate similar crosswise structures by unknown reasons. For example, the connection between the hemispheres of human brain and the halves of human body possesses the similar crosswise character: the left hemisphere serves the right half of the body and the right hemisphere (Figure 6) (Annett, 1985, 1992; Gazzaniga, 1995; Hellige, 1993). The system of optic cranial nerves from two eyes possesses the crosswise structures as well: the optic nerves transfer information about the right half of field of vision into the left hemisphere of brain, and information about the left half of field of vision into the right hemisphere. The same is held true for the hearing system (Penrose, 1989, Chapter 9). One can suppose that these inherited physiological phenomena are connected with genetic crosswise structures, which include, in particular, crosswise matrices of dyadic shifts, of hyperbolic matrions and octet Yin-Yang-numbers from Section 3 to provide noise-immunity properties of genetic systems.

FUTURE TRENDS AND CONCLUSION

The described matrix approach shows first examples of usefulness of utilizing symmetrical features of ensembles of genetic elements for development of new mathematical tools of genetic investigations. Such an approach permits one to represent all sets of genetic multiplets in the well-ordered general

Figure 6. The crosswise schemes of some morpho-functional structures in human organism. On the left side: the crosswise connections of brain hemispheres with the left and the right halves of a human body. In the middle: the crosswise structure of optic nerves from eyes in brain. On the right side: a chromosome

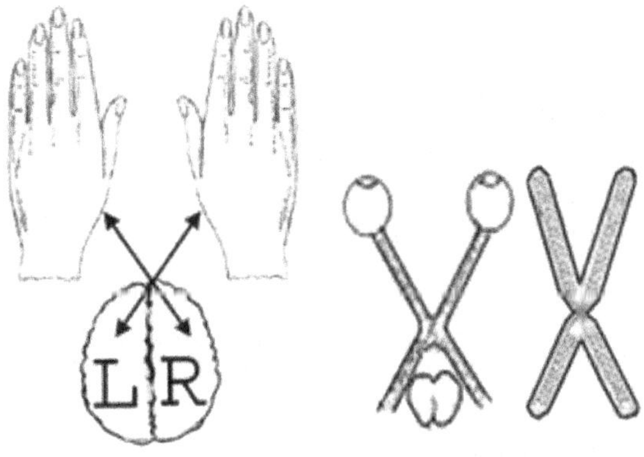

form of matrices of the Kronecker family. Each multiplet obtains its individual number in the proposed natural system of numbering the genetic multiplets. It obtains also its own individual disposition in an appropriate genetic matrix of the Kronecker family. The described natural system of numbering the multiplets is recommended for utilizing in computerized catalogs of genetic sequences. This Kronecker family of genetic matrices is the new cognitive form of presentation and analysis of ensembles of elements of the genetic code, which is utilized intensively in the next chapters of the book and which leads to many significant results.

Revealing the connection between the genetic matrices of the Kronecker family and matrices of diadic shifts, which are known in computer informatics, gives us ability to use the mathematical ideology of diadic spaces and diadic metrics for genetic systems.

The first described constructions in the field of matrix genetics gave us new abilities for investigations of genetic systems in the future. One of them is the creation of catalogs of matrices with all possible multiplets for various tasks. For example, such catalogs permit the investigation of how introns and exons are disposed in these genetic matrices; what kinds of matrix mosaics appear for them; and how these mosaics are related to components of multi-dimensional numeric systems described in Section 3. It can lead to reveal new appropriate regularities.

We also note that revealing the connection between the genetic matrices $[C\ A; G\ U]^{(n)}$ and the matrices of diadic shifts $P^{(n)}_{DIAD}$ (Figure 5) leads one to utilize the notions and formalisms of "diadic spaces", "diadic metrics", etc. (Harmut, 1989), which are known in the field of computer informatics, in new fields of matrix genetics and bioinformatics.

The conception by Stent (1969) and Jacob (1974) about possible relation between the genetic code and the symbolic system of the ancient Chinese "I Ching" obtains new materials for further examinations. Additional discussions along this direction will be described in the next chapters of the book.

Investigations of ensembles of elements of the genetic code with their symmetrical features have led to the construction of the Kronecker family of the genetic matrices. This matrix family presents all sets of genetic multiplets in the well-ordered general form, where each multiplet obtains its own individual number in binary notation on the basis of molecular characteristics of the genetic letters A, C, G,

U/T. Such a general form is connected with important principles and methods of computer informatics and of the noise-immunity in digital technique. It gives us new mathematical ability to study genetic systems and their connections with computer informatics and algebraic theory of coding. For example, first evidences were obtained that the logics of structures of the genetic code are related to the logical modulo-2 addition.

REFERENCES

Ahmed, N., & Rao, K. (1975). *Orthogonal transforms for digital signal processing*. New York: Springer-Verlag Inc.

Annett, M. (1985). *Left, right, hand, and brain: The right shift theory*. New Jersey: Lawrence Erlbaum.

Annett, M. (1992). Spatial ability in subgroupt of left- and right-handers. *The British Journal of Psychology, 83*, 493–962.

Arques, D., & Michel, C. (1996). A complementary circular code in the protein coding genes. *Journal of Theoretical Biology, 182*, 45–56. doi:10.1006/jtbi.1996.0142

Arques, D., & Michel, C. (1997). A circular code in the protein coding genes of mitochondria. *Journal of Theoretical Biology, 189*, 45–58. doi:10.1006/jtbi.1997.0513

Arques, D., & Michel, C. J. (1994). Analytic expression of the purine/pyrimidine autocorrelation function after and before random mutations. *Mathematical Biosciences, 123*, 103–125. doi:10.1016/0025-5564(94)90020-5

Bakhtiarov, K. I. (2001). Logical structure of genetic code. *Symmetry: Culture and Science, 12*(3-2), 401-406.

Bashfold, Y. D., Tsohantjis, I., & Yarvis, P. D. (1997). Codon and nucleotide assignments in a supersymmetric model of the genetic code. *Physics Letters. [Part A], 233*, 481. doi:10.1016/S0375-9601(97)00475-1

Bellman, R. (1960). *Introduction to matrix analysis*. New York, Toronto, London: McGraw-Hill.

Bernal, I., Hamilton, W. C., & Ricci, J. S. (1972). *Symmetry: A stereoscopic guide for chemists*. San Francisco: Freeman.

Birss, R. R. (1964). *Symmetry and magnetism*. Amsterdam: Norht Holland Publ. Co.

Blahut, R. E. (1985). *Fast algorithms for digital signal processing*. Massachusetts: Addison-Wesley Publishing Company.

Cantor, C. R., & Schimmel, P. R. (1980). *Biophysical chemistry*. San Francisco: W. H. Freeman and Company.

Capra, F. (2000). *The Tao of physics: An exploration of the parallels between modern physics and eastern mysticism*. New Jersey: Shambhala Publications, Inc, Chapevillle, F., & Haenni, A.-L. (1974). *Biosynthese des proteins*. Paris: Hermann Collection.

Chernavskiy, D. S. (2000). The problem of origin of life and thought from the viewpoint of the modern physics. [in Russian]. *Uspehi Phizicheskih Nauk, 170*(2), 157–183. doi:10.3367/UFNr.0170.200002c.0157

Darvas, G. (2007). *Symmetry*. Basel: Birkhäuser Book.

Dawkins, R. (1991). *The blind watchmaker*. New York: Longman Scientific & Technical.

Dragovich, B., & Dragovich, A. (2007). *P-adic modelling of the genome and the genetic code*. Retrieved on July 7, 2007, from http://arXiv:0707.3043

Eingorin, M. Y. (2001). *Bases of coding and management in molecular biology*. Nizhniy Novgorod: State Medical Academy (in Russian).

Eingorin, M. Y. (2003). Grammar of coding of genetic texts. In I. Ivanov (Ed.), *Sixth ISTC Scientific Advisory Committee Seminar "Science and Computing," Science and Computing* (Vol. 1, pp. 173-178). Moscow: ISTC&RAS.

Eingorin, M. Y. (2006). About grammar of coding of hidden levels of genes. *Vestnik Nizhegorodskogo Universiteta Imeni N. I. Lobachevskogo, Series* " [in Russian]. *Mathematical Modeling and Optimal Management, 1*(30), 128–141.

Engstrom, A., & Strandberg, B. (Eds.). (1968). *Nobel symposium, "aymmetry and function of biological systems at the macromolecular level."* Oslo: Plenum Press.

Fickett, J., & Burks, C. (1989). Development of a database for nucleotide sequences. In M. S. Waterman (Ed.), *Mathematical methods in DNA|sequences* (pp.1-34). Florida: CRC Press.

Forger, M., Hornos, Y. M. M., & Hornos, Y. E. M. (1997). Global aspects in the algebraic approach to the genetic code. *Physical Review E: Statistical Physics, Plasmas, Fluids, and Related Interdisciplinary Topics, 56*, 7078–7082. doi:10.1103/PhysRevE.56.7078

Forger, M., Hornos, Y. M. M., & Hornos, Y. E. M. (1999). Symmetry and symmetry breaking: An algebraic approach to the genetic code. *Intern. Journal Mod. Phys, 1313*, 2795–2885.

Frank-Kamenetskiy, M. D. (1988). *The most principal molecule*. Moscow: Nauka (in Russian).

Frappat, L., Sciarrino, A., & Sorba, P. (1998). A crystal base for the genetic code. *Physics Letters. [Part A], 250*, 214. doi:10.1016/S0375-9601(98)00761-0

Frey, G., & Michel, C. (2003). Circular codes in archaeal genomes. *Journal of Theoretical Biology, 223*, 413–431. doi:10.1016/S0022-5193(03)00119-X

Frey, G., & Michel, C. J. (2006). Identification of circular codes in bacterial genomes and their use in a factorization method for retrieving the reading frames of genes. *Computational Biology and Chemistry, 30*, 87–101. doi:10.1016/j.compbiolchem.2005.11.001

Fujita, S. (1991). *Symmetry and combinatorial enumeration in chemistry*. Berlin: Springer.

Gamov, G. (1954). Possible relation between deoxyribonucleic acid and protein structures. *Nature, 173*, 318. doi:10.1038/173318a0

Gamov, G., & Metropolis, N. (1954). Numerology of polypeptide chains. *Science, 120*, 779–780.

Gardner, M. (1991). *The new ambidextrous universe: Symmetry and asymmetry, from mirror reflections to superstrings*. USA: W.H. Freeman & Company.

Gazale, M. J. (1999). *Gnomon. From pharaons to fractals*. New Jersey: Princeton University Press.

Gazzaniga, M. S. (1995). Principles of human brain organization derived from split brain studies. *Neuron, 14*, 217–228. doi:10.1016/0896-6273(95)90280-5

Geadah, Y. A., & Corinthios, M. J. (1977). Natural, dyadic, and sequency order algorithms and processors for the Walsh-Hadamard transform. *IEEE Transactions on Computers, C-26*, 435–442. doi:10.1109/TC.1977.1674860

Gell-Mann, M., & Ne'eman, Y. (2000). *The eightfold way*. New York: Westview Press.

Ginzburg, I. F., Mihailov, M. Y., & Pokrovskiy, V. L. (2001). Yuriy borisovich rumer. [in Russian]. *Uspehi Physicheskih Nauk, 171*(10), 1131–1136. doi:10.3367/UFNr.0171.200110n.1131

Hahn, W. (1989). *Symmetrie als entwicklungsprinzip in natur und kunst*. Königstein: Langewiesche.

Hahn, W. (1998). *Symmetry as a developmental principle in nature and art*. Singapore: World Scientific.

Hamming, R. W. (1980). *Coding and information theory*. New Jersey: Princeton-Hall, Inc.

Hann, W. (1998). *Symmetry, as a developmental principle in nature and art*. Singapore: World Scientific.

Hargittai, I. (Ed.). (1986). *Symmetry: Unifying human understanding*. New York: Pergamon Press.

Hargittai, I. (Ed.). (1989). *Symmetry: Unifying human understanding 2*. Oxford: Pergamon Press.

Hargittai, I. (2001). Double symmetry of the double helix. *Symmetry: Culture and Science, 12*(3-4), 247–254.

Hargittai, I., & Hargittai, M. (1994). *Symmetry: A unifying concept*. Bolinas, CA: Shelter Publications.

Harmut, H. F. (1989). *Information theory applied to space-time physics*. Washington: The Catholic University of America, D.C.

He, M. (2001). On double helical sequences and doubly stochastic matrices. *Symmetry: Culture and Science, 12*(3-4), 307–330.

He, M., Narasimhan, G., & Petoukhov, S. (Eds.). (2005). *Advances in bioinformatics and its applications, series in mathematical biology and medicine, 8*. New Jersey, London, Singapore: World Scientific.

He, M., Petoukhov, S., & Ricci, P. (2004). Genetic code, hamming distance and stochastic matrices. *Bulletin of Mathematical Biology, 66*, 1405–1421. doi:10.1016/j.bulm.2004.01.002

He, M., & Petoukhov, S. V. (2007). Harmony of living nature, symmetries of genetic systems, and matrix genetics. *International journal of integrative biology, 1*(1), 41-43.

Hellige, J. B. (1993). *Hemispheric asymmetry: What's right and what's left.* Cambridge, M: Harvard University Press.

Hornos, J. E., Braggion, L., & Magini, M. (2001). Symmetry and the genetic code: The symplectic model. *Symmetry: Culture and Science, 12*(3-4), 349–370.

Jacob, F. (1974). Le modele linguistique en biologie. *Critique, Mars, 30*(322), 197-205.

Jacob, F. (1977). The linguistic model in biology. In D. Armstrong & C. H. van Schooneveld (Ed.). *Roman Jakobson: Echoes of his scholarship* (pp. 185-192). Lisse: Peter de Ridder.

Jean, R. V. (1994). *Phyllotaxis: A systematic study in plant morphogenesis.* Cambridge: Cambridge University Press.

Jean, R. V., & Barabe, D. (Eds.). (2001). *Symmetry in plants.* New Jersey: World Scientific.

Jimenes-Montano, M. A. (2005). Applications of hyper genetic code to bioinformatics. In M. He, G. Narasimhan, & S. Petoukhov (Ed.), *Advances in Bioinformatics and its Applications, Series in Mathematical Biology and Medicine*, 8 (pp. 473-481). New Jersey, London, Singapore: World Scientific.

Kappraff, J. (2002). *Beyond measure: Essays in nature, myth, and number.* Singapore: World Scientific.

Karasev, V. A. (2003). *The genetic code: New horizons.* Sankt-Petersburg: Tessa (in Russian).

Karasev, V. A., Luchinin, V. V., & Stefanov, V. E. (2005). A dodecahedron-based model of spatial representation of the canonical set of amino acids. In M. He, G. Narasimhan & S. Petoukhov (Ed.), *Advances in bioinformatics and its applications, aeries in mathematical biology and medicine, 8* (pp. 482-493). New Jersey, London, Singapore: World Scientific.

Kargupta, H. (2001). A striking property of genetic code-like transformations. *Complex systems, 11*, 43-50.

Khrennikov, A. Y., & Kozyrev, S. V. (2007). *Genetic code on the dyadic plane.* Archiv. Retrieved in January 2007, from http://arXiv:q-bio/0701007

Konopelchenko, B. G., & Rumer, Y. B. (1975a). Classification of the codons of the genetic code, I & II. *Preprints 75-11 and 75-12 of the Institute of Nuclear Physics of the Siberian department of the USSR Academy of Sciences.* Novosibirsk: Institute of Nuclear Physics.

Konopelchenko, B. G., & Rumer, Y. B. (1975b). On the correlation between the structure of amino acids and the degeneracy of the genetic code. *Preprint 75-25 of the Institute of Nuclear Physics of the Siberian Department of the USSR Academy of Sciences.* Novosibirsk: Institute of Nuclear Physics.

Konopelchenko, B. G., & Rumer, Y. B. (1975c). The wobble hypothesis and the sequence of nucleotides. *Preprint 75-26 of the Institute of Nuclear Physics of the Siberian Department of the USSR Academy of Sciences.* Novosibirsk: Institute of Nuclear Physics.

Konopelchenko, B. G., & Rumer, Y. B. (1975d). Classification of the codons in the genetic code. [in Russian]. *Doklady Akademii Nauk SSSR, 223*(2), 145–153.

Landau, L. D., & Rumer, Y. B. (2003). *What is relativity?* London: Dover Publications.

Laubenbacher, R., & Sturmfels, B. (2008). *Computer algebra in systems biology* (pp. 1-12).

Laurent, C. (Ed.). *Symmetry2000* (pp. 265-278). Oxford: Pergamon Press.

Lee, M. H., & Kaveh, M. (1986). Fast Hadamard transform based on a simple matrix factorization. *IEEE Transactions on Acoustics, Speech, and Signal Processing, ASSSP-34*(6), 1666–1667. doi:10.1109/TASSP.1986.1164972

Leyton, M. (1992). *Symmetry, causality, mind.* Cambridge, MA: MIT Press.

Lochak, G. (1994). *La geometrisation de la physique*. France: Flammarion.

Loeb, A. L. (1971). *Color and symmetry*. New York: Wiley.

Loeb, A. L. (1993). *Concepts & images: Visual mathematics*. Boston: Birkhäuser.

MacDonaill, D. A. (2003). Why nature chose A, C, G, and U/T: An error-coding perspective of nucleotide alphabet composition. *Origins of Life and Evolution of the Biosphere, 33*, 433–455. doi:10.1023/A:1025715209867

MacDonaill, D. A. (2005). Molecular mappings: Group theory, coding theory, and the emergence of replication. In M. He, G. Narasimhan & S. Petoukhov (Ed.), *Advances in bioinformatics and its applications, series in mathematical biology and medicine, 8* (pp. 494-501). New Jersey, London, Singapore: World Scientific.

Mainzer, K. (1988). *Symmetrien der natur: Ein handbuch zur natur- und wissenschaftsphilosophie.* Berlin: de Gruyter.

Makovskiy, M. M. (1992). *Linquistic genetics*. Moscow: Nauka (in Russian).

Mandelbrot, B. B. (1983). *The fractal geometry of nature.* New York: Freeman.

Marcus, S. (1990). *Algebraic linguistics; analytical models*. Bucharest: Academic.

Marcus, S. (2001). Internal and external symmetry in genetic information. *Symmetry: Culture and Science, 12*(3-2), 395-400.

Marcus, S. (2006). At the roots of the symmetry phenomena. *Symmetry: Culture and Science, 17*(1-2), 89–90.

Marcus, S. (2007). *Words and languages everywhere.* Bucharest: Polimetrica.

Miller, W., Jr. (1972). *Symmetry groups and their applications*. New York: Academic Press.

Moller, A. P., & Swaddle, J. P. (1997). *Asymmetry, developmental stability, and evolution.* Oxford: Oxford University Press.

Ne'eman, Y. (1999). Symmetry as the leitmotif at the fundamental level in the twentieth century physics. *Symmetry: Culture and Science, 10*, 143–162.

Ne'eman, Y. (2002). Pythagoreanism in atomic, nuclear, and particle physics. In I. Hargittai & T.

Ne'eman, Y., & Kirsh, Y. (1986). *The particle hunters*. Cambridge: Cambridge University Press.

Negadi, T. (2001). Symmetry and proportion in the genetic code, and genetic information from the basic units of life. *Symmetry: Culture and Science, 12*(3-4), 371–393.

Negadi, T. (2005). Symmetry and information in the genetic code. In M. He, G. Narasimhan & S. Petoukhov (Ed.), *Advances in bioinformatics and its applications, series in mathematical biology and medicine, 8* (pp. 502-511). New Jersey, London, Singapore: World Scientific.

Negadi, T. (2006). The genetic code structure, from inside. *Symmetry: Culture and Science, 17*(1-2), 317–340.

Nielsen, M. A., & Chuang, I. L. (2001). *Quantum computation and quantum information*. Cambridge: Cambridge University Press.

Paun, G., Rozenberg, G., & Salomaa, A. (2006). *DNA computing: New computing paradigms*. New Jersey: Prentice Hall.

Penrose, R. (1989). *The emperor's new mind*. Oxford: Oxford University Press.

Penrose, R. (2004). *The road to reality. A complete guide to the laws of the universe*. London: Jonathan Cape.

Peterson, W. W., & Weldon, E. J. (1972). *Error-correcting codes*. Cambridge: MIT Press.

Petoukhov, S. V. (1981). *Biomechanics, bionics, and symmetry*. Moscow: Nauka.

Petoukhov, S. V. (1989). Non-Euclidean geometries and algorithms of living bodies. In I. Hargittai (Ed.), *Computers & Mathematics with Applications, 17*(4-6), 505-534. Oxford: Pergamon Press.

Petoukhov, S. V. (1990). Symmetric-algorithmic properties of regular biostructures. *Symmetry: Culture and Science, 1*(3), 295–312.

Petoukhov, S. V. (1999a). *Biosolitons. The basis of solitonic biology*. Moscow: GPKM (in Russian).

Petoukhov, S. V. (1999b). Genetic code and the ancient Chinese "book of changes". *Symmetry: Culture and Science, 10*, 211–226.

Petoukhov, S. V. (2001). *The bi-periodic table of genetic code and the number of protons*. Moscow: MKC (in Russian).

Petoukhov, S. V. (2001). Genetic codes 1: Binary sub-alphabets, bi-symmetric matrices, and golden section. Genetic codes 2: numeric rules of degeneracy and the chronocyclic theory. *Symmetry: Culture and Science, 12*(1), 255–306.

Petoukhov, S. V. (2003-2004). Attributive conception of genetic code, its bi-periodic tables, and problem of unification bases of biological languages. *Symmetry: Culture and Science, 14-15*(part 1), 281–307.

Petoukhov, S. V. (2003). The biperiodic table and attributive conception of genetic code. A problem of unification bases of biological languages. In F. Valafar & H. Valafar (Ed.), *Proceedings of "The 2003 International Conference on Mathematics and Engineering Techniques in Medicine and Biological Sciences," session "Bioinformatics 2003."* Las Vegas, NV: CSREA Press.

Petoukhov, S. V. (2005a). The rules of degeneracy and segregations in genetic codes. The chronocyclic conception and parallels with Mendel's laws. In M. He, G. Narasimhan & S. Petoukhov (Ed.), *Advances in Bioinformatics and its Applications, Series in Mathematical Biology and Medicine,* 8, (pp. 512-532). Singapore: World Scientific.

Petoukhov, S. V. (2006). Bioinformatics: Matrix genetics, algebras of the genetic code, and biological harmony. *Symmetry: Culture and Science, 17*(1-4), 251–290.

Petoukhov, S. V. (2008a). *Matrix genetics, algebras of the genetic code, noise-immunity*. Moscow: RCD (in Russian).

Petoukhov, S. V. (2008b). The degeneracy of the genetic code and Hadamard matrices. 1-8.

Ratner, V. A. (2002). *Genetics, molecular cybernetics*. Novosibirsk: Nauka.

Retrieved on December 27, 2007, from http://arXiv:0712.4248

Retrieved on February 8, 2008, from http://arXiv:0802.3366

Roller, A. (1974). *Discovering the basis of life. An introduction to molecular biology*. New York: McGraw-Hill Book Company.

Rosen, J. (1983). *Symmetry primer for scientists*. New York: Wiley.

Rosen, J. (1995). *Symmetry in science: An introduction to the general theory*. New York: Springer.

Rumer, Y. B. (1936). *Spinor analysis*. Moscow: Nauka.

Rumer, Y. B. (1956). *Investigations on 5-optics*. Moscow: GITTL.

Rumer, Y. B. (1966). About systematization of the codons in the genetic code. [in Russian]. *Doklady Akademii Nauk SSSR, 167*(6), 1393–1394.

Rumer, Y. B. (1968). Systematization of the codons of the genetic code. [in Russian]. *Doklady Akademii Nauk SSSR, 183*(1), 225–226.

Rumer, Y. B., & Fet, A. I. (1970). *The theory of unitary symmetry*. Moscow: Nauka.

Rumer, Y. B., & Fet, A. I. (1977). *The group theory and quantified fields*. Moscow: Nauka.

Rumer, Y. B., & Ryvkin, M. S. (1972). *Thermodynamics, statistical physics, and kinetics*. Moscow: Nauka.

Sanchez, R., & Grau, R. (2008). An algebraic hypothesis about the primeval genetic code. 1-25. Retrieved on May 8, 2008, from http://arXiv:0805.1128

Schrodinger, E. (1955). *What is life? The physical aspect of the living cell.* Cambridge: University Press.

Schutskiy, Y. K. (1997). *The Chinese classical "book of changes."* Moscow: Vostochnaya literatura (in Russian).

Seeman, N. (2004). Nanotechnology and the double helix. *Scientific American, 290*(6), 64–75.

Shapiro, E., & Benenson, Y. (2007). Bringing DNA computers to life. *Scientific American, 17*(3), 40–47.

Shcherbak, V. I. (1988). The cooperative symmetry of the genetic code. *Journal of Theoretical Biology, 132*, 121–124. doi:10.1016/S0022-5193(88)80196-6

Shechtman, D., Blech, I., Gratias, D., & Cahn, J. (1984). Metallic phase with long-range orientational order and no translational symmetry. *Physical Review Letters, 53*, 1951–1954. doi:10.1103/PhysRevLett.53.1951

Shubnikov, A. V., & Koptsik, V. A. (1974). *Symmetry in science and art.* New York: Plenum Press.

Shultz, G. E., & Schirmer, R. H. (1979). *Principles of protein structure.* Berlin: Springer-Verlag.

Sklar, B. (2001). *Digital communication. Fundamentals and applications.* New York: Prentice Hall.

Stambuk, N. (1999). Circular coding properties of gene and protein sequences. *Croatica Chemica Acta, 72*(4), 999–1008.

Stambuk, N., Konyevoda, P., & Gotovac, N. (2005). Symbolic coding of amino acid and nucleotide properties. In M. He, G. Narasimhan & S. Petoukhov (Ed.), *Advances in Bioinformatics and its Applications, Series in Mathematical Biology and Medicine, 8* (pp. 512-532). New Jersey, London, Singapore: World Scientific.

Stent, G. S. (1969). *The coming of the golden age.* New York: The Natural History Press.

Stent, G. S. (1971). *Molecular genetics.* San Francisco: W. H. Freeman and Company.

Stewart, I., & Golubitsky, M. (1992). *Fearful symmetry: Is God a geometer?* Oxford: Blackwell.

Szabo, R., & He, M. (2006). Statistical analyses of patterns in tripetides. *Symmetry: Culture and Science, 17*(1-2), 293–316.

Szabo, R., He, M., Burnham, E., & Jurani, J. (2005). Analyzing patterns of tripeptids using statistical approach and neural network paradigms. In M. He, G. Narasimhan & S. Petoukhov (Ed.), *Advances in Bioinformatics and its Applications, Series in Mathematical Biology and Medicine, 8* (pp. 544-553). New Jersey, London, Singapore: World Scientific.

Thompson, d'Arcy W. (1942). *On growth and form.* Cambridge: Cambridge University Press.

Trahtman, A. M. (1972). *Introduction in generalized spectral theory of signals.* Moscow: Sovetskoie Radio (in Russian).

Trahtman, A. M., & Trahtman, V. A. (1975). *The foundations of the theory of discrete signals on finite intervals.* Moscow: Sovetskoie Radio (in Russian).

Vavilov, N. I. (1922). The law of homologous series in variation. *Journal of Genetics, 12*(1), 47–89. doi:10.1007/BF02983073

Vernadsky, V. I. (1965). *Chemical structure of the earth and its surrounding.* Moscow: Nauka.

Volkenstein, M. V., & Rumer, Y. B. (1966). On the regularities in codons. *Preprint T-20 of the Institute of Mathematics of the Siberian department of the USSR Academy of Sciences*. Novosibirsk: Institute of Mathematics.

Waterman, M. S. (Ed.). (1999). *Mathematical methods for DNA sequences*. Florida: CRC Press, Inc.

Watson, J. D. (1968). *The double helix; a personal account of the discovery of the structure of DNA*. New York: Atheneum.

Watson, J. D., & Crick, F. H. C. (1953). Molecular structure of nucleic acids: A structure of deoxyribose nucleic acid. *Nature*, *171*, 737–738. doi:10.1038/171737a0

Weyl, H. (1931). *The theory of groups and quantum mechanics*. New York: Dover.

Weyl, H. (1946). *The classical groups, their invariants, and representations*. New Jersey: Princeton University Press.

Weyl, H. (1952). *Symmetry*. New Jersey: Princeton University Press.

Wigner, E. P. (1965). Violations of symmetry in physics. *Scientific American*, *213*(6), 28–36.

Wigner, E. P. (1967). *Symmetries and reflections, scientific essays of Eugene P. Wigner*. Bloomington: Indiana University Press.

Wigner, E. W. (1970). *Symmetries and reflections*. Bloomington-London: Indiana University Press.

Wittmann, H. G. (1961). Ansatze zur entschlusselung des genetishen codes. *Naturwissenschaften*, *48*(24), 55. doi:10.1007/BF00590622

Yang, C. M. (2001). Chemistry and the 28-gon polyhedral symmetry of the genetic code. *Symmetry .Cultura e Scuola*, *12*(3-4), 331–347.

Yang, C. M. (2005). Molecular vs. atomic information logic behind the genetic coding contents constrained by two evolutionary axes and the Fibonacci-Lucas sequence. In M. He, G. Narasimhan, & S. Petoukhov (Ed.), *Advances in Bioinformatics and its Applications, Series in Mathematical Biology and Medicine, 8* (pp. 554-564). New Jersey, London, Singapore: World Scientific.

Yarlagadda, R., & Hershey, J. (1997). *Hadamard matrix analysis and synthesis with applications to communications and signal/image processing*. New York: Kluwer Academic Publ.

Ycas, M. (1969). *The biological code*. Amsterdam, London: North-Holland Publishing Company.

Chapter 2
Symmetries of the Degeneracy of the Vertebrate Mitochondrial Genetic Code in the Matrix Form

ABSTRACT

Symmetries of the degeneracy of the vertebrate mitochondrial genetic code in the mosaic matrix form of its presentation are described in this chapter. The initial black-and-white genomatrix of this code is reformed into a new mosaic matrix when internal positions in all triplets are permuted simultaneously. It is revealed unexpectedly that for all six variants of positional permutations in triplets (1-2-3, 2-3-1, 3-1-2, 1-3-2, 2-1-3, 3-2-1) the appropriate genetic matrices possess symmetrical mosaics of the code degeneracy. Moreover the six appropriate mosaic matrices in their binary presentation have the general non-trivial property of their "tetra-reproduction," which can be utilized in particular for mathematical modeling of the phenomenon of the tetra-division of gametal cells in meiosis. Mutual interchanges of the genetic letters A, C, G, U in the genomatrices lead to new mosaic genomatrices, which possess similar symmetrical and tetra-reproduction properties as well.

INTRODUCTION AND BACKGROUND

Chapter 1 described the construction of genomatrices of the Kronecker family, including the genomatrix $P^{CAUG}_{123} = [C\ A; U\ G]^{(3)}$, which contain 64 triplets in the well-ordered form. But how are amino acids and stop-codons, which are encoded by these triplets, disposed in this genomatrix? Does the genetic code possess any features which may give the symmetrical character for this genomatrix? Such questions are investigated in this chapter. Really, the degeneracy of the genetic code has lead to a symmetrical black-and-white mosaic of the genomatrix in the case of the vertebrate mitochondrial genetic code, which is the most symmetrical dialect of the genetic code.

DOI: 10.4018/978-1-60566-124-7.ch002

By analogy of the theory of digital signals, where permutations of signal elements play significant role, we study two kinds of permutations of elements of the genetic code, which transform initial mosaic genomatrices into new mosaic genomatrices. The first of these kinds of permutations is permutations of three positions inside all triplets: 1-2-3, 2-3-1, 3-1-2, 1-3-2, 2-1-3, 3-2-1. The second kind is mutual interchanges of the genetic letters A, C, G, U. Both of these kinds lead unexpectedly to such new genomatrices, which possess symmetrical black-and-white mosaics and the binary forms of which possess the mathematical property of tetra-self-reproducing. This chapter sets out results of these investigations.

The main objectives of this chapter are the following:

1. In-depth study of matrix symmetries of the degeneracy of the vertebrate mitochondrial genetic code in the matrix form of its presentation;
2. Investigations of reforming these matrix symmetries under some kinds of permutations of elements of the genetic code;
3. Demonstrating new phenomenological materials in the field of matrix genetics to develop algebraic models of the genetic code.

PECULIARITIES OF DEGENERACY OF THE GENETIC CODE

Modern science knows many dialects of the genetic code, data about which are shown on the NCBI's website http://www.ncbi.nlm.nih.gov/Taxonomy/Utils/wprintgc.cgi. According to general traditions, theory of symmetry studies initially those natural objects which possess the most symmetrical character, and then it constructs a theory for cases of violations of this symmetry in other kindred objects. Correspondingly the authors of this book investigate initially the vertebrate mitochondrial genetic code which is the most symmetrical code among dialects of the genetic code. One can also note that some authors consider this dialect not only as the most "perfect" but also as the most ancient dialect (Frank-Kamenetskiy, 1988) while the last aspect is a debatable one. The vertebrate mitochondrial code is used as a basic dialect in some other mathematical works where a presentation of the 64 triplets exists in a form of square tables (Dragovich & Dragovich, 2006, 2007; Khrennikov & Kozyrev, 2007). Figure 1 shows the correspondence between the set of 64 triplets and the set of 20 amino acids with stop-signals (Stop) of protein synthesis in this code.

The set of 64 triplets contains such 16 subfamilies of triplets, every one of which contains 4 triplets with the same two letters on the first positions of each triplet (an example of such subsets is the case of the four triplets CAC, CAA, CAU, CAG with the same two letters CA on their first positions). We shall name such subfamilies as the subfamilies of *NN*-triplets. In the case of the vertebrate mitochondrial code, the set of these 16 subfamilies of *NN*-triplets is divided into two equal subsets from the viewpoint of degeneration properties of the code (Figure 1). The first subset contains 8 subfamilies of so called "two-position" *NN*-triplets, a coding value of which is independent of a letter on their third position. An example of such subfamilies is the four triplets CGC, CGA, CGU, CGC, all of which encode the same amino acid Arg, though they have different letters on their third position. All members of such subfamilies of *NN*-triplets are marked by black color in the genomatrix [C A; U G]$^{(3)}$ on the Figure 2.

The second subset contains 8 subfamilies of "three-position" *NN*-triplets, a coding value of which depends on a letter on their third position. An example of such subfamilies is the four triplets CAC, CAA, CAU, CAC, two of which (CAC, CAU) encode the amino acid His and other two (CAA, CAG) encode

Figure 1. The case of the vertebrate mitochondrial genetic code; The initial data were taken from the NCBI's web-site http://www.ncbi.nlm.nih.gov/Taxonomy/Utils/wprintgc.cgi

8 subfamilies of the "two-position" NN-triplets and the amino acids, which are encoded by them	**8 subfamilies of the "three-position" NN-triplets and the amino acids, which are encoded by them**
CCC, CCA, CCU, CCG ➔ Pro	CAC, CAA, CAU, CAG ➔ His, Gln
CUC, CUA, CUU, CCG ➔ Leu	AAC, AAA, AAU, AAG ➔ Asn, Lys
CGC, CGA, CGU, CGG ➔ Arg	AUC, AUA, AUU, AUG ➔ Ile, Met
ACC, ACA, ACU, ACG ➔ Thr	AGC, AGA, AGU, AGG ➔ Ser, Stop
UCC, UCA, UCU, UCG ➔ Ser	UAC, UAA, UAU, UAG ➔ Tyr, Stop
GCC, GCA, GCU, GCG ➔ Ala	UUC, UUA, UUU, UUG ➔ Phe, Leu
GUC, GUA, GUU, GUG ➔ Val	UGC, UGA, UGU, UGG ➔ Cys, Trp
GGC, GGA, GGU, GGG ➔ Gly	GAC, GAA, GAU, GAG ➔ Asp, Glu

another amino acid Gln. All members of such subfamilies of *NN*-triplets are marked by white color in the genomatrix $P^{(3)}$ = [C A; U G]$^{(3)}$ on the Figure 2. So the genomatrix [C A; U G]$^{(3)}$ has 32 black triplets and 32 white triplets. Each subfamily of four *NN*-triplet is disposed in an appropriate (2x2)-subquadrant of the genomatrix [C A; U G]$^{(3)}$ due to the Kronecker algorithm of construction of genomatrix [C A; U G]$^{(3)}$ of triplets from the alphabet genomatrix *P* (Figure 3 of the previous chapter).

Here we recall the work by Rumer (1968) that a combination of letters on the first two positions of each triplet was named as a "root" of this triplet. A set of 64 triplets contains 16 possible variants of such roots. Taking into account of the properties of triplets, Rumer has divided the set of 16 possible roots into two subsets with eight roots in each. Roots CC, CU, CG, AC, UC, GC, GU, and GG form the first of such octets. They were called by Rumer as "strong roots". Other eight roots CA, AA, AU, AG, UA, UU, UG, and GA form the second octet and they were called as weak roots. When Rumer published his works, the vertebrate mitochondrial genetic code was unknown. But one can easily check that the set of 32 black (white) triplets, which we described for the case of the vertebrate mitochondrial genetic code (Figures 1 and 2), is identical to the set of 32 triplets with strong (weak) roots described by Rumer. So, using notions proposed by Rumer, the black triplets can be named as triplets with the strong roots and the white triplets can be named as triplets with the weak roots. Rumer believed that this symmetrical

Figure 2. The representation of the genomatrix $P^{(3)} = P^{CAUG}_{123} = [C\ A;\ U\ G]^{(3)}$ (Figure 3 in Chapter 1) for the case of the vertebrate mitochondrial genetic code. The matrix contains 64 triplets and 20 amino acids with their traditional abbreviations. Stop-codons are marked as "Stop". Numeration of columns and rows in decimal notation is shown

	0 (000)	1 (001)	2 (010)	3 (011)	4 (100)	5 (101)	6 (110)	7 (111)
0	CCC Pro 0	CCA Pro 1	CAC His 2	CAA Gln 3	ACC Thr 4	ACA Thr 5	AAC Asn 6	AAA Lys 7
1	CCU Pro 8	CCG Pro 9	CAU His 10	CAG Gln 11	ACU Thr 12	ACG Thr 13	AAU Asn 14	AAG Lys 15
2	CUC Leu 16	CUA Leu 17	CGC Arg 18	CGA Arg 19	AUC Ile 20	AUA Met 21	AGC Ser 22	AGA Stop 23
3	CUU Leu 24	CUG Leu 25	CGU Arg 26	CGG Arg 27	AUU Ile 28	AUG Met 29	AGU Ser 30	AGG Stop 31
4	UCC Ser 32	UCA Ser 33	UAC Tyr 34	UAA Stop 35	GCC Ala 36	GCA Ala 37	GAC Asp 38	GAA Glu 39
5	UCU Ser 40	UCG Ser 41	UAU Tyr 42	UAG Stop 43	GCU Ala 44	GCG Ala 45	GAU Asp 46	GAG Glu 47
6	UUC Phe 48	UUA Leu 49	UGC Cys 50	UGA Trp 51	GUC Val 52	GUA Val 53	GGC Gly 54	GGA Gly 55
7	UUU Phe 56	UUG Leu 57	UGU Cys 58	UGG Trp 59	GUU Val 60	GUG Val 61	GGU Gly 62	GGG Gly 63

division into two binary-oppositional categories of roots is very important for understanding the nature of the genetic code systems.

Let us introduce the symbol of the genomatrix $P^{(3)} = [C\ A;\ U\ G]^{(3)}$ by the symbol P^{CAUG}_{123}, which is more comfortable for a comparative analyses of this (8x8)-genomatrix with other (8x8)-genomatrices below. Here the bottom index "123" shows the appropriate queue of positions 1-2-3 in triplets; the upper index shows the kind of the kernel [C A; U G] of the Kronecker family of genomatrices. The exponent (3) is not written because the bottom index is enough for understanding that this symbol means the (8x8)-genomatrix of triplets. This change of the symbol is useful because we shall consider later the genomatrices with permutations of positions in triplets (2-3-1, 3-1-2, etc.) and with another kernels of Kronecker families of genomatrices ([G C; A U], [C A; G U], etc.).

Below we will demonstrate the phenomenological fact of a symmetric character of dispositions of the 32 white triplets and the 32 black triplets in the genomatrix $[C\ A;\ U\ G]^{(3)}$. We will also analyze the genetic matrices, which are produced from the genomatrix $[C\ A;\ U\ G]^{(3)}$ as a result of positional and alphabetic permutations in all triplets. One should note here that permutations of elements play an important role in the theory of digital signals processing (Ahmed & Rao, 1975; Blahut, 1985; Trahtman, 1972; Trahtman & Trahtman, 1975). It was the reason of the special interest to investigate the genomatrices with such permutations. On these way analogies between the famous fact of the tetra-division of gametal cells and some properties of these genomatrices with permutations are revealed.

SYMMETRICAL PROPERTIES OF GENETIC MATRICES OF TRIPLETS UNDER PERMUTATIONS OF POSITIONS INSIDE TRIPLETS

The specifics of the degeneracy of the genetic code provoke many questions. One of them is the following: was the code degeneracy an accidental choice of nature or not? Deep investigations of symmetries in a matrix map of the code degeneracy can give many useful materials for such questions.

We use the genomatrices from the Kronecker family, which was described in Chapter 1, to present the degeneracy of the genetic code in a special cognitive form. We investigate possibilities of this matrix form of presentation step by step to obtain evidences of its adequacy and usefulness. This form gives new viewpoints for the question about specifics of the code degeneracy. It gives us new results, which are much unexpected sometimes, about relations of matrix symmetries in sets of elements of the genetic code. One of the ways to study these symmetries is connected with permutations of three positions inside all triplets, which leads to new genomatrices with symmetrical peculiarities. The second way, which leads to new symmetrical genomatrices as well, is connected with a mutual replacing of the genetic letters A, C, G, U in the kernel of the Kronecker family of the genomatrices. Both of these ways and the appropriate results are described in this chapter.

The genomatrix $[C\ A;\ U\ G]^{(3)} = P^{CAUG}{}_{123}$ of Figure 2 shows all triplets together with amino acids and stop-codon, which are encoded by the triplets in the case of the vertebrate mitochondrial genetic code (compare with Figure 1). Black cells of the genomatrix contain the triplets, which belong to the set of the two-position *NN*-triplets, and white cells contain the triplets, which belong to the set of the three-position *NN*-triplets.

So, the black-and-white mosaic of the genomatrix $P^{CAUG}{}_{123}$ on Figure 2 reflects the specificity of the degeneracy of this basic dialect of the genetic code. Unexpectedly it has a few interesting symmetrical peculiarities as follows.

The left and right halves of the matrix mosaic are mirror-anti-symmetric to each other in its colors: any pair of cells, disposed by mirror-symmetrical manner in these halves, possesses the opposite colors.

The genomatrix $P^{CAUG}{}_{123}$ consists of the four pairs of neighbor rows with even and odd numeration numbers in each pair: 0-1, 2-3, 4-5, 6-7. The rows of each pair are equivalent to each other from the viewpoint of a disposition of the same amino acids in their appropriate cells.

The black-and-white matrix mosaic has a symmetric figure of a diagonal cross: diagonal quadrants of the matrix are equivalent to each other from the viewpoint of their mosaic.

Mosaics of all rows have a meander-line character, which is connected with Rademacher functions from the theory of discrete signals processing.

The turning of the genomatrix $P^{CAUG}{}_{123}$ into a cylinder with an agglutination of its upper and lower borders reveals an ornamental pattern of a cyclic shift. This pattern has the character of cyclic shifts that permits one to think about a possible genetic meaning of cyclic codes, which play a significant role in the theory of digital signal processing. This pattern is demonstrated more clearly by a tessellation of a plane with this mosaic genomatrix (Figure 3, at the left). The plane with this tessellation possesses the ornamental pattern with two pattern units which are identical in their forms, but contrary in their colors (black and white) and orientations (left and right).

This symmetrical character of the degeneracy of the genetic code, which is presented by the matrix mosaic, is the key for many secrets of the genetic code. Let us note the following peculiarity of the presented "black-and-white" degeneracy of the genetic code on the Figure 2. The black triplets encode 8 amino acids, each of which is encoded by 4 triplets or more: Ala, Arg, Gly, Leu, Pro, Ser, Thr, Val. We

Figure 3. At the left: the tessellation of a plane with the mosaic of genomatrix P^{CAUG}_{123} from Figure 2. At the right: the tessellation of a plane with the mosaic of genomatrix P^{CAUG}_{231} from Figure 4

will name them as high-degeneracy amino acids. Another 12 amino acids form the sub-set, any member of which is encoded by less quantity of triplets: Asn, Asp, Cys, Gln, Glu, His, Ile, Lys, Met, Phe, Trp, Tyr. We will name them as low-degeneracy amino acids. The fact of the existence of these two sub-sets will be utilized in Chapter 3 for a comparative analysis of dialects of the genetic code.

The phenomenon of existence of cyclic shifts in the genetic pattern on Figure 3 has led to the investigation of a possible meaning of cyclic shifts of three positions in all triplets. If one changes the initial order 1-2-3 in all triplets by the cyclic shift into the new order 2-3-1, then many cells of the initial genomatrix P^{CAUG}_{123} are occupied by new triplets. For example, the matrix cell with the triplet CAU is occupied by the triplet AUC, etc. As a result the whole genomatrix P^{CAUG}_{123} is reconstructed into the new genomatrix P^{CAUG}_{231} (Figure 2).

It is unexpected that this "cyclic-generated" genomatrix P^{CAUG}_{231} with new matrix dispositions of triplets and amino acids possesses similar symmetric characteristics (Petoukhov, 2006, 2008a,c):

1. All its (4x4)-quadrants are identical to each other by its mosaics;
2. The upper and the lower halves of P^{CAUG}_{231} are identical to each other from the viewpoint of dispositions of all amino acids and stop-signals;
3. All rows of the (8x8)-genomatrix and its (4x4)-quadrants have a meander-line character again, which is connected with Rademacher functions;
4. The genomatrix P^{CAUG}_{231} possesses 4 pairs of identical rows as well: 0-1, 2-3, 4-5, 6-7 (but the rows with these numbers are disposed in new matrix positions on Figure 2 and they differ from the rows with the same numbers on Figure 2).

Note, that the mosaic of the initial (8x8)-genomatrix P^{CAUG}_{123} is reproduced in (4x4)-quadrants of this P^{CAUG}_{231} in a fractal manner: the coefficient of fractal ranging of areas is equal to 4. The tessellations of a plane by the mosaics of P^{CAUG}_{123} and of P^{CAUG}_{231} demonstrate their fractal correspondence very clearly (Figure 2). Such scale transformation of areas in the mosaics of the code degeneracy can be named "tetra-reproduction" transformation. Due to this tetra-reproduction, the cyclic-generated genomatrix P^{CAUG}_{231} has the quantity of the pattern units 4 times more than the initial genomatrix P^{CAUG}_{123} (Figures 2-4).

This fact is interesting because an analogical tetra-reproduction (or a tetra-division) exists in the living nature in the course of division of gametal cells, which are transmitters of genetic information. In this

Figure 4. The representation of the genomatrix P^{CAUG}_{231}, which is produced from the genomatrix P^{CAUG}_{123} (Figure 2) by the cyclic shift of positions in triplets (1-2-3 → 2-3-1)

	0 (000)	2 (010)	4 (100)	6 (110)	1 (001)	3 (011)	5 (101)	7 (111)
0	CCC Pro	CAC His	ACC Thr	AAC Asn	CCA Pro	CAA Gln	ACA Thr	AAA Lys
2	CUC Leu	CGC Arg	AUC Ile	AGC Ser	CUA Leu	CGA Arg	AUA Met	AGA Stop
4	UCC Ser	UAC Tyr	GCC Ala	GAC Asp	UCA Ser	UAA Stop	GCA Ala	GAA Glu
6	UUC Phe	UGC Cys	GUC Val	GGC Gly	UUA Leu	UGA Trp	GUA Val	GGA Gly
1	CCU Pro	CAU His	ACU Thr	AAU Asn	CCG Pro	CAG Gln	ACG Thr	AAG Lys
3	CUU Leu	CGU Arg	AUU Ile	AGU Ser	CUG Leu	CGG Arg	AUG Met	AGG Stop
5	UCU Ser	UAU Tyr	GCU Ala	GAU Asp	UCG Ser	UAG Stop	GCG Ala	GAG Glu
7	UUU Phe	UGU Cys	GUU Val	GGU Gly	UUG Leu	UGG Trp	GUG Val	GGG Gly

mysterious act of meiosis, one gamete is divided into four new gametes (this fact was mentioned specially by Erwin Schrodinger in his famous book (Schrodinger, 1955, §13)). The described tetra-reproduction of the mosaics of the genomatrices can be utilized, in particular, in formal models of meiosis.

Materials of the matrix genetics lead us to questions of biological meaning. Really, we revealed unexpectedly that a simple algorithmic re-packing (re-arrangement) of elements in triplets by the cyclic shift is sufficient to receive new genomatrix with the fractal tetra-reproducing the mosaics of the code degeneration. It seems that a similar re-packing of molecular elements in biological object can be sufficient also to provide foundations of a process of tetra-reproducing in some cases, first of all, in the case of meiosis. These and other considerations permit us to put forward the hypothesis of molecular re-packing. According to this hypothesis, the mysterious process of meiosis is based on a mechanism of algorithmic re-arrangement of molecular elements of gametes with a participation of algorithms of cyclic and dyadic shifts. In our opinion, the principle of re-packing of biological molecules and of their ensembles is an important general principle of biological self-organization. It is interesting also that one can compare the tetra-division of material gametes with the tetra-division of the code genomatrices, which are information objects. These materials show that meiosis is not an accidental material process but it is coordinated with more ancient information structures of the genetic code in their matrix form.

Permutations of elements play an important role in the theory of signals processing (Ahmed & Rao, 1975; Trahtman & Trahtman, 1975). Six variants of permutations of positions in triplets are possible only: 1-2-3, 2-3-1, 3-1-2, 1-3-2, 2-1-3, 3-2-1. The genomatrices P^{CAUG}_{123} and P^{CAUG}_{231} for the first two of these permutations were considered above (Figures 2 and 4). Let us consider other four variants which lead to genomatrices P^{CAUG}_{312}, P^{CAUG}_{132}, P^{CAUG}_{213}, P^{CAUG}_{321}. They are presented in Figure 5. It is an unexpected phenomenological fact, that all of these genomatrices have symmetrical peculiarities, which are similar to symmetrical peculiarities of P^{CAUG}_{123} and P^{CAUG}_{231}. The whole considered genetic code seems to be in agreement with these permutations and corresponding symmetries in the mosaics of all these 6 genomatrices.

Really, one can note that all genomatrices $P^{CAUG}{}_{213}{}^{(3)}$, $P^{CAUG}{}_{321}{}^{(3)}$, $P^{CAUG}{}_{312}{}^{(3)}$, $P^{CAUG}{}_{132}{}^{(3)}$ on Figure 5 have symmetric features as well. For example:

1. their mosaics have the mirror-antisymmetry between their left half and their right half;
2. each of these genomatrices has 4 pairs of identical rows again: 0-1, 2-3, 4-5, 6-7 (see their decimal numeration on Figure 5), which are not adjacent rows in these matrices;
3. all rows of the (8x8)-genomatrix and its (4x4)-quadrants have a meander-line character again, which is connected with Rademacher functions, etc.

Figure 5. The genomatrices $P^{CAUG}{}_{213}$, $P^{CAUG}{}_{321}$, $P^{CAUG}{}_{312}$, $P^{CAUG}{}_{132}$; Each matrix cell has a triplet and an amino acid (or a stop-signal) coded by this triplet. The black-and-white mosaic reflects the specificity of the degeneracy of this code

$P^{CAUG}{}_{213}$:

	0 (000)	1 (001)	4 (100)	5 (101)	2 (010)	3 (011)	6 (110)	7 (111)
0	CCC Pro	CCA Pro	ACC Thr	ACA Thr	CAC His	CAA Gln	AAC Asn	AAA Lys
1	CCU Pro	CCG Pro	ACU Thr	ACG Thr	CAU His	CAG Gln	AAU Asn	AAG Lys
4	UCC Ser	UCA Ser	GCC Ala	GCA Ala	UAC Tyr	UAA Stop	GAC Asp	GAA Glu
5	UCU Ser	UCG Ser	GCU Ala	GCG Ala	UAU Tyr	UAG Stop	GAU Asp	GAG Glu
2	CUC Leu	CUA Leu	AUC Ile	AUA Met	CGC Arg	CGA Arg	AGC Ser	AGA Stop
3	CUU Leu	CUG Leu	AUU Ile	AUG Met	CGU Arg	CGG Arg	AGU Ser	AGC Ser
6	UUC Phe	UUA Leu	GUC Val	GUA Val	UGC Cys	UGA Trp	GGC Gly	GGA Gly
7	UUU Phe	UUG Leu	GUU Val	GUG Val	UGU Cys	UGG Trp	GGU Gly	GGG Gly

$P^{CAUG}{}_{312}$:

	0 (000)	4 (100)	1 (001)	5 (101)	2 (010)	6 (110)	3 (011)	7 (111)
0	CCC Pro	ACC Thr	CCA Pro	ACA Thr	CAC His	AAC Asn	CAA Gln	AAA Lys
4	UCC Ser	GCC Ala	UCA Ser	GCA Ala	UAC Tyr	GAC Asp	UAA Stop	GAA Glu
1	CCU Pro	ACU Thr	CCG Pro	ACG Thr	CAU His	AAU Asn	CAG Gln	AAG Lys
5	UCU Ser	GCU Ala	UCG Ser	GCG Ala	UAU Tyr	GAU Asp	UAG Stop	GAG Glu
2	CUC Leu	AUC Ile	CUA Leu	AUA Met	CGC Arg	AGC Ser	CGA Arg	AGA Stop
6	UUC Phe	GUC Val	UUA Leu	GUA Val	UGC Cys	GGC Gly	UGA Trp	GGA Gly
3	CUU Leu	AUU Ile	CUG Leu	AUG Met	CGU Arg	AGU Ser	CGG Arg	AGG Stop
7	UUU Phe	GUU Val	UUG Leu	GUG Val	UGU Cys	GGU Gly	UGG Trp	GGG Gly

$P^{CAUG}{}_{321}$:

	0 (000)	4 (100)	2 (010)	6 (110)	1 (001)	5 (101)	3 (011)	7 (111)
0	CCC Pro	ACC Thr	CAC His	AAC Asn	CCA Pro	ACA Thr	CAA Gln	AAA Lys
4	UCC Ser	GCC Ala	UAC Tyr	GAC Asp	UCA Ser	GCA Ala	UAA Stop	GAA Glu
2	CUC Leu	AUC Ile	CGC Arg	AGC Ser	CUA Leu	AUA Met	CGA Arg	AGA Stop
6	UUC Phe	GUC Val	UGC Cys	GGC Gly	UUA Leu	GUA Val	UGA Trp	GGA Gly
1	CCU Pro	ACU Thr	CAU His	AAU Asn	CCG Pro	ACG Thr	CAG Gln	AAG Lys
5	UCU Ser	GCU Ala	UAU Tyr	GAU Asp	UCG Ser	GCG Ala	UAG Stop	GAG Glu
3	CUU Leu	AUU Ile	CGU Arg	AGU Ser	CUG Leu	AUG Met	CGG Arg	AGG Stop
7	UUU Phe	GUU Val	UGU Cys	GGU Gly	UUG Leu	GUG Val	UGG Trp	GGG Gly

$P^{CAUG}{}_{132}$:

	0 (000)	2 (010)	1 (001)	3 (011)	4 (100)	6 (110)	5 (101)	7 (111)
0	CCC Pro	CAC His	CCA Pro	CAA Gln	ACC Thr	AAC Asn	ACA Thr	AAA Lys
2	CUC Leu	CGC Arg	CUA Leu	CGA Arg	AUC Ile	AGC Ser	AUA Met	AGA Stop
1	CCU Pro	CAU His	CCG Pro	CAG Gln	ACU Thr	AAU Asn	ACG Thr	AAG Lys
3	CUU Leu	CGU Arg	CUG Leu	CGG Arg	AUU Ile	AGU Ser	AUG Met	AGG Stop
4	UCC Ser	UAC Tyr	UCA Ser	UAA Stop	GCC Ala	GAC Asp	GCA Ala	GAA Glu
6	UUC Phe	UGC Cys	AUU Ile	AGU Ser	CUG Leu	CGG Arg	GUA Val	GGA Gly
5	UCU Ser	UAU Tyr	UCG Ser	UAG Stop	GCU Ala	GAU Asp	GCG Ala	GAG Glu
7	UUU Phe	UGU Cys	UUG Leu	UGG Trp	GUU Val	GGU Gly	GUG Val	GGG Gly

Now let us also consider on Figure 5 the genomatrix $P^{CAUG}{}_{321}{}^{(3)}$ with the inverse order of positions in all triplets (3-2-1 instead of 1-2-3). One can compare its mosaic with the mosaic of the $P^{CAUG}{}_{213}{}^{(3)}$ based on the cyclic shift of positions in all triplets: 2-1-3 instead of 3-2-1. In this case the similar phenomenon of the tetra-self-reproduction of these mosaics becomes apparent again but with a new pattern (Figure 6).

In addition, one can note that all six genomatrices on Figures 2, 4, and 5 are interconnected by special permutations of their columns and rows. The same genomatrices can be obtained from the initial genomatrix $P^{CAUG}{}_{123}{}^{(3)}$ by appropriate permutations of positions in binary 3-digit numbering their columns and rows. In other words, the "local" permutations of positions in triplets give the same results as the "global" permutations of positions in binary 3-digit numbering the columns and the rows. All six genomatrices on these Figures are connected with Hadamard matrices on the basis of the U-algorithm described in Chapter 6. The presented permutations gave interesting results in their application to genomatrices. It seems that applications of similar permutations to genetic sequences of triplets can give interesting results as well because each gene belongs to a group of six genetic sequences, which are differed from each other by orders of positions in their triplets: 1-2-3, 2-3-1, 3-1-2, 1-3-2, 2-1-3, 3-2-1.

The revelation of the permutation group of the six symmetric genomatrices $P^{CAUG}{}_{123}{}^{(3)}$, $P^{CAUG}{}_{231}{}^{(3)}$, $P^{CAUG}{}_{213}{}^{(3)}$, $P^{CAUG}{}_{321}{}^{(3)}$, $P^{CAUG}{}_{312}{}^{(3)}$, $P^{CAUG}{}_{132}{}^{(3)}$ seems to be the essential fact because of heuristic associations with the mathematical theory of digital signal processing, where similar permutations are utilized for a long time as the useful tool. For example, the book (Ahmed & Rao, 1975, § 4.6) gives the example of the important role of the method of data permutations and of the binary inversion for one of variants of the algorithm of a fast Fourier transformation. In this example the numeric sequence 0, 1, 2, 3, 4, 5, 6, 7 is reformed into the sequence 0, 4, 2, 6, 1, 5, 3, 7. But the same change of the numeration of the columns and the rows takes a place in our case (Figure 5) where the genomatrix $P^{CAUG}{}_{123}$ is reformed into the genomatrix $P^{CAUG}{}_{321}$ as a result of the inversion of the binary numbering the columns and the rows (or of the inversion of the positions in the triplets). These and other facts permit one to think that the genetic system has a connection with a fast Fourier transformation (or with a fast Hadamard transformation) (Petoukhov, 2006, 2008a,b).

GENOMATRICES WITH THE PROPERTY OF THE TETRA-SELF-REPRODUCING

Why has nature chosen this variant of degeneration of genetic code, which gives such mosaics? Do these six "triplets-permutations" genomatrices $P^{CAUG}{}_{123}{}^{(3)}$, $P^{CAUG}{}_{231}{}^{(3)}$, $P^{CAUG}{}_{213}{}^{(3)}$, $P^{CAUG}{}_{321}{}^{(3)}$, $P^{CAUG}{}_{312}{}^{(3)}$, $P^{CAUG}{}_{132}{}^{(3)}$ possess such mutual mathematical property that can be associated with famous biological facts of genetic inheritance? Yes, such a mutual property exists and it is connected with the tetra-reproduction by

Figure 6. The tessellations of a plane by the mosaics of the genomatrices $P^{CAUG}{}_{213}$ (at the left) and $P^{CAUG}{}_{321}$ (on the right)

analogy with meiosis again. This property is a non-trivial one and it does not exist in the most variants of arbitrary dispositions of 32 black triplets and 32 white triplets in (8*8)-matrices.

Let us represent the black-and-white mosaic of each of the six genomatrices as a binary mosaic of numbers "+1" and "-1" by means of replacing black (white) color of each matrix cell by an element "+1" ("-1") correspondingly. As a result, these genomatrices $P^{CAUG}{}_{123}{}^{(3)}$, $P^{CAUG}{}_{231}{}^{(3)}$, $P^{CAUG}{}_{213}{}^{(3)}$, $P^{CAUG}{}_{321}{}^{(3)}$, $P^{CAUG}{}_{312}{}^{(3)}$, $P^{CAUG}{}_{132}{}^{(3)}$ are reformed into the genomatrices B_{123}, B_{231}, B_{312}, B_{132}, B_{213}, B_{321} (Figure 7).

The unexpected mutual property of these six binary genomatrices is the following one. The multiplication of each genomatrix with itself (the square of each genomatrix) gives a phenomenon of its tetra-reproduction: the four duplicates of the genomatrix appeared in Figure 8. Really the following formulas take place:

$$(B_{123})^2 = 4*B_{123};\ (B_{231})^2 = 4*B_{231};\ (B_{312})^2 = 4*B_{312}$$

$$(B_{132})^2 = 4*B_{132};\ (B_{213})^2 = 4*B_{213};\ (B_{321})^2 = 4*B_{321} \qquad (1)$$

This fact is interesting because the genetic code is destined by nature for reproduction of biological structures, and matrices of the genetic code in their binary representation possess the non-trivial algebraic property of their own self-reproduction. The set of these six binary genomatrices has many other interesting properties (for instance, $B_{123}*B_{321}+B_{123}*B_{132} = 4*B_{123}$), which generate heuristic associations with

Figure 7. The binary numeric genomatrices B_{123}, B_{231}, B_{312}, B_{132}, B_{213}, B_{321}, in which each black cell means the element "+1"; and each white cell means the element "-1"

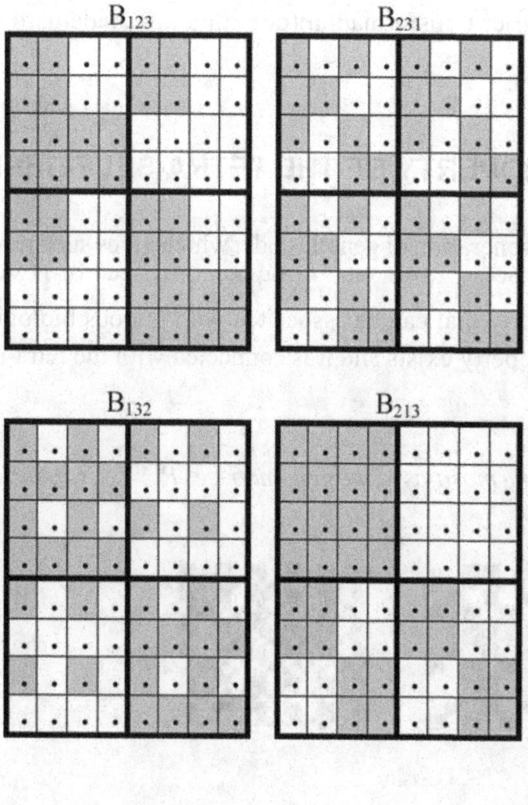

Figure 8. Tetra-self-reproduction of each of the binary numeric genomatrices B_{123}, B_{231}, B_{312}, B_{132}, B_{213}, B_{321} from Figure 7 due to operation of its multiplication with itself

genetic phenomena and which can be utilized to model the meiosis process of the tetra-self-reproduction of gametal cells with a specific behavior of chromosomes to some extent (Figures 8 and 9). These matrix properties are connected with the octave Yin-Yang algebra of the genetic code (see Chapter 7).

It can be also mentioned that one can consider those "complementary" variants of the genomatrices $P^{CAUG}_{123}{}^{(3)}$, $P^{CAUG}_{231}{}^{(3)}$, $P^{CAUG}_{213}{}^{(3)}$, $P^{CAUG}_{321}{}^{(3)}$, $P^{CAUG}_{312}{}^{(3)}$, $P^{CAUG}_{132}{}^{(3)}$, which are achieved by the replacement of each triplet by its complementary triplet (the example of the complementary triplets is CAG and GUC). In each case the "complementary" matrix is identical to 180-degree turn of the initial matrix. The "complementary" genomatrices in similar binary presentations possess the same properties as their tetra-self-reproduction.

One can ask, why nature did not chose the more simple variant of the mosaic of the degeneracy of the genetic code, for example, such a variant where the left half of the matrix is occupied by black triplets and the right half is occupied by white triplets (Figure 10)? This variant and many other possible variants of (8x8)-matrices with 32 black cells and 32 white cells do not possess those interesting properties, which natural genomatrices possess: the properties of the tetra-self-reproduction; the algorithmic relation to Hadamard matrices; the connection with hyperbolic matrions and the octet Yin-Yang algebra (Chapters 6-8), etc. For example, the matrix with the black-and-white mosaic of the signs "+1" and "-1" in Figure 10 does not possess the described property of the tetra-self-reproduction because its square is equal to the null matrix.

Figure 9. The illustration of the process of the tetra-self-reproduction of a gametal cell in a course of meiosis

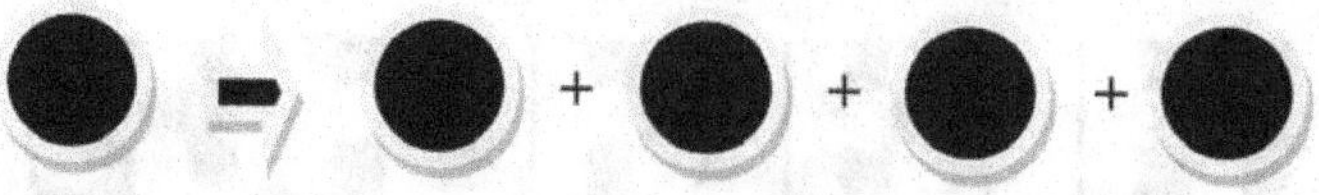

SYMMETRICAL PROPERTIES OF GENETIC MATRICES UNDER ALPHABETIC PERMUTATIONS IN THE SET OF 64 TRIPLETS

Until now we considered the Kronecker family of the genomatrices with the kernel [C A; U G] (Figure 3 of Chapter 1) and obtained some interesting properties of the mosaic genomatrices $[C\ A;\ U\ G]^{(3)}$. This paragraph demonstrates that analogical properties exist for the other mosaic genomatrices with various kernels: $[C\ A;\ G\ U]^{(3)}$, $[G\ C;\ A\ U]^{(3)}$, etc. These new variants of kernels of the Kronecker families of genomatrices are produced by permutations of the four letters C, A, U, G on positions in the (2x2)-matrix, for example by mutual interchanges C↔G, A↔U (such permutations produce a change of letter compositions of triplets in matrix cells in comparison with the described genomatrix P^{CAUG}_{123}).

We continue to utilize the upper index in a symbols of each (8x8)-genomatrix to show the kind of the kernel of the Kronecker family of this genomatrix. Such upper indexes can be CAGU, CGUA, ACUG, ACGU, UACG, UGCA, GAUC, GAUC, GUAC, etc. For example, the symbol P^{GCAU}_{123} means the genomatrix $[G\ C;\ A\ U]^{(3)}$, which differs from the described genomatrix $P^{CAUG}_{123} = [C\ A;\ U\ G]^{(3)}$, of course.

The 24 variants of such (2x2)-genomatrix exist, which differ from each other by dispositions of the letters inside the matrix (Figure 11). We will pay attention to a disposition of the particular letter U, which is replaced by the letter T in the course of transfer from RNA to DNA for unknown reason and which differs from other letters C, G, A by this feature. Such attention to the letter U is explained in Chapter 6 which deals with a connection between genomatrices and Hadamard matrices. This letter U can occupy one of the four positions in the alphabetic (2x2)-matrix.

Correspondingly one can divide the whole set of 24 variants of such (2x2)-matrices into the four categories (Figure 11). The first category contains (2x2)-matrices with the letter U in their left lower corner. The second category possesses the letter U in the right upper corner. The third category possesses the letter U in the left upper corner. The fourth category possesses the letter U in the right lower corner.

Figure 10. One of hypothetic variants of octet matrices with 32 black triplets and 32 white triplets (explanation in the text)

CCC	CCA	CAC	CAA	ACC	ACA	AAC	AAA
CCU	CCG	CAU	CAG	ACU	ACG	AAU	AAG
CUC	CUA	CGC	CGA	AUC	AUA	AGC	AGA
CUU	CUG	CGU	CGG	AUU	AUG	AGU	AGG
UCC	UCA	UAC	UAA	GCC	GCA	GAC	GAA
UCU	UCG	UAU	UAG	GCU	GCG	GAU	GAG
UUC	UUA	UGC	UGA	GUC	GUA	GGC	GGA
UUU	UUG	UGU	UGG	GUU	GUG	GGU	GGG

Figure 11. The four categories of possible 24 variants of alphabetic (2x2)-matrices; the left column shows the number of each of the four categories

$$1)\ \begin{vmatrix} C & A \\ U & G \end{vmatrix}\ \begin{vmatrix} C & G \\ U & A \end{vmatrix}\ \begin{vmatrix} A & C \\ U & G \end{vmatrix};\ \begin{vmatrix} A & G \\ U & C \end{vmatrix};\ \begin{vmatrix} G & A \\ U & C \end{vmatrix};\ \begin{vmatrix} G & C \\ U & A \end{vmatrix}$$

$$2)\ \begin{vmatrix} U & G \\ C & A \end{vmatrix}\ \begin{vmatrix} U & A \\ C & G \end{vmatrix}\ \begin{vmatrix} U & G \\ A & C \end{vmatrix}\ \begin{vmatrix} U & C \\ A & G \end{vmatrix}\ \begin{vmatrix} U & C \\ G & A \end{vmatrix}\ \begin{vmatrix} U & A \\ G & C \end{vmatrix}$$

$$3)\ \begin{vmatrix} G & U \\ A & C \end{vmatrix}\ \begin{vmatrix} A & U \\ G & C \end{vmatrix}\ \begin{vmatrix} G & U \\ C & A \end{vmatrix}\ \begin{vmatrix} C & U \\ G & A \end{vmatrix}\ \begin{vmatrix} C & U \\ A & G \end{vmatrix}\ \begin{vmatrix} A & U \\ C & G \end{vmatrix}$$

$$4)\ \begin{vmatrix} A & C \\ G & U \end{vmatrix}\ \begin{vmatrix} G & C \\ A & U \end{vmatrix}\ \begin{vmatrix} C & A \\ G & U \end{vmatrix}\ \begin{vmatrix} G & A \\ C & U \end{vmatrix}\ \begin{vmatrix} A & G \\ C & U \end{vmatrix}\ \begin{vmatrix} C & G \\ A & U \end{vmatrix}$$

We will use the name "mirror-coupling" for a reconstruction of any matrix (Figure 11) from one category into a matrix of another category by means of a permutation of its columns or rows. Each matrix of any category has one mirror-coupling matrix in each of the other categories. Such mirror-coupling matrices of various categories are disposed in one column on the Figure 11. For example, the genomatrix [C A; U G], which is disposed in the first column of the first category, is reconstructed into the genomatrix [U G; C A] of the second category; etc. The relation of such mirror-coupling for these matrices is conserved at their rising in Kronecker powers. For example, the matrices $[U\ G; C\ A]^{(n)}$ & $[G\ U; A\ C]^{(n)}$ are mirror-coupling to each another: one of them can be reconstructed from the second matrix by mirror permutations of columns relative to the middle vertical line of the matrix.

Genomatrices of the first category and the third category contain the particular letter U on their second diagonal. They are connected with the Hadamard matrices [1 1; -1 1] or [1 -1; 1 1] and with the matrix form of representation of complex numbers and multi-dimensional generalization of complex numbers (see Chapter 8). Genomatrices of the second category and of the fourth category contain the letter U

Figure 12. The genomatrix P^{GCAU}_{123} and the mosaic of the code degeneracy (explanation in the text)

GGG Gly	GGC Gly	GCG Ala	GCC Ala	CGG Arg	CGC Arg	CCG Pro	CCC Pro
GGA Gly	GGU Gly	GCA Ala	GCU Ala	CGA Arg	CGU Arg	CCA Pro	CCU Pro
GAG Glu	GAC Asp	GUG Val	GUC Val	CAG Gln	CAC His	CUG Leu	CUC Leu
GAA Glu	GAU Asp	GUA Val	GUU Val	CAA Gln	CAU His	CUA Leu	CUU Leu
AGG Stop	AGC Ser	ACG Thr	ACC Thr	UGG Trp	UGC Cys	UCG Ser	UCC Ser
AGA Stop	AGU Ser	ACA Thr	ACU Thr	UGA Trp	UGU Cys	UCA Ser	UCU Ser
AAG Lys	AAC Asn	AUG Met	AUC Ile	UAG Stop	UAC Tyr	UUG Leu	UUC Phe
AAA Lys	AAU Asn	AUA Met	AUU Ile	UAA Stop	UAU Tyr	UUA Leu	UUU Phe

on the main diagonal. They are connected with Hadamard matrices [1 1; 1 -1] or [-1 1; 1 1] and their Kronecker powers. These Hadamard matrices in various Kronecker powers play an important role in the theory of quantum computers, spectral methods of discrete signals processing, etc. (see Chapter 6).

Taking into account the mirror-coupling among genomatrices of the four categories, it is sufficient to consider examples of the genomatrices of the fourth category (Figure 11). For this reason, let us consider the genomatrices P^{GCAU}_{123} = [G C; A U]$^{(3)}$ and P^{CAGU}_{123} =[C A; G U]$^{(3)}$. Other genomatrices of the fourth category can be considered analogically.

The example of the genomatrix P^{GCAU}_{123}. Figure 12 shows the genomatrix P^{GCAU}_{123} = [G C; A U] with its black-and-white mosaic of the degeneracy of the vertebrate mitochondrial genetic code.

This genomatrix P^{GCAU}_{123} possesses the following symmetric features:

1. The left half and the right half of the genomatrix are symmetric each to the other in the sense of translation symmetry of their mosaics;
2. Two quadrants along each matrix diagonal are inversion-anti-symmetric each to the other in their mosaics;
3. The neighboring rows in four pairs of the rows are identical each to the other from the viewpoint of a disposition of amino acids and stop-codons;
4. Four pairs of neighboring rows are identical.

Permutations of positions inside triplets, which were described above, produce the five genomatrices P^{GCAU}_{231}, P^{GCAU}_{312}, P^{GCAU}_{132}, P^{GCAU}_{213}, P^{GCAU}_{321}, which are shown in Figures 13, 14, 15, 16, and 17. One can see without additional explanations that all of them possess similar symmetric features as well.

The example of the genomatrix P^{CAGU}_{123}. The Figure 18 shows the genomatrix P^{CAGU}_{123} = [C A; G U]$^{(3)}$ with its black-and-white mosaic of the degeneracy of the vertebrate mitochondrial genetic code. The Figures 19, 20, 21, 22, and 23 demonstrate the genomatrices P^{CAGU}_{231}, P^{CAGU}_{312}, P^{CAGU}_{132}, P^{CAGU}_{213}, P^{CAGU}_{321}, which are produced from P^{CAGU}_{123} by all possible permutations of positions inside all triplets. One can see without additional explanations that all these six genomatrices possess symmetrical features as well. We do not show amino acids and stop-codons in some of these genomatrices to decrease tabular materials.

Figure 13. The genomatrix P^{GCAU}_{231} and the mosaic of the code degeneracy (explanation in the text)

GGG Gly	GCG Ala	CGG Arg	CCG Pro	GGC Gly	GCC Ala	CGC Arg	CCC Pro
GAG Glu	GUG Val	CAG Gln	CUG Leu	GAC Asp	GUC Val	CAC His	CUC Leu
AGG Stop	ACG Thr	UGG Trp	UCG Ser	AGC Ser	ACC Thr	UGC Cys	UCC Ser
AAG Lys	AUG Met	UAG Stop	UUG Leu	AUC Ile	AUC Ile	UAC Tyr	UUC Phe
GGA Gly	GCA Ala	CGA Arg	GCA Ala	GGU Gly	GCU Ala	CGU Arg	CCU Pro
GAA Glu	GUA Val	CAA Gln	CUA Leu	GAU Asp	GUU Val	CAU His	CUU Leu
AGA Stop	ACA Thr	UGA Trp	UCA Ser	AGU Ser	ACU Thr	UGU Cys	UCU Ser
AAA Lys	AUA Met	UAA Stop	UUA Leu	AAU Asn	AUU Ile	UAU Tyr	UUU Phe

Figure 14. The genomatrix P^{GCAU}_{312} and the mosaic of the code degeneracy (explanation in the text)

GGG Gly	CGG Arg	GGC Gly	CGC Arg	GCG Ala	CCG Pro	GCC Ala	CCC Pro
AGG Stop	UGG Trp	AGC Ser	UGC Cys	ACG Thr	UCG Ser	ACC Thr	UCC Ser
GGA Gly	CGA Arg	GGU Gly	CGU Arg	GCA Ala	CCA Pro	GCU Ala	CCU Pro
AGA Stop	UGA Trp	AGU Ser	UGU Cys	ACA Thr	UCA Ser	ACU Thr	UCU Ser
GAG Glu	CAG Gln	GAC Asp	CAC His	GUG Val	CUG Leu	GUC Val	CUC Leu
AAG Lys	UAG Stop	AAC Asn	UAC Tyr	AUG Met	UUG Leu	AUC Ile	UUC Phe
GAA Glu	CAA Gln	GAU Asp	CAU His	GUA Val	CUA Leu	GUU Val	CUU Leu
AAA Lys	UAA Stop	AAU Asn	UAU Tyr	AUA Met	UUA Leu	AUU Ile	UUU Phe

Figure 15. The genomatrix P^{GCAU}_{132} and the mosaic of the code degeneracy (explanation in the text)

GGG Gly	GCG Ala	GGC Gly	GCC Ala	CGG Arg	CCG Pro	CGC Arg	CCC Pro
GAG Glu	GUG Val	GAC Asp	GUC Val	CAG Gln	CUG Leu	CAC His	CUC Leu
GGA Gly	GCA Ala	GGU Gly	GCU Ala	CGA Arg	CCA Pro	CGU Arg	CCU Pro
GAA Glu	GUA Val	GAU Asp	GUU Val	CAA Gln	CUA Leu	CAU His	CUU Leu
AGG Stop	ACG Thr	AGC Ser	ACC Thr	UGG Trp	UCG Ser	UGC Cys	UCC Ser
AAG Lys	AUG Met	AAC Asn	AUC Ile	UAG Stop	UUG Leu	UAC Tyr	UUC Phe
AGA Stop	ACA Thr	AGU Ser	ACU Thr	UGA Trp	UCA Ser	UGU Cys	UCU Ser
AAA Lys	AUA Met	AAU Asn	AUU Ile	UAA Stop	UUA Leu	UAU Tyr	UUU Phe

Figure 16. The genomatrix P^{GCAU}_{213} and the mosaic of the code degeneracy (explanation in the text)

GGG Gly	GGC Gly	CGG Arg	CGC Arg	GCG Ala	GCC Ala	CCG Pro	CCC Pro
GGA Gly	GGU Gly	CGA Arg	CGU Arg	GCA Ala	GCU Ala	CCA Pro	CCU Pro
AGG Stop	AGC Ser	UGG Trp	UGC Cys	ACG Thr	ACC Thr	UCG Ser	UCC Ser
AGA Stop	AGU Ser	UGA Trp	UGU Cys	ACA Thr	ACU Thr	UCA Ser	UCU Ser
GAG Glu	GAC Asp	CAG Gln	CAC His	GUG Val	GUC Val	CUG Leu	CUC Leu
GAA Glu	GAU Asp	CAA Gln	CAU His	GUA Val	GUU Val	CUA Leu	CUU Leu
AAG Lys	AAC Asn	UAG Stop	UAC Tyr	AUG Met	AUC Ile	UUG Leu	UUC Phe
AAA Lys	AAU Asn	UAA Stop	UAU Tyr	AUA Met	AUU Ile	UUA Leu	UUU Phe

Figure 17. The genomatrix P^{GCAU}_{321} *and the mosaic of the code degeneracy (explanation in the text)*

GGG Gly	CGG Arg	GCG Ala	CCG Pro	GGC Gly	CGC Arg	GCC Ala	CCC Pro
AGG Stop	UGG Trp	ACG Thr	UCG Ser	AGC Ser	UGC Cys	ACC Thr	UCC Ser
GAG Glu	CAG Gln	GUG Val	CUG Leu	GAC Asp	CAC His	GUC Val	CUC Leu
AAG Lys	UAG Stop	AUG Met	UUG Leu	AAC Asn	UAC Tyr	AUC Ile	UUC Phe
GGA Gly	CGA Arg	GCA Ala	CCA Pro	GGU Gly	CGU Arg	GCU Ala	CCU Pro
AGA Stop	UGA Trp	ACA Thr	UCA Ser	AGU Ser	UGU Cys	ACU Thr	UCU Ser
GAA Glu	CAA Gln	GUA Val	CUA Leu	GAU Asp	CAU His	GUU Val	CUU Leu
AAA Lys	UAA Stop	AUA Met	UUA Leu	AAU Asn	UAU Tyr	AUU Ile	UUU Phe

Figure 18. The genomatrix P^{CAGU}_{123} *and the mosaic of the code degeneracy (explanation in the text)*

CCC Pro	CCA Pro	CAC His	CAA Gln	ACC Thr	ACA Thr	AAC Asn	AAA Lys
CCG Pro	CCU Pro	CAG Gln	CAU His	ACG Thr	ACU Thr	AAG Lys	AAU Asn
CGC Arg	CGA Arg	CUC Leu	CUA Leu	AGC Ser	AGA Stop	AUC Ile	AUA Met
CGG Arg	CGU Arg	CUG Leu	CUU Leu	AGG Stop	AGU Ser	AUG Met	AUU Ile
GCC Ala	GCA Ala	GAC Asp	GAA Glu	UCC Ser	UCA Ser	UAC Tyr	UAA Stop
GCG Ala	GCU Ala	GAG Glu	GAU Asp	UCG Ser	UCU Ser	UAG Stop	UAU Tyr
GGC Gly	GGA Gly	GUC Val	GUA Val	UGC Cys	UGA Trp	UUC Phe	UUA Leu
GGG Gly	GGU Gly	GUG Val	GUU Val	UGG Trp	UGU Cys	UUG Leu	UUU Phe

Figure 19. The genomatrix P^{CAGU}_{231} *and the mosaic of the code degeneracy (explanation in the text)*

CCC	CAC	ACC	AAC	CCA	CAA	ACA	AAA
CGC	CUC	AGC	AUC	CGA	CUA	AGA	AUA
GCC	GAC	UCC	UAC	GCA	GAA	UCA	UAA
GGC	GUC	UGC	UUC	GGA	GUA	UGA	UUA
CCG	CAG	ACG	AAG	CCU	CAU	ACU	AAU
CGG	CUG	AGG	AUG	CGU	CUU	AGU	AUU
GCG	GAG	UCG	UAG	GCU	GAU	UCU	UAU
GGG	GUG	UGG	UUG	GGU	GUU	UGU	UUU

Figure 20. The genomatrix P^{CAGU}_{312} and the mosaic of the code degeneracy (explanation in the text)

CCC	ACC	CCA	ACA	CAC	AAC	CAA	AAA
GCC	UCC	GCA	UCA	GAC	UAC	GAA	UAA
CCG	ACG	CCU	ACU	CAG	AAG	CAU	AAU
GCG	UCG	GCU	UCU	GAG	UAG	GAU	UAU
CGC	AGC	CGA	AGA	CUC	AUC	CUA	AUA
GGC	UGC	GGA	UGA	GUC	UUC	GUA	UUA
CGG	AGG	CGU	AGU	CUG	AUG	CUU	AUU
GGG	UGG	GGU	UGU	GUG	UUG	GUU	UUU

Figure 21. The genomatrix P^{CAGU}_{132} and the mosaic of the code degeneracy (explanation in the text)

CCC	CAC	CCA	CAA	ACC	AAC	ACA	AAA
CGC	CUC	CGA	CUA	AGC	AUC	AGA	AUA
CCG	CAG	CCU	CAU	ACG	AAG	ACU	AAU
CGG	CUG	CGU	CUU	AGG	AUG	AGU	AUU
GCC	GAC	GCA	GAA	UCC	UAC	UCA	UAA
GGC	GUC	GGA	GUA	UGC	UUC	UGA	UUA
GCG	GAG	GCU	GAU	UCG	UAG	UCU	UAU
GGG	GUG	GGU	GUU	UGG	UUG	UGU	UUU

One can check that the fractal property of the tetra-scaling, which was described above in connection with Figures 3 and 6, between mosaics of certain pairs of genomatrices exists for all categories of considered (8x8)-genomatrices. Those pairs of genomatrices possess this property, which are connected by cyclic shifts in their lower indexes: (123)-(231) and (321)-(213).

All of the described mosaic genomatrices are connected with appropriate Hadamard matrices by means of the same U-algorithm, which is presented in Chapter 6.

FUTURE TRENDS AND CONCLUSION

The described investigations demonstrate that the degeneracy of the genetic code is connected with the system of genomatrix symmetries and with the system of invariants relative to some kinds of permutations in triplets. The described results show that the degeneracy of the genetic code is not the accidental choice of nature at all. The matrix genetics proposes the effective cognitive form of the matrix presentation of ensembles of the genetic code elements. This cognitive form should be utilized in future investigations of genetic systems as well.

The aim of each scientific theory is an explanation of phenomenological facts. The more phenomenological facts exist, the more bases exist to create theories. The results described in this chapter give us new interesting phenomenological facts about some permutation properties of the genetic code. They should be explained theoretically and they can be a prompting a new mathematical theory of genetic code systems. For example, these results lead to the idea that the genetic code in its matrix form of presentation can be connected with algebraic multi-dimensional constructions, which possess matrix forms of presentation also. A confirmation of this idea is described in Chapter 7.

Figure 22. The genomatrix P^{CAGU}_{213} and the mosaic of the code degeneracy (explanation in the text)

CCC	CCA	ACC	ACA	CAC	CAA	AAC	AAA
CCG	CCU	ACG	ACU	CAG	CAU	AAG	AAU
GCC	GCA	UCC	UCA	GAC	GAA	UAC	UAA
GCG	GCU	UCG	UCU	GAG	GAU	UAG	UAU
CGC	CGA	AGC	AGA	CUC	CUA	AUC	AUA
CGG	CGU	AGG	AGU	CUG	CUU	AUG	AUU
GGC	GGA	UGC	UGA	GUC	GUA	UUC	UUA
GGG	GGU	UGG	UGU	GUG	GUU	UUG	UUU

Figure 23. The genomatrix P^{CAGU}_{321} and the mosaic of the code degeneracy (explanation in the text)

CCC	ACC	CAC	AAC	CCA	ACA	CAA	AAA
GCC	UCC	GAC	UAC	GCA	UCA	GAA	UAA
CGC	AGC	CUC	AUC	CGA	AGA	CUA	AUA
GGC	UGC	GUC	UUC	GGA	UGA	GUA	UUA
CCG	ACG	CAG	AAG	CCU	ACU	CAU	AAU
GCG	UCG	GAG	UAG	GCU	UCU	GAU	UAU
CGG	AGG	CUG	AUG	CGU	AGU	CUU	AUU
GGG	UGG	GUG	UUG	GGU	UGU	GUU	UUU

The unexpected properties of tetra-reproducing and tetra-scaling in the set of the mosaic genomatrices (Figures 3, 6 and 8) can be utilized for mathematical modeling the phenomenon of the tetra-division of gametal cells in meiosis; they can be useful for the theory of self-development systems and self-organizing systems as well.

The results of this chapter show that the permutations of various kinds are important not only for the theory of digital signals processing but also for genetic code systems. One should investigate further similar analogies between the genetic field and the advanced theory of digital informatics. Why does the degeneracy of the genetic code possess the permutation properties described in this chapter? Why has nature chosen such a variant of the degeneracy of the genetic code? What kind of algebraic numeric structures possess such matrix features and can be a mathematical model of the genetic code with its degeneracy? Many of such questions should be answered from the viewpoint of a general theory in the future. The proposed cognitive forms of matrix genetics can be useful to create such a theory. The described results permit one to search algebraic multi-dimensional constructions as a genetic code model with analogical matrix properties. In the case of a success of such algebraic searching, the problem of algebraization of bioinformatics can draw attention to the fundamental role of the genetic code. The theory of self-development systems and self-organizing systems can utilize the described data about the properties of the tetra-reproducing and the tetra-scaling of the genomatrices.

The presentation of the vertebrate mitochondria genetic code in the form of the genomatrices of Kronecker family reveals unexpectedly a set of symmetries in matrix mosaics of its degeneracy. Possible permutations of positions in triplets produce new genomatrices, which possess similar matrix symmetries as well. Mutual interchanges of alphabetic letters A, C, G, U in matrix kernels of Kronecker families produce new genomatrices, which also possess similar matrix symmetries. These phenomenological facts show the prospect that the genetic code and its degeneracy are not accidental choices of nature at all. These facts are bases for searching algebraic multi-dimensional systems with similar properties, which can serve as a model of the genetic code.

REFERENCES

Ahmed, N., & Rao, K. (1975). *Orthogonal transforms for digital signal processing.* New York: Springer-Verlag Inc.

Blahut, R. E. (1985). *Fast algorithms for digital signal processing.* Massachusetts: Addison-Wesley Publishing Company.

Dragovich, B., & Dragovich, A. (2006). *A p-adic model of DNA sequence and genetic code.* Retrieved on July 13, 2006, from http://arXiv:q-bio/0607018

Dragovich, B., & Dragovich, A. (2007). *P-adic modelling of the genome and the genetic code.* Retrieved on July 7, 2007, from http://arXiv:0707.3043

Frank-Kamenetskiy, M. D. (1988). *The most principal molecule.* Moscow: Nauka (in Russian).

Khrennikov, A. Y., & Kozyrev, S. V. (2007). Genetic code on the dyadic plane. *Archiv.* Retrieved on January 3, 2007, from http:// arXiv:q-bio/0701007

Petoukhov, S. V. (2006). Bioinformatics: Matrix genetics, algebras of the genetic code, and biological harmony. *Symmetry: Culture and Science, 17*(1-4), 251–290.

Petoukhov, S. V. (2008a). *Matrix genetics, algebras of the genetic code, noise-immunity.* Moscow: RCD (in Russian).

Petoukhov, S. V. (2008b). *The degeneracy of the genetic code and Hadamard matrices* (pp. 1-8). Retrieved on February 8, 2008, from http://arXiv:0802.3366

Petoukhov, S. V. (2008c). *Matrix genetics, part 1: Permutations of positions in triplets and symmetries of genetic matrices* (pp. 1-12). Retrieved on March 8, 2008, from http://arXiv:0803.0888

Schrodinger, E. (1955). *What is life? The physical aspect of the living cell.* Cambridge: University Press.

Trahtman, A. M. (1972). *Introduction in generalized spectral theory of signals.* Moscow: Sovetskoie Radio (in Russian).

Trahtman, A. M., & Trahtman, V. A. (1975). *The foundations of the theory of discrete signals on finite intervals.* Moscow: Sovetskoie Radio (in Russian).

Chapter 3
Biological Evolution of Dialects of the Genetic Code

ABSTRACT

This chapter demonstrates results of a comparative investigation of characteristics of degeneracy of all known dialects of the genetic code. This investigation is conducted on the basis of the results of symmetrological analysis, which were described in Chapter 2, about the division of the set of the 20 amino acids into the two canonical subsets: the subset of the 8 high-degeneracy acids and the subset of the 12 low-degeneracy acids. The existence of numerical and structural invariants in the set of these dialects is shown. The derived results from the comparative investigation permit one to formulate some phenomenological rules of evolution of these dialects. These numeric invariants and parameters of code degeneracy draw attention to the formal connection of this evolution with famous facts of chrono-biology and chrono-medicine. The chronocyclic conception of the functioning of molecular-genetic systems is proposed on this basis. The biophysical basis of this conception provides connection to the genetic code structures with mechanisms of photosynthesis which produce living substance by means of utilization of solar energy. And the solar energy comes cyclically on the surface of the Earth. The revealed numeric invariants of evolution of the genetic code give new approaches to the fundamental question, why do 20 amino acids exist? We will demonstrate new patterns of the genetic code systems.

INTRODUCTION AND BACKGROUND

Beginning with the level of the code correspondence between 64 triplets and 20 amino acids, some evolutional changes take place, which lead to many different dialects of the genetic code. Each amino acid is encoded in a concrete dialect by a certain quantity of triplets. This quantity of its triplets is called

DOI: 10.4018/978-1-60566-124-7.ch003

"number of degeneracy" of the genetic code. For example, the amino acid Thr is encoded by 4 triplets in one genetic dialect; the number of degeneracy of this amino acid in this dialect is equal to 4. But this amino acid is encoded by 8 triplets in another dialect of the genetic code, where its number of degeneracy is equal to 8, etc. Structures of the set of such dialects reflect features of biological evolution on very basic levels. It seems that the comparative analysis of these dialects can give important information about essence and mechanisms of biological organisms. The symmetry analysis of phenomenological data is useful for answering these questions as well.

One direction, where such information can be useful, is connected with knowledge about physiological rhythms in organisms. The statement that biological organisms exist in accordance with cyclic processes of environment and with their own cyclic physiological processes is one of the most classical statements of biology and medicine from ancient times. Many branches of ancient and modern medicine take into account the time of day especially, when diagnostic, pharmacological and therapeutic actions should be made for individuals. The set of this medical and biological knowledge is usually united under the names chrono-medicine and chrono-biology. But is it possible to spread this chrono-biological viewpoint from the usual level of macro-physiological systems into the molecular-genetic level? This chapter analyzes this problem.

The second direction, where results of the comparative analysis of the dialects of the genetic code can be useful, is connected with the question of internal structure of the set of 20 amino acids. This question is considered in the last paragraph of this chapter.

The third direction is related to algebraic foundations of the genetic code, which will be considered in Chapter 7.

So, the objectives of this chapter are, firstly, the comparative analysis of all known dialects of the genetic code, secondly, the utilization of its results to develop appropriate thoughts about chrono-biology at the molecular-genetic level and about the internal structure of the set of 20 amino acids.

The various dialects of the genetic code exist in different kinds of organisms or of their subsystems (first of all, in mitochondria, which play a role of factories of energy in biological cells). For this book all initial data about the dialects of the genetic code were taken by the authors from the website of the National Center for Biotechnology Information http://www.ncbi.nlm.nih.gov/Taxonomy/Utils/wprintgc.cgi. These dialects differ one from another through their specifics of the degeneracy (through concrete relations between 20 amino acids and 64 triplets). Based on these data, one can find that 17 dialects are known only which differ one from another by the numbers of the degeneracy of the amino acids (see these 17 dialects in Table 1). A small quantity of the dialects from the website differ one from another by their start-codons only but not by the numbers of the degeneracy of the amino acids; we consider these dialects as the same dialect in our investigation.

Concerning chrono-biology and chrono-medicine, literature sources have many brilliant words about the great importance of biological rhythms for organisms. For example, the famous Russian physiologist A. Bogomolets wrote about "universal rhythmic movement in biology": "*The world exists in rhythms, cosmic processes follow the law of rhythmic movement ... The day replaces night, the time of activity replaces the dream ... The vital processes work in an organism rhythmically ... A heart works rhythmically, and lungs breathe rhythmically, and processes of feeding of an organism are worked rhythmically, and nervous system follows the law of a rhythm, creating a rhythm of mental life*" (Vogralik & Vogralik, 1978, p. 11).

According to the famous concepts of Ancient Oriental medicine about the cyclic nature of biological processes, "*each organ has more or less a definite time interval for its culmination (its own time interval), when its activity is maximal, ... each organ has a maximum sensitiveness to pathogenic and medicinal influences just in this special time interval*" (Vogralik & Vogralik, 1978, p.11). This phenomenological knowledge about the chronocyclic essence of biological organisms was used and tested during several thousand years by generations of oriental doctors, which were specially selected from many candidates in accordance with the criteria of their talents and of their brains. Many effective methods were constructed on the basis of this knowledge. (for example see (Cheng Xinnong, 1987; Needham, 1956)). One of them is the pulse diagnostics of Tibetan medicine. This pulse diagnostics was a universal method of diagnostics for an experienced doctor, who could determine not only many kinds of diseases, but report sometimes about physiological past and future of his patient. It is known that a doctor traditionally examines the state of 12 main organs during a session of pulse diagnostics (Tsydypov, 1988, p. 7). This method shows additionally, that chronocyclic processes (pulse processes, etc.) in biological organisms carry astonishingly complete information about organism on the whole.

Modern medicine and biology agree with many views of the Ancient Oriental medicine in questions of chrono-medicine and internal biological clock of organisms (see for example (Dubrov, 1989; Wright, 2002)). Many diseases are connected with disturbances of natural biological rhythms in organisms. The problem of internal clocks of organisms, which participate in coordination of all interrelated processes of any organism, is one of the main physiological problems.

From ancient times, medicine connects chronocyclic processes of biological organisms with chronocycles of the surrounding world, first of all, with the solar cycles of the changing of days and nights. It was found that the duration of such solar cycles could be divided comfortably for many practical tasks into 24 equal parts ("hours" by their modern name). For example, this division was comfortable in connection with the periodical activity of human organs. Ancient Oriental doctors divided 24 hours into 12 equal parts with a two-hour duration for each part. Each part was considered traditionally as a time interval of culmination activity of one of 12 main physiological organs. The other 11 main organs work in this time interval as well, but without their culmination activity. This division of 24 hours into 12 equal parts is used intensively in recipes of acupuncture, in methods of pulse-diagnostics and in other branches of Oriental medicine (see, for example (Vogralik & Vogralik, 1978)). It is very interesting that many of these branches of Oriental medicine, including acupuncture and pulse-diagnostics, recommend time intervals of application of their recipes and methods in accordance with a table of 64 hexagrams and other symbolic structures of "I Ching" ("The Book of Changes") (see, for example (Falev, 1991)). In these applications, the table of 64 hexagrams (which is connected with the genetic matrices of 64 triplets, as Chapter 11 of our book demonstrates) has an interpretation and meanings in terms of chronocycles.

It should be noted that a set of biological organisms consists of two main categories of organisms: autotrophic and heterotrophic organisms. Autotrophic organisms obtain carbon, which is needed to build their bodies, from CO_2 of the surrounding world only by means of their mechanisms of photosynthesis, based on the use of solar energy. But the sun shines from morning till night only. Intervals of cyclic activity of autotrophic mechanisms of photosynthesis are dependent on solar cycles "day-night". It is well known that "*autotrophic organisms with their photosynthesis mechanisms play a decisive role in nature because they generate a main mass of organic material in the biosphere... The existence of all other organisms and the course of biogeochemical cycles are determined by activities of autotrophic organisms*" (Giliarov, 1989, p. 9). It seems to be obvious that the solar cycle with its form "day-night" is the most important for autotrophic organisms. This solar 24-hour cycle can be considered as a main cycle of the outer world for biological objects. Is it possible that structural evolution of genetic code dialects was realized without a connection with chronocycles of the whole organism and of the outer world and, in particular, without a connection with this solar 24-hour cycle?

Due to the reasons described above, one may conclude that genetic codes of autotrophic organisms are the most significant ones for the investigation of a possible connection between genetic structures and the solar 24-hour cycle. Heterotrophic organisms are less interesting for this task. They obtain carbon for their bodies not from CO_2 and photosynthesis, but from exogenous organic materials. And heterotrophic organisms can be adapted to secondary chronocycles of those biological organisms, from which they obtain their organic food. So, one should differentiate cases of autotrophic and heterotrophic organisms in investigations of the dialects of the genetic code.

PHENOMENOLOGICAL RULES OF EVOLUTION OF KNOWN DIALECTS OF GENETIC CODES

Chapter 2 described the applications of methods of symmetry to analyze internal structure of the set of the 64 triplets and the set of the 20 amino acids. In the case of the vertebrate mitochondrial genetic code it was revealed that the set of the 20 amino acids is divided into the two sub-sets: the sub-set of the 8 high-degeneracy amino acids (Ala, Arg, Gly, Leu, Pro, Ser, Thr, Val) and the sub-set of the 12 low-degeneracy amino acids (Asn, Asp, Cys, Gln, Glu, His, Ile, Lys, Met, Phe, Trp, Tyr).

As we mentioned in Chapter 2, the vertebrate mitochondrial genetic code is the most ancient and "perfect" (symmetrical) dialect of the genetic code. We consider this dialect, which is shown in Table 1 under number 1 on the first column, as the basic dialect to compare with other dialects. Let us analyze the 17 dialects of the genetic code to reveal the possible phenomenological rules and numeric invariants of evolution of the genetic code.

The table on Table 1 demonstrates the 17 dialects of the genetic code with their numbers of degeneracy. Numbers of degeneracy (ND), which are observed in the dialects, are equal to numbers from 1 to 8. For example, the first dialect of the genetic code in the table on Table 1 possesses 12 amino acids, which number of degeneracy is equal to 2 (Asn, Asp, Cys, Gln, Glu, His, Ile, Lys, Met, Phe, Trp, Tyr); 6 amino acids, which number of degeneracy is equal to 4 (Ala, Arg, Gly, Pro, Thr,Val), and 2 amino acids, which number of degeneracy is equal to 6 (Leu, Ser). At first it seems, that the distribution of numbers of degeneracy in a set of the 17 dialects of the genetic codes on Table 1 is chaotic on the whole. But this impression disappears, if one divides the set of 20 amino acids into the two subsets, which were mentioned above: the subset of low-degeneracy amino acids, each of which is encoded by 3 triplets or less in the dialect of the vertebrate mitochondrial genetic code, and the subset of high-degeneracy amino acids, each of which is encoded by 4 triplets or more in the same basic dialect. Such division reveals hidden regularities. Other kinds of the division of the set of 20 amino acids into two subsets do not reveal hidden regularities.

The numbers of the dialects of the genetic code on Table 1 correspond to the following dialects: 1) The Vertebrate Mitochondrial Code; 2) The Standard Code; 3) The Mold, Protozoan, and Coelenterate Mitochondrial Code and the Mycoplasma/Spiroplasma Code; 4) The Invertebrate Mitochondrial Code; 5) The Echinoderm and Flatworm Mitochondrial Code; 6) The Euplotid Nuclear Code; 7) The Bacterial and Plant Plastid Code; 8) The Ascidian Mitochondrial Code; 9) The Alternative Flatworm Mitochondrial Code; 10) Blepharisma Nuclear Code; 11) Chlorophycean Mitochondrial Code; 12) Trematode Mitochondrial Code; 13) Scenedesmus obliquus mitochondrial Code; 14) Thraustochytrium Mitochondrial

Table 1. The 17 dialects of the genetic code and distributions of their numbers of degeneracy (ND) among 20 amino acids (AA). The two right columns show quantities of the low-degenerate and high-degenerate acids (ΣAA). Bold frames mark two categories of numbers of the degeneracy: from 1 to 3 and from 4 to 8 (Petoukhov, 2001a). Initial data were taken from the NCBI's website http://www.ncbi.nlm.nih.gov/Taxonomy/Utils/wprintgc.cgi

Dialects	Distribution of numbers of degeneracy from 1 to 8 among 20 AA								ΣAA with ND from 1 to 3	ΣAA with ND from 4 to 8
	1	2	3	4	5	6	7	8		
1		12		6		2			**12**	**8**
2	2	9	1	5		3			**12**	**8**
3	1	10	1	5		3			**12**	**8**
4		12		6		1		1	**12**	**8**
5	2	8	2	6		1		1	**12**	**8**
6	2	8	2	5		3			**12**	**8**
7	2	9	1	5		3			**12**	**8**
8		12		5		3			**12**	**8**
9	2	7	3	6		1		1	**12**	**8**
10	2	8	2	5		3			**12**	**8**
11	2	9	1	5		2	1		**12**	**8**
12	1	10	1	6		1		1	**12**	**8**
13	2	9	1	5	1	1	1		**12**	**8**
14	2	9	1	5	1	2			**12**	**8**
15	2	9	1	5	1	1	1		**12**	**8**
16		13		5		1		1	**13**	**7**
17	2	8	1	6		3			**11**	**9**

Code; 15) The Alternative Yeast Nuclear Code; 16) The Yeast Mitochondrial Code; 17) The Ciliate, Dasycladacean and Hexamita Nuclear Code.

The data on the Table 1 permit us to formulate the following phenomenological rule (Petoukhov, 2001a):

The phenomenological rule № 1: in genetic codes, the set of 20 amino acids contains two opposite subsets: the first subset consists of 12 low-degeneracy amino acids (with their numbers of degeneracy from 1 to 3), and the second subset consists of 8 high-degeneracy amino acids (with their numbers of degeneracy from 4 to 8).

As the authors can conclude, this rule about the canonical ratio 12:8 for two categories of amino acids is held true in nature without any exceptions for dialects of the genetic code of autotrophic organisms. These types of organisms play the main role in biogeochemical cycles. But this rule has small exceptions in two cases of heterotrophic organisms in a form of minimal numeric shifting from the regular ratio "12:8" to the nearest integers ratios: The "Yeast Mitochondrial Code" possesses the ratio "13:7" for these two categories of amino acids, and the "Ciliate, Dasycladacean and Hexamita Nuclear Code" possesses the ratio "11:9". These non-standard ratios encircle the canonical ratio "12:8" from the contrary sides of numeric axis. These non-standard ratios demonstrate additionally the main role of the canonical ratio 12:8 as that centre, around which minimal numeric fluctuations exist.

The data about evolution of the genetic code also demonstrates the existence of the following rule about canonical sub-sets of the low-degeneracy and high-degeneracy amino acids.

The phenomenological rule № 2: if a triplet encodes different amino acids in different genetic codes, then these amino acids belong to the same canonical subset of amino acids. In other words, it is practically forbidden for those triplets, which encode amino acids from one canonical subset of degeneracy, to pass into the group of triplets during biological evolution, which encode amino acids from another canonical subset.

A single exception to this rule exists: the triplet UAG can encode amino acids Leu or Gln in the different canonical subsets. The rule says nothing about stop-codons, and so it does not consider those evolutionary cases, when triplets which encode stop-codons (or amino acids) in one genetic code begin to encode amino acids (or stop-codons respectively) in another code.

Phenomenological rules described above testify that two independent branches of evolution of the genetic code exist at billions biological species: one branch – for canonical subset of high-degeneracy amino acids, and another branch - for canonical subset of low-degeneracy amino acids. These evolutionary branches within the consolidated code system can be compared with a parallel evolution of male and female organisms within a frame of one biological species. It reveals simultaneously that nature realizes an association of two very different subsets of 8 and 12 amino acids in the set of 20 amino acids. Thereby the matrix genetics reveal the existence of such internal structure in the set of 20 amino acids, which possesses the invariant properties in evolution of the genetic code. One can find additional details about such phenomenological rules of the dialects in the article (Petoukhov, 2001b).

THE CHRONOCYCLIC CONCEPTION AND THE DEGENERACY IN THE DIALECTS OF THE GENETIC CODE

One can note two numerical peculiarities of the natural system of the degeneracy numbers of amino acids in the set of the dialects of the genetic code:

1. Number **24** is the least divisible integer for numbers 8 and 12;
2. Main numbers of degeneracy of amino acids in all dialects codes are 1, 2, 3, 4, 6, 8; all of them are divisors of number **24** (four dialects have a single amino acid with its number of degeneracy 5 or 7; a rate of each of these non-typical numbers of degeneracy in the whole set of the dialects is equal to 0.88%).

Number **24** can be considered as the hidden constant of coordination among numbers of degeneracy in the dialects of the genetic code. But number **24** is well known in chrono-biology and chrono-medicine since ancient time as we mentioned above.

Chrono-medicine, which has thousand-year history, asserts that physiological systems of any individual organism undergo regular changes of their physiological activity and passivity within the limits of the certain time intervals, which are connected with division of day on 24 equal parts. Oriental chrono-medicine are related to the "day-night" cycle of entrance of solar energy on the surface of the Earth. In the field of chrono-medicine, the number 24 represents not an arbitrary division of day into some parts, but phenomenological concordance of duration of physiological cycles with the duration of the day-night cycle.

Modern molecular biology knows that existence of proteins, structures of which are encoded genetically, possesses a cyclic character as well. Really, it is the well-known fact that proteins in biological organisms are re-built (re-created) by systematic cyclic processes. It means that a set of physicochemical factors inside biological organisms disintegrates proteins into amino acids permanently and then it rebuilds them from amino acids again in a cyclic manner. A half-life period (a duration of renovation of half of a set of molecules) for proteins of human organisms is approximately equal to 80 days in most cases; for proteins of the liver and blood plasma – 10 days; for the mucilaginous cover of bowels – 3-4 days; for insulin – 6-9 minutes. Such permanent rebuilding of proteins provides a permanent cyclic renovation of human organisms. These known facts are described in biological encyclopedias (for example, see (Aksenova, 1998, v. 2, p. 19)). Such cyclic processes at the molecular-genetic level should be investigated from various theoretical viewpoints. One possible viewpoint is given by the chronocyclic conception (Petoukhov, 2001b, 2008), which is described below.

This chronocyclic conception interprets separate groups of amino acids (or groups of triplets) as special "organs", which have their culmination time intervals of their cyclic activity in 24-hour solar cycle by analogy with time intervals of culmination activity of macro-physiological organs from the above-mentioned conception of Oriental medicine. Of course, the cyclic activity of such genetic "organs" is coordinated with a cyclic activity of physico-chemical factors, which provides their work, including a necessary activation of amino acids. It is well known that "*the necessary condition of proteins synthesis, which is expressed by polymerization of amino acids, is the existence of non-free, but so called activated (!) amino acids in the system, which amino acids have their own resource of energy. Activation of free amino acids is realized by means of specific ferments*" (Berezov & Korovkin, 1990, p. 409).

From the viewpoint of the chronocyclic conception, the 12 low-degenerated amino acids can be interpreted conditionally as a certain interrelated ensemble of "organs", which divides a 24-hour cycle into a sequence of 12 equal parts with a 2-hour duration of each part. And each part corresponds to a time interval of a culmination activity of one of these amino acids (together with physicochemical factors, which serves this amino acid). The idea of chronocyclic culmination activities of the considered amino acids (with their teams of servicing) is placed here in a parallel with the phenomenological knowledge of Oriental medicine about the chronocyclic culmination activities of physiological macro-systems. It is essential that one can examine experimentally the existence of cyclic culmination activity of each amino acid. In our opinion, this experimental task of investigation of chronocyclic activities of amino acids in vivo is very important for understanding the genetic coding system.

Another group with the 8 high-degenerated amino acids can be interpreted in such a model as a certain interrelated ensemble of "organs", which divides a solar 24-hour cycle into a sequence of 8 equal parts with the 3-hour duration of each part.

By the way, for those readers, who are interested in Oriental medicine and in Ancient Oriental culture, it could be mentioned that there is a special meaning to the numeric pair "12 and 8", which is one of the distinquished pairs there: "*8 and 12 are a standard measure of alternative separations of space-time in Chinese chronotopograms ... The symbol of the Earth – a square – is characterized by number 8, and the symbol of heaven – a circle - is characterized by number 12*" (Kobzev, 1994, p. 39, 40).

In the modelling approach, each amino acid receives a new theoretical parameter, connected with chronocyclic processes: the duration of its time interval of the culmination activity. More precisely, the 12 low-degenerated amino acids receive a relatively shorter duration (2 hours). The 8 high-degenerated amino acids receive a relatively greater duration (3 hours). It permits to introduce comfortable and heuristic terminology from linguistics for two considered categories of amino acids.

The set of 20 amino acids is the genetic alphabet for proteins. An analogy between genetic code and linguistics has been widely used in science for a long time by many authors. Moreover the famous conception exists for many years, that all linguistic languages were formed not on an empty place, but they are a continuation of the genetic language or, anyway, are closely connected to it, confirming the idea of unification of information bases of organisms (for example, see (Baily, 1982; Jacob, 1974)). The book "Linguistic genetics" marks: "*The opinion about the language as about living organism, which submitted to the natural laws of nature, ascend to a deep antiquity... Research of a nature, of character and of reasons of isomorphism between genetic and linguistic laws is one of the most important cardinal problems for linguistics of our time*" (Makovskiy, 1992, p. 15).

But alphabets of linguistic languages always consist of consonant letters and vowel letters, which differ phonetically in terms of their time durations and relative quantities in each alphabet (the quantity of consonant letters is greater than the quantity of vowel letters). The alphabet of 20 amino acids with the two canonical categories of amino acids, which differ in terms of their time durations in the described modelling approach, has a new obvious parallel with the alphabets of human languages relative to their two categories of consonant letters and of vowel letters. Due to this parallel, one can name 12 considered amino acids with the shorter time duration (2 hours) as "consonant" amino acids, and 8 other amino acids with the relative greater duration (3 hours) as "vowel" amino acids. The quantity of consonant amino acids is greater than the quantity of vowel amino acids in concordance with the relative quantities of consonant letters and of vowel letters in linguistic alphabets.

Human speech and writing are constructed on the basis of alternating change of vowel and consonant elements, and chained sequence of proteins is based on alternating changes of vowel and consonant amino acids. It is probable that numerous number of physiological processes is constructed in a similar chained pattern with alternating changes of their "vowel" and "consonant" elements, which differ typically by their time duration and which are produced there by nature. In this context about binary-opposite categories of physiological sub-processes, vowel element is a representative from a category of more prolonged sub-processes, and consonant element is a representative from a category of shorter sub-processes. For example, the human cardio cycle lasts 1 second approximately at rest. In rest this cardio cycle consists of a more prolonged activity phase in 0.6 sec and a shorter repose phase in 0.4 sec. The ratio of duration of these phases is equal to 6:4 = 12:8 = 3:2. These two phases of cardio cycles can be correlated to two categories of durations ("vowel" and "consonant"). It should be noted that this cardio ratio 6:4 = 12:8 = 3:2 is equal to the described ratio 12:8 between the quantity of the consonant and vowel amino acids. This ratio 3:2 is named the quint (or the fifth) in the field of musical harmony. This quint ratio underlies the harmony of ancient Chinese music and the Pythagorean musical scale as Chapter 4 of our book describes in connection with materials of matrix genetics.

From an informational viewpoint, all physiological processes in an organism can be represented as information messages to interchange by information among different subsystems of organisms (or all physiological processes have their information components additionally). In our opinion, this information interchanging is realized in more or less uniform languages, which are coordinated with genetic languages, and their alphabets can possess vowel and consonant elements as well. Due to this reason, one should investigate all physiological processes to find representatives from two binary-opposite categories of durations ("vowel" and "consonant") there by analogy with linguistic alphabets.

By the way, computer informatics does not use such ordinary alphabetic symbols, which are differed by their time durations. The reason is that a trigger technology provides equal times for trigger transitions into "on" or "off" states. So, computer informatics and human languages have important differences in this aspect, which is connected with deep physiological mechanisms of biological informatics including human speech on the whole.

The very important hypothesis in the frame of the chronocyclic conception is the hypothesis about the connection between the structure of the genetic code and mechanisms of photosynthesis. The mechanisms of photosynthesis play the role of the initial mechanisms, which produce in cyclic manner the living substance of autotrophic organisms in accordance with cyclic arrivals of solar energy to organisms. By this hypothesis, the genetic code structures are connected by means of mechanisms of photosynthesis with the 24-hour cycle of arrival of solar energy on the surface of the Earth. From this viewpoint, those cyclic processes of macro-physiological systems of organisms, which are co-ordinated with phases of the day-night cycle, have molecular-genetic forerunners, which are co-ordinated with these phases as well. And secrets of structures of the genetic code are related to secrets of biological phenomenon of photosynthesis. The efficiency of photosynthesis is not reproduced in modern laboratories till now. But its mechanisms produce cyclically the living substance, which exists cyclically and which is encoded genetically by means of adequate cyclic forms. In other words, one can think that photosynthesis is primary in relation to the genetic code which promotes the coded inheritance of already photosynthesized primary matter. And structures of the genetic code (for example, the phenomenon of division into sub-sets of the 8 and 12 acids, the specifics of numbers of degeneracy of amino acids) depend on mechanisms of photosynthesis and the 24-hour cycle of day-night.

One of additional indirect arguments of chrono-biological dependence of structures of the genetic code is the coincidence of matrix structures of the genetic code with tables of the ancient Chinese "I Ching", which underlie the Oriental chrono-medicine (see Chapter 12 of our book). By the way, G. Stent (1969), who is the famous specialist in the field of molecular genetics, has put forward the hypothesis about a possible connection between the set of 64 genetic triplets and the table of 64 hexagrams from "I Ching". As far as we know, it was the first publication on this theme, and so Stent should be considered as a pioneer in this field of analyzing of parallels between modern molecular genetics and mysterious knowledge of Ancient civilizations.

The chronocyclic theory of genetic codes considers molecular-genetic processes as chronocycles, included in a mutual chorus of chronocycles of nature. It has been known for a long time, that processes of synthesis of proteins have a cyclic character. From an ordinary viewpoint, structures of genetic code are destined to code amino acids in their space sequence in proteins. From the viewpoint of the chronocyclic theory, it is likely that these genetic structures are coding simultaneously time parameters of cyclic processes of amino acids and of protein's synthesis. Moreover, one can think that genetic structures are coding, first of all, these chronocycles exactly, due to which the coding of amino acids (and of proteins) is realized in a secondary manner. In other words, DNA and RNA are carriers of information not only about primary composition of proteins, but about the chronocyclic organization of amino acids and proteins as well. From this position, those biological rhythms, which are observed at very different physiological levels so widely, should be derived not from peculiarities of final ensembles of proteins only, but also from peculiarities of pre-protein's genetic structures, which carry chronocyclic information in a long train of biological generations. The proposed chronocyclic conception includes the thesis that internal genetic clocks exist, which are distributed inside many parts of individual organism to participate in coordination of the whole chorus of cyclic physiological processes at different levels.

It is very likely that universal nitrogenous bases of the genetic code have one more hidden attribute (or the trait) – chronocyclic (time) attribute. For example, complementary nitrogenous bases can be characterized by equal typical time of a process of their junction during the formation of DNA (and of their separation during the splitting of DNA). Two pairs of complementary bases with their 3 and 2 hydrogen bonds can have the appropriate ratio 3:2 of their typical times in some sense. Appropriate genetic matrices, which include a factor of time, can be written for mono- and multiplets of genetic systems. It is very probable that genetic structures are coding not only the synthesis of proteins, which is the first stage in life of proteins, but also the whole cycle of their life including their disintegration phase. One can think that a future theory of genetic systems will include a theoretic consideration of these cyclic phenomena of protein's life in connection with other cycles of nature.

Concerning the unity of a biological organisms, one should emphasize that structures of all physiological systems, which have a chronocyclic character of their work, should be coordinated with structural-cyclic peculiarities of genetic coding system to provide the evolutionary survival of these physiological systems by means of their reproduction in next generations.

WHY 20 AMINO-ACIDS?

Many attempts to answer on this fundamental question are known. On the basis of the described phenomenological rules of evolution of dialects of the genetic code, one can propose the new possible answer: the set of 20 acids are presented in genetic code, that is formed by two alternative subsets of 8 and 12 amino acids. Therefore the initial question comes down to the deeper question: why dose the two alternative sub-sets of the 8 high-degeneracy acids and the 12 low-degeneracy acids exist in the set of 20 amino acids?

A possible answer on this new fundamental question is related to the revealed fact, that these two sub-sets constitute two independent branches of evolution within a genetic code as it was described above. These numbers 8 and 12 have their tetra-presentation: 8 = 4x**2**, 12 = 4x**3**. In this presentation, the number 8 contains the number 2 as its modular block, and the number 12 contains the number 3 in the analogical role. Just a biological mechanism of tetra-segregations can be responsible for the realization of such two sub-sets of amino acids. Each modular block of the sub-set of the 8 high-degeneracy acids consists of two amino acids, and each modular block of the sub-set of the 12 low-degeneracy acids consists of three amino acids. One can note the formal analogy between these 2-part and 3-part blocks and the famous fundamental hypothesis from the quite different field of physics of elementary particles: according to the quark hypothesis, baryons consist of three quarks, and mesons consist of two quarks (the quark and the anti-quark). Is it possible for formal elements of the quark theory to be transferred into the field of the theory of the genetic structures? The future will show.

Additional investigations have revealed that the considered pair of alternative attributes ("high-degenerative and low-degenerative") is not a single pair for a division of the set of 20 amino acids into subsets of 8 and 12 amino acids. The genetic code is constructed so, that such division is a typical one for many other pairs of real binary-opposite attributes, in relation to which such division is considered. Similar divisions, but with different sub-sets of 8 and 12 amino acids, are given by such binary-opposite attributes as "complementary-uncomplementary" amino acids, "high-carbon or low-carbon" amino acids, "hydrophobic or hydrophilic" amino acids, "eightfold or non-eightfold quantity of protons" in amino acids (Petoukhov, 2001b, 2005). One can think that such multichoice phenomenon of the typical segregation of the set of 20 amino acids into the two sub-sets with their ratio 12:8 is connected with providing the parallel channels of biological information, which work with different binary-oppositional attributes (He & Petoukhov, 2007).

For the proposed viewpoint about the principle of the tetra-segregation, Figure 1 demonstrates a confirmative example with sub-sets of complementary and uncomplimentary amino acids: the existence of the sub-sets of the 8 and 12 amino acids is provided by the principle of their tetra-construction from typical modular blocks with 2 units and with 3 units (8=4x2 and 12=4x3). Complementary amino acids are those, which are encoded by groups of codons and their anti-codons. One can see from Figure 2 of Chapter 2, that the 8 amino acids form the four pairs of complementary amino acids (Pro-Gly, Arg-Ala, Lys-Phe, Met-Tyr), but other 12 amino acids are uncomplementary ones. The sub-set of the 8 complementary amino acids is divided into those four pairs (or four modular blocks with two amino acids), each of which is encoded by triplets from a separate family of N-triplets. And the sub-set of the 12 uncomplementary amino acids is divided into those four triples (or four modular blocks with three amino acids), each of which is encoded by triplets from a separate families of N-triplets (in the case of each of the amino acids Ser and Leu, we take here into account those family of N-triplets, all four triplets of which encode it in Figure 2 of Chapter 2). So, each of four families of N-triplets encodes 2 complementary amino acids and 3 uncomplementary ones.

FUTURE TRENDS AND CONCLUSION

The analysis of symmetries in numbers of degeneracy of many kinds of the dialects of the genetic code have led to discoveries of some phenomenological rules about numeric invariants and regularities in this evolution. The obtained results produced new concepts about chronocyclic aspects of the molecular-genetic system and about the fundamental question, why do 20 amino acids exist. These results will be used in Chapter 7, where the 8-dimensional algebra of the genetic code is described.

The phenomenological data about evolution of the genetic code needs to be investigated further. Why do only some triplets change their coding meaning in the course of biological evolution? In what aspects do these variable triplets differ from conservative triplets? Is it possible to propose such adequate mathematical model of the genetic code, which reflects the evolutionary peculiarities of the dialects? One can think that methods of symmetry analysis will be useful to solve these and other similar questions as well.

The proposed chronocyclic conception gives some new approaches to investigate functional features of the molecular-genetic systems experimentally and theoretically. The heuristic research of internal genetic clocks, which are distributed along all parts of the whole organism, has an important meaning to understand a coordinated organization of various biological rhythms at different biological levels. Obvi-

Figure 1. An example of the presentation of the set of 20 amino acids (AA) as a sum of two subsets with 8 = 4 x 2 and 12 = 4 x 3 acids of complementary and uncomplementary types according to Figure 2 of Chapter 2. The first tetra-subset has the four pairs of amino acids of the complementary type. The second tetra-subset has the four triples of the uncomplementary amino acids correspondingly. Triplets from the same four families of N-triplets encode these two tetra-subsets of amino acids

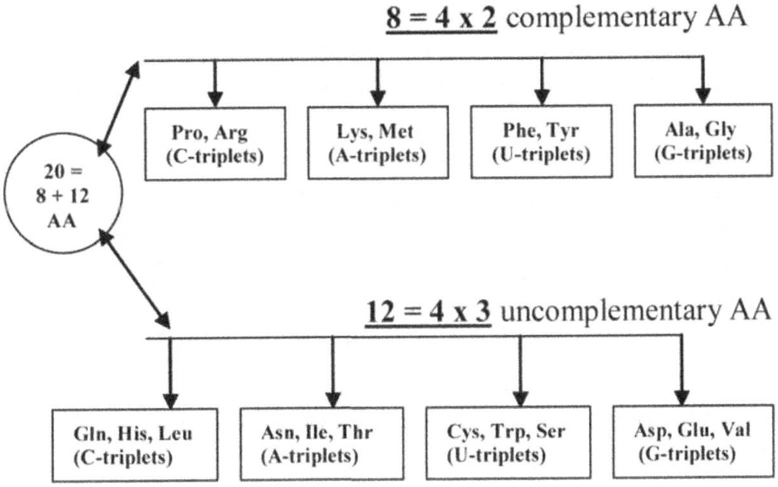

ously, such research has not only theoretical, but practical aims also. For example, at what time and how frequently should we give pharmacological medicines? It is well-known that the same pharmacological medicines have very different effects depending on a time of its taking. A knowledge about internal genetic clocks of organisms is very valuable for pharmacological and physiotherapeutic influences, for conducting of morphogenetic and growth processes, for ergonomic stimulation in man-machine systems and for many other tasks.

The chronocyclic conception extends additionally the traditional field of investigations of parallels between genetic and linguistic languages by introducing the reasoned notions of the vowel and consonant amino acids. The conception produces arguments also to study typical binary-oppositional kinds of time durations (short and long durations with typical ratios between them, for example, with the quint ratio 3:2) in various physiological processes on different biological levels.

The proposed answer on the fundamental question about the set of 20 amino acids brings down this question to the deeper one about the 8 high-degeneracy acids and the 12 low-degeneracy acids. Investigations of various aspects of this answer revealed the phenomenon of existence of many variants of division of this set into two sub-sets of 8 and 12 acids depending on a choice of a few kinds of molecular binary-oppositional attributes.

In our opinion, an additional comparative analysis of dialects of the genetic code will give many essential results to understand specifics of the genetic code systems and their evolution. Methods of symmetry and of matrix genetics will be utilized extensively in these researches. They will permit to discover not only phenomenological rules of molecular-genetic evolution but also to develop appropriate mathematical models of genetic systems (see Chapter 7).

The chronocyclic conception will unite many isolated facts and details of cyclic processes in molecular-genetic systems and will facilitate understanding the connection between cyclic processes at molecular-genetic and macro-physiological levels. The progress in investigations of genetic clocks, which are distributed inside many parts of the whole organism, can help to overcome those diseases, which are connected with disturbances of biological rhythms. One can hope that this progress will be useful for solving the problem of ageing of human organisms, which is related to violation of physiological cycles to some extent as well. The science knows examples of biological organisms, which are immortal practically and which utilize endless cycles.

A bridge between the famous theory of hypercycles by Eigen (1971, 1979, 1988, 1992, 1993) and hierarchies of cyclic processes at molecular-genetic level is possible to some extent for tasks of modeling.

The investigation field of parallels between genetic and linguistic languages will be extended by utilizing the notions "vowel" and "consonant". Possible researches of physiological processes with typical binary-oppositional kinds of their durations (with the quint ratio 3:2 and others) will demonstrate new fields of a specific coordination of processes in biological organism at its various levels.

The proposed answer to the question of the set of 20 amino acids stimulates new ideas about structures of the genetic code and leads to new researches about the phenomenon of many variants of division of this set into two sub-sets with 8 and 12 acids depending on concerned binary-oppositional attributes of genetic molecules. A possible meaning of this new typical pattern of the division with the ratio 8:12 should be investigated from various viewpoints.

The analysis of symmetries in internal structures of the sets of the 64 triplets and of the 20 amino acids leads to useful results, which help in investigating the evolution of the genetic code. Experimental data of molecular genetics about many dialects of the genetic code can be utilized by means of the comparative symmetrical analysis to discover phenomenological rules of evolution of the genetic code. The described phenomenological rules draw attention to existence of numeric invariants in evolution of these dialects and to phenomenon of the division of the set of the 20 amino acids into two sub-sets of the 8 high-degeneracy acids and the 12 low-degeneracy acids. This division is the new typical pattern in molecular genetics.

The proposed chronocyclic conception leads to new ideas for experimental and theoretical researches to understand the general chorus of cyclic processes in each biological organism.

The results of investigations, which were described in this chapter, will be utilized to develop the mathematical model of the genetic code and to extend applications of methods of symmetry in the field of molecular genetics.

REFERENCES

Aksenova, M. (Ed.). (1998). *Encyclopedia of biology*. Moscow: Avanta+ (in Russian).

Baily, C. (1982). *On the yin and yang nature of language*. London: Ann Arbor.

Berezov, T. T., & Korovkin, B. F. (1990) *Biological chemistry*. Moscow: Nauka (in Russian).

Cheng, X. (Ed.). (1987). *Chinese acupuncture and moxibustion.* Beijing: Foreign Languages Press.

Eigen, M. (1971) *Self-organization of matter and the evolution of biological macromolecules*. Berlin, New York: Springer-Verlag.

Eigen, M. (1979). *The hypercycle-a principle of natural self-organization.* Berlin: Springer-Verlag.

Eigen, M. (1988). *Perspektiven der wissenschaft.* Stuttgart: Deutsche Verlagsanstalt.

Eigen, M. (1992). *Steps towards life. A perspective of evolution.* Oxford: Oxford University Press.

Entsiklopedia (in Russian). Dubrov, A. P. (1989) *Symmetry of biorhythms and reactivity.* New York, London, Tokyo: Gordon & Breach Science Publishers.

Falev, A. I. (1991). *Classic methodology of traditional Chinese tzhen-tsiu therapy.* Moscow: Prometei (in Russian).

Giliarov, M. S. (Ed.). (1989). *Biological encyclopedic dictionary.* Moscow: Sovetskaia

He, M., & Petoukhov, S. V. (2007). Harmony of living nature, symmetries of genetic systems and matrix genetics. *International Journal of Integrative Biology, 1*(1), 41–43.

Jacob, F. (1974). Le modele linguistique en biologie. *Critique, Mars, 30*(322), 197-205.

Kobzev, A. I. (1994). *Studies on symbols and numbers in Chinese classical philosophy.* Moscow: Vostochnaia literature (in Russian).

Makovskiy, M. M. (1992). *The linguistic genetics.* Moscow: Nauka.

Needham, J. (1956). *Science and civilization in China.* In 3 volumes. Cambridge: University Press.

Petoukhov, S. V. (2001a). Genetic codes I: Binary sub-alphabets, bi-symmetric matrices, and golden section. *Symmetry: Culture and Science, 12*(1), 255–274.

Petoukhov, S. V. (2001b). Genetic codes II: Numeric rules of degeneracy and the chronocyclic theory. *Symmetry: Culture and Science, 12*(1), 275–306.

Petoukhov, S. V. (2005). The rules of degeneracy and segregations in genetic codes. The chronocyclic conception and parallels with Mendel's laws. In M. He, G. Narasimhan & S. Petoukhov (Ed.), *Advances in Bioinformatics and its Applications, Series in Mathematical Biology and Medicine, 8*, (pp. 512-532). Singapore: World Scientific.

Petoukhov, S. V. (2008). *Matrix genetics, algebras of the genetic code, noise-immunity.* Moscow: RCD (in Russian).

Stent, G. S. (1969) *The coming of the golden age.* New York: The Naural Histore Press.

Tsydypov, C. Ts. (Ed.). (1988). *Pulse diagnostics of Tibetan medicine.* Novosibirsk: Nauka (in Russian).

Vogralik, V. G., & Vogralik, M. V. (1978). *Acupuncture.* Gor'kii: Meditsina (in Russian).

Wright, K. (2002). Times of our lives. *Scientific American, 287*(3), 58–65.

Section 2
Symmetrical Analysis Techniques and Numeric Matrices of the Genetic Code

Section 2 is organized into three chapters. This section discusses numeric matrices of genetic code and establishes the relationships between genetic code, stochastic matrices, and Hadamard matrices. The noise immunity, encoding and principle of molecular economy in genetic informatics, and Fibonacci numbers and phyllotaxis laws are presented in this section.

Chapter 4
Numeric Genomatrices of Hydrogen Bonds, the Golden Section, Musical Harmony, and Aesthetic Feelings

ABSTRACT

This chapter is devoted to a consideration of the Kronecker family of the genetic matrices, but in the new numerical form of their presentation. This numeric presentation gives opportunities to investigate ensembles of parameters of the genetic code by means of system analysis including matrix and symmetric methods. In this way, new knowledge is obtained about hidden regularities of element ensembles of the genetic code and about connections of these ensembles with famous mathematical objects and theories from other branches of science. First of all, this chapter demonstrates the connection of molecular-genetic system with the golden section and principles of musical harmony.

INTRODUCTION AND BACKGROUND

Till this moment we analyzed the symbolic genetic matrices. Now we begin to analyze numeric genetic matrices, which are produced from the symbolic genomatrices. What are initial reasons to pay attention to numeric genomatrices?

The previous chapters demonstrated that the Kronecker product of matrices is useful for analysis of genetic code and is adequate for its structure. But the Kronecker product possesses some distinctive properties, which are connected with eigenvalues of matrices: eigenvalues of the Kronecker product A⊗B for two matrices A and B, which have their eigenvalues α_i and β_k, are equal to the products $\alpha_i*\beta_k$ of these eigenvalues. This property gives an additional opportunity to introduce the notion of the Kronecker product into mathematics (Bellman, 1960). But if eigenvalues are so important for the theme of Kronecker products, one should investigate numeric genomatrices, which possess eigenvalues (symbolic matrices do not possess eigenvalues).

DOI: 10.4018/978-1-60566-124-7.ch004

Also we will try to investigate genetic sequences from the viewpoint of the theory of digital signal processing. This theory presents a signal in the form of a sequence of its numeric values in points of references. Discrete signals are interpreted as vectors of multi-dimensional spaces: a value of the signal in each time (a moment of reference) is interpreted as the value of one of the coordinates of multi-dimensional space of signals (Trahtman, 1972). The theory of discrete signals processing is the geometrical science about multidimensional spaces in some extent. The number of dimensions of such space is equal to the quantity of moments of references for the signal. Appropriate metric notions and other necessary things for providing the reliability, velocity and economy of information transfer are introduced in these multi-dimensional vector spaces. For example, important information notions of the energy and of the power of a discrete signal are correspondingly the square of the length of the vector-signal and the same square of the length of the vector-signal, which is divided by the number of dimensions. Various signals and their ensembles are compared as geometrical objects of such metric multi-dimensional spaces.

These methods underline technologies of signal intelligence and pattern recognition, detections and corrections of information mistakes, artificial intellect and robot learning, etc. If we wish to use the methods of the theory of discrete signals processing for analyzing the genetic structures, we should learn to turn from the symbolic genetic matrices and genetic sequences to their numeric analogies.

The method, which is utilized in this book for such a turn, replaces the letter symbols A, C, G, U(T) of the genetic alphabet by quantitative parameters of these nitrogenous bases, which determine their physical-chemical role (Petoukhov, 2001a). First of all, these symbols are replaced in this chapter by numbers of the hydrogen bonds, which are suspected long ago as important participants of transferring of genetic information. Each molecular element of the genetic code is a component part of a harmonic system of genetic coding. Its molecular parameters are coordinated with quantitative parameters of other elements of this system. Quantitative characteristics of separate elements should be investigated as a part of the set of quantitative characteristics of system ensemble of elements. The matrix approach is known in science long ago as very effective for system investigations, for example, in the fields of quantum mechanics, physics of elementary particles, etc. In the field of matrix genetics, this approach unites parameters of a set of separate elements not only in a general matrix, but in the whole family of genetic matrices, which embraces sets of multiplets of different lengths (Figure 3 of Chapter 1). In this way hidden connections between parameters of separate parts of the united genetic system can be revealed together with their relations to famous physical and mathematical constants and other objects.

One of such famous constant is the golden section or "the divine proportion" φ, which is equal to $(1+5^{0.5})/2 = 1.618\ldots$. This chapter demonstrates the connection of the genetic code parameters with the golden section in particular. The golden section is related to the famous series of Fibonacci numbers F_n, where n = 0, 1, 2, 3,.... This Fibonacci series F_n (Figure 1) begins with the numbers 0 and 1. Each next member of this series is equal to the sum of two previous members: $F_{n+2} = F_n + F_{n+1}$. Fibonacci numbers are used widely in the theory of optimization and in many other fields. One can find a rich collection of data about the golden section and the Fibonacci numbers on the web-site of "The museum of harmony and the golden section" by A.Stakhov (www.goldenmuseum.com) and in works (Jean, 2006; Kappraff, 1990, 1992).

Another hidden connection, which is revealed by means of the matrix approach, is the connection of the genetic code parameters with the Pythagorean musical scale. It is known that thoughts about the key significance of harmonious vibrations in the organization of the world exist from ancient time. For example, Pythagoreans thought about musical intervals in the planetary system and in all around. J. Kepler wrote the famous book "Harmonices Mundi", etc. Modern atomic physics found the harmonic

Figure 1. The Fibonacci series

n	0	1	2	3	4	5	6	7	8	9	10	11	...
F_n	0	1	1	2	3	5	8	13	21	34	55	89	...

ratios in spectral series by T. Lyman in the atom of hydrogen, which has been named "music of atomic spheres" by A. Einstein and A. Sommerfeld (Voloshinov, 2000). The importance of Pythagorean ideas about a role of musical harmony was emphasized also by the Nobel winner in physics R. Feynman (1963, v. 4, Chapter 50).

The scientific studies of physiological mechanisms of musical perception took place long ago. One can find the review on this topic in the article (Weinberger, 2004). Beginning with 4-months old, infants turn to a source of pleasant sounds (consonances) and turn aside a source of unpleasant sounds (dissonances). The human brain does not possess a special center of music. The feeling of love to music seems to be dispersed in the whole organism. The musical sound addresses to all in the person, or to person's archetypes. Data are known that the first shout of the baby, who has been born, corresponds to sounds on frequency of the music note "la" (440 Hz) irrespective of its timbre and of loudness, as a rule. (http://www.rods.ru/Html/Russian/MoreResonance.html). This frequency is used traditionally for tuning musical instruments by means of a tuning fork. This speaks certain biological unification of musical sounds. According to statistics, physical reactions to music (in the form of skin reactions, tears, laugh, etc.) arise in 80% of adult people. Animals also are not indifferent to human music. All such data show that the perception of music has biological essence and that the feeling of musical harmony is based on inborn mechanisms. Therefore it is necessary to search for connections of the genetic system with musical harmony. In particular this chapter presents such a search.

It is known for a long time in the field of mechanics that harmonious vibrations are capable of structurally forming and ordering influences leading, for example, to the formation of so-called figures by Chladni. The book (Jenni, 1972) presents the scientific field, which studies ordering action of harmonious vibrations on many shapeless free-flowing and liquid substances. This book demonstrates through many photos how these vibrations produce in these substances beautiful morphological patterns including five-symmetrical patterns, which are forbidden in classical crystallography. Questions about symmetries in music and poetry were investigated traditionally (see for example (Bruhn, 1992, 1996; Darvas, 2007; Goldman, 1992; Lendvai, 1993; Shubnikov & Koptsik, 1974; Tusa, 1994)).

The living substance is compared with crystals frequently. For example, E. Schrodinger (1955) named it "aperiodic crystal". Whether annals of modern science contain any data about a connection of musical harmony with crystals? Yes, such data exist. The book (Berger, 1997, p. 270-281) gives the following historical data about a few prominent crystallographers, which emphasized a connection of crystal structures with musical harmony.

In 1818, C.S. Weiss, who discovered crystallographic systems and who was one of founders of crystallography, emphasized a musical analogy in crystallographic systems. He investigated ratios among segments, which are formed by faces of crystals of the cubic system. Weiss has shown that these ratios are identical absolutely to ratios between musical tones.

In 1829, J. Grassman, who wrote a well-known book "Zur physischen Kristallonomie und geometrishen Combinationslehre" and developed many mathematic methods in crystallography, noted impressive musical analogies in the field of crystallography. The statement is about many analogies described by him between ratios of musical tones and segments, formed by faces of the same zone of crystals. According to his figurative expression, "*crystal polyhedron is a fallen asleep chord - a chord of the molecular fluctuations made in time of its formation*" (from (Berger, 1997, p. 270)).

At the end of 1890's the outstanding crystallographer V. Goldschmidt returned to the same ideas. The prominent Russian mineralogist and geochemist A. E. Fersman wrote about his thematic publications: "*These works represent the historical page in crystallography, which has lead Goldschmidt to revealing by him laws of harmonic ratios. Goldschmidt has extended these laws logically from the world of crystals into the world of other correlations in the regions of paints, colors, sounds and even biological correlations. It has become one of the most favourite themes of philosophical researches by Goldschmidt*" (from (Berger, 1997, p. 270)). This list of such historical examples can be continued.

Taking into account, that Shrodinger named a living substance as aperiodic crystal and that the classicists of crystallography emphasized a connection between crystal structures and musical harmony, it seems natural to try to find traces of musical harmony in living substance as well. This idea about a possible participation of musical harmony in the organization of biological organisms is not new for modern biophysics. For example, the famous Russian biophysics Simon Shnoll (1989) wrote: "*From possible consequences of interaction of macromolecules of enzymes, which are carrying out conformational (cyclic) fluctuations, we shall consider pulsations of pressure - sound waves. The range of numbers of turns of the majority of enzymes corresponds to acoustic sound frequencies. We shall consider ... a fantastic picture of "musical interactions" among biochemical systems, cells, bodies, and a possible physiological role of these interactions. ...... It leads to pleasant thoughts about nature of hearing, about an origin of musical perception and about many other things, which belong to area of biochemical aesthetics already*". This term "biochemical aesthetics" proposed by Schnoll reflects many materials of this chapter.

Let us recall some fundamental notions of the theory of musical harmony. Each musical note is characterized by its certain frequency of sounding. For musical melody, a ratio between frequencies of neighboring notes is important, but not absolute values of frequencies of separate notes. For this reason the melody is easily distinguished irrespective of what acoustic range of frequencies it is produced in, for example, by child, woman or adult man with quite different voices. An aggregate of frequency values between sounds in musical system is named a musical scale.

The same note, for example, the note "do" is distinguished by the person as the same if its frequency is increased or reduced twice i.e. if it belongs to another octave. The interval of frequencies from some note frequency f_0 up to frequency $2*f_0$ is named an octave. Each note "do" is considered usually as the beginning of the appropriate octave. For example, the first octave reaches from frequency 260 Hz approximately (the note "do" of the first octave) up to the double frequency 520 Hz (the note "do" of the second octave).

Small quantity of frequencies of the octave diapason is traditionally used for musical notes only but not the whole infinite set of its frequencies. The notes, which correspond to these frequencies, form the certain sequence in ascending order of frequencies. A musical scale represents a sequence of numerical values ("interval values") between frequencies of the adjacent notes (musical tones).

For Europeans the idea of musical harmony of a universe is connected basically with the name Pythagoras and his school. After ancient thinkers (first of all, ancient Chinese thinkers) Pythagoreans considered that the world is arranged by principles of musical harmony. The Pythagorean musical scale, which is based on the quint ratio 3:2, played the main role in these views. One should note that this musical scale was known in Ancient China long before Pythagoras, who has presumably got acquainted within his life in Egypt and Babylon (the analysis of these questions is presented in detail in the book (Needham, v.4, 1962)). In Ancient China this quint music scale had a cosmic meaning connected with "The Book of Changes" ("I Ching"): numbers 2 and 3 were named "numbers of Earth and Heaven" there. After Ancient China, Pythagoreans considered numbers 2 and 3 as the female and male numbers which can give birth to new musical tones in their interconnection. According to some data, the quint system of the musical scale is the most ancient among known systems in the history of musical scales (http://www.arbuz.uz/t_octava.html).

Ancient Greeks attached an extraordinary significance to the search of the quint 3:2 in natural systems because of their thoughts about musical harmony in the organization of the world. For example, the great mathematician and the mechanician Archimedes considered as the best result of his life the detection of the quint 3:2 between volumes and areas of a cylinder and a sphere entered in it (Voloshinov, 2000). Just these geometrical figures with the quint ratio were pictured on his gravestone according to Archimedes testament. And due to these figures Cicero has found Archimedes's grave later, 200 years after his death. This chapter demonstrates, in particular, the connection of the Kronecker family of the genomatrices of hydrogen bonds with the Pythagorean musical scale based on the quint ratio 3:2.

NUMERIC GENOMATRICES OF HYDROGEN BONDS

As we mentioned above, numeric genomatrices are derived from the replacement of each symbol A, C, G, U/T of the nitrogenous bases in the symbolic genomatrixes $P^{(n)}$=[C A; U G]$^{(n)}$ (Figure 3 of Chapter 1) by quantitative parameters of these bases. For example, let us consider the genomatrices of hydrogen bonds of these nitrogenous bases. The hydrogen bonds of complementary letters of the genetic alphabet are suspected for a long time for their important information meaning. In addition hydrogen plays the main role in the composition of our Universe, where 93 atoms of hydrogen are presented among each 100 atoms and where "chemical influence of omnipresent hydrogen is the defining factor" (Ponnamperuma, 1972). Thus the investigation of a possible meaning of hydrogen bonds in genetic information has a special interest.

The complementary letters C and G have 3 hydrogen bonds (C = G = 3) and the complementary letters A and U have 2 hydrogen bonds (A = U = 2). Let us replace each multiplet in the Kronecker family of the genomatrices $P^{(n)}$ = [C A; U G]$^{(3)}$ by the product of these numbers of its hydrogen. In this case, we get the Kronecker family of the multiplicative matrices marked as $P_{MULT}^{(n)}$ = [3 2; 2 3]$^{(n)}$ conditionally (another family of additive matrices was considered in the works (Petoukhov 1999, 2001, 2003-2004)). For example, the triplet CAU will be replaced by number 12 (=3*2*2) in the genomatrix $P_{MULT}^{(3)}$. Figure 2 demonstrates the three initial genomatrices from this Kronecker family of genomatrices [3 2; 2 3]$^{(n)}$ constructed in this way. Numeric characteristics of each genomatrix [3 2; 2 3]$^{(n)}$ are connected with the quint ratio 3:2; for this reason we name such genomatrices as quint genomatrices conditionally.

All matrices $P_{MULT}^{(n)}$ are nonsingular. They are symmetrical relative to both diagonals and can be named "bi-symmetric matrices". All rows and all columns of this matrix differ from each other by the sequences of their numbers. But the sums of all numbers in the cells of each row and of each column in any matrix $P_{MULT}^{(n)}$ are identical to each other. For example, in the case of the matrix $P_{MULT}^{(3)}$, these sums are equal to $125 = 5^3$ and the total sum of numbers inside the matrix is equal to 1000. A rank of this matrix is equal to 8. Its determinant is equal to 5^{12}. Eigenvalues of $P_{MULT}^{(3)}$ are 1, 5, 5, 5, 5^2, 5^2, 5^2, 5^3. The matrix $P_{MULT}^{(3)}$ has four kinds of numbers only: 8, 12, 18 and 27. The certain laws are observed

Figure 2. The beginning of the family of the quint multiplicative genomatrices $P_{MULT}^{(n)} = [3\ 2;\ 2\ 3]^{(n)}$, which are based on product of numbers of hydrogen bonds (C=G=3, A=U=2)

$$P_{\text{MULT}}^{(1)} = \begin{vmatrix} 3 & 2 \\ 2 & 3 \end{vmatrix}; P_{\text{MULT}}^{(2)} = \begin{vmatrix} 9 & 6 & 6 & 4 \\ 6 & 9 & 4 & 6 \\ 6 & 4 & 9 & 6 \\ 4 & 6 & 6 & 9 \end{vmatrix}; P_{\text{MULT}}^{(3)} = \begin{vmatrix} 27 & 18 & 18 & 12 & 18 & 12 & 12 & 8 \\ 18 & 27 & 12 & 18 & 12 & 18 & 8 & 12 \\ 18 & 12 & 27 & 18 & 12 & 8 & 18 & 12 \\ 12 & 18 & 18 & 27 & 8 & 12 & 12 & 18 \\ 18 & 12 & 12 & 8 & 27 & 18 & 18 & 12 \\ 12 & 18 & 8 & 12 & 18 & 27 & 12 & 18 \\ 12 & 8 & 18 & 12 & 18 & 12 & 27 & 18 \\ 8 & 12 & 12 & 18 & 12 & 18 & 18 & 27 \end{vmatrix}$$

in their disposition, which are connected with a few interesting properties of this matrix, including the property of invariance of its numeric mosaic under many mathematical operations with this matrix (see below).

THE NUMERIC GENOMATRICES AND THE GOLDEN SECTION

In biology, a genetic system provides the self-reproduction of biological organisms in their generations. In mathematics, the "golden section" (or the "divine proportion") and its properties were a mathematical symbol of self-reproduction from the Renaissance and they were studied by Leonardo da Vinci, J. Kepler and many other prominent thinkers (see details in the website "Museum of Harmony and Golden Section" by A. Stakhov, www.goldenmuseum.com). Is there any connection between these two systems? Yes, and this paragraph demonstrates such unexpected connection.

The golden section is the value $\varphi = (1+5^{0.5})/2 = 1.618\ldots$ (Sometimes the inverse of this value is called the golden section in literature). If the simplest genetic matrix $P_{MULT}^{(1)}$ is raised to the power ½ in the ordinary sense (that is, if we take the square root), the result is the bi-symmetric matrix $\Phi = (P_{MULT}^{(1)})^{1/2}$, the matrix elements of which are equal to the golden section and to its inverse value. And if any other genomatrix $P_{MULT}^{(n)} = [3\ 2;\ 2\ 3]^{(n)}$ is raised to the power ½ in the ordinary sense, the result is the bi-symmetric matrix $\Phi^{(n)} = (P_{MULT}^{(n)})^{1/2}$, the matrix elements of which are equal to the golden section in various integer powers with elements of symmetry among these powers (Figure 3). For instance, the matrix $\Phi_{MULT}^{(3)} = (P_{MULT}^{(n)})^{1/2}$ has only two pairs of inverse numbers: φ^1 and φ^{-1}, φ^3 and φ^{-3} (Figure 3). Matrices with matrix elements, all of which are equal to golden section φ in different powers only, can be referred to as "golden matrices". Figuratively speaking, the quint genomatrices have the secret substrate from the golden matrices. The product of all numbers in any row and in any column of these golden matrices is equal to 1.

The mentioned matrix elements of the matrix $\Phi^{(n)} = (P_{MULT}^{(n)})^{1/2}$ can be constructed from a combination of φ and φ^{-1} directly by the following algorithm. We take a corresponding multiplet of the matrix $P^{(n)} = [C\ A;\ U\ G]^{(n)}$ and change its letters C and G to φ. Then we take letters A and U in this multiplet and change each of them to φ^{-1}. As a result, we obtain a chain with "n" links, where each link is φ or φ^{-1}. The product of all such links gives the value of corresponding matrix elements in the matrix $\Phi^{(n)}$.

Figure 3. The beginning of the Kronecker family of the golden matrices $\Phi^{(n)} = (P_{MULT}^{(n)})^{1/2}$, where $\varphi = (1+5^{0.5})/2 = 1,618...$ is the golden section

$$(P_{MULT})^{1/2} = \Phi = \begin{vmatrix} \varphi & \varphi^{-1} \\ \varphi^{-1} & \varphi \end{vmatrix}; \quad (P_{MULT}^{(2)})^{1/2} = \Phi^{(2)} = \begin{vmatrix} \varphi^{2} & \varphi^{0} & \varphi^{0} & \varphi^{-2} \\ \varphi^{0} & \varphi^{2} & \varphi^{-2} & \varphi^{0} \\ \varphi^{0} & \varphi^{-2} & \varphi^{2} & \varphi^{0} \\ \varphi^{-2} & \varphi^{0} & \varphi^{0} & \varphi^{2} \end{vmatrix}$$

$$(P_{MULT}^{(3)})^{1/2} = \Phi^{(3)} = \begin{vmatrix} \varphi^{3} & \varphi^{1} & \varphi^{1} & \varphi^{-1} & \varphi^{1} & \varphi^{-1} & \varphi^{-1} & \varphi^{-3} \\ \varphi^{1} & \varphi^{3} & \varphi^{-1} & \varphi^{1} & \varphi^{-1} & \varphi^{1} & \varphi^{-3} & \varphi^{-1} \\ \varphi^{1} & \varphi^{-1} & \varphi^{3} & \varphi^{1} & \varphi^{-1} & \varphi^{-3} & \varphi^{1} & \varphi^{-1} \\ \varphi^{-1} & \varphi^{1} & \varphi^{1} & \varphi^{3} & \varphi^{-3} & \varphi^{-1} & \varphi^{-1} & \varphi^{1} \\ \varphi^{1} & \varphi^{-1} & \varphi^{-1} & \varphi^{-3} & \varphi^{3} & \varphi^{1} & \varphi^{1} & \varphi^{-1} \\ \varphi^{-1} & \varphi^{1} & \varphi^{-3} & \varphi^{-1} & \varphi^{1} & \varphi^{3} & \varphi^{-1} & \varphi^{1} \\ \varphi^{-1} & \varphi^{-3} & \varphi^{1} & \varphi^{-1} & \varphi^{1} & \varphi^{-1} & \varphi^{3} & \varphi^{1} \\ \varphi^{-3} & \varphi^{-1} & \varphi^{-1} & \varphi^{1} & \varphi^{-1} & \varphi^{1} & \varphi^{1} & \varphi^{3} \end{vmatrix}$$

For example, in the case of the matrix $\Phi^{(3)} = (P_{MULT}^{(3)})^{1/2}$, let us calculate a matrix element, which is disposed at the same place as the triplet CAU in the matrix $[C\ A;\ U\ G]^{(3)} = P^{(3)}$. According to the described algorithm, one should change the letter C to φ and the letters A and U to φ^{-1}. In the considered example, we obtain the following product: $(\varphi * \varphi^{-1} * \varphi^{-1}) = \varphi^{-1}$. This is the desired value of the considered matrix element for the matrix $\Phi^{(3)}$ on Figure 3.

A ratio between adjacent numbers in numerical sequences inside each of such matrices $\Phi^{(n)}$ (for example, $...\boldsymbol{\varphi^{-3}}, \boldsymbol{\varphi^{-1}}, \boldsymbol{\varphi^{1}}, \boldsymbol{\varphi^{3}}$ ...) is equal to $\boldsymbol{\varphi^{2}}$ always. The same ratio $\boldsymbol{\varphi^{2}}$ exists in regular 5-stars (Figure 4) as a ratio between sides of the adjacent stars entered in each other (this pentagram is the ancient symbol of health).

The golden section is presented in 5-symmetrical objects of biological bodies (flowers, etc.), which are presented widely in the living nature but which are forbidden in classical crystallography. It exists as well in many figures of modern generalized crystallography: quasi-crystals by D. Shechtman, R. Penrose's mosaics (Gardner, 1988; Penrose, 1989), dodecahedrons of ensembles of water molecules, icosahedron figures of viruses, biological phyllotaxis laws, etc.

One can propose the new principal - "matrix-genetic" - definition of the golden section on the basis of the matrix specifics of genetic code systems: the golden section φ and its inverse value φ^{-1} are single matrix elements of a bi-symmetrical matrix Φ_{MULT}, which is the square root from such a bi-symmetrical (2x2)-matrix P_{MULT}, the elements of which are genetic numbers of hydrogen bonds (C = G = 3, A = U = 2) and which has a positive determinant.

This matrix-genetic definition does not use traditional elements of definition of the golden section: line segments, their proportions, *etc.* Probably, many realizations of the golden section in nature are related to its matrix essence and with its matrix representation. It should be investigated especially and systematically, where in natural systems and phenomena we have the bi-symmetric matrix P_{MULT} with its matrix elements 3 and 2 in a direct or masked form (for example, in a form of pairs of numbers 6 and

Figure 4. Sizes of pentagrams, which are entered in each other, differ by scale factor φ^2

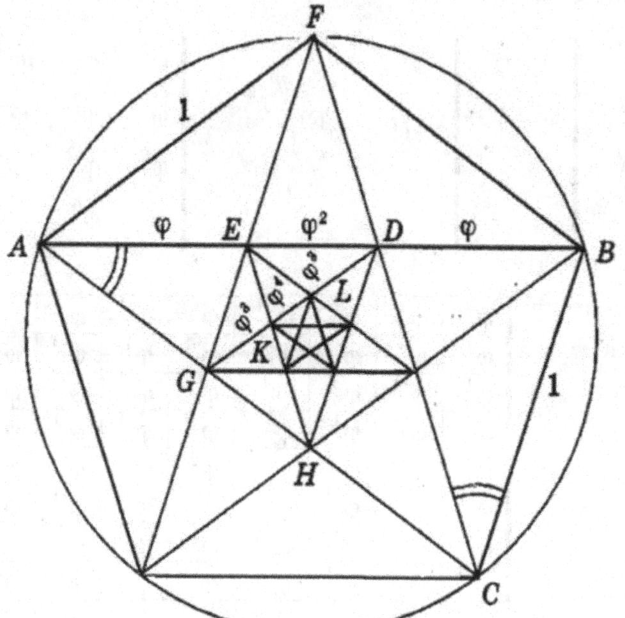

4, or 9 and 6, or 12 and 8, etc. with the same proportion 3:2, which is so frequent for ratios of elements in genetic codes). One can hope to discover many new system phenomena and connections between them in nature in this way.

The new theme of the golden section in genetic matrices seems to be important because many physiological systems and processes are connected with it. It is known that proportions of a golden section characterize many physiological processes: cardio-vascular processes, respiratory processes, electric activities of brain, locomotion activity, etc. The golden section is described and is investigated for a long time in phenomena of aesthetic perception as well. Taking into account these facts, the golden section should be considered as the candidate for the role of one of base elements in an inherited interlinking of the physiological subsystems, which provides unity of an organism. The matrix relation between the golden section φ and significant parameters of genetic codes testifies in a favor of a molecular-genetic providing such physiological phenomena. One can hope that the algebra of bi-symmetric genetic matrices, which are connected with the theme of the golden section, will be useful for explanation and the numeric forecast of separate parameters in different physiological sub-systems of biological organisms with their cooperative essence and golden section phenomena.

The Kronecker families of the golden genomatrices and of the quint genomatrices are connected with the famous triangle by Pascal by means of quantities of equal numbers, which are presented in sequences of the matrices of the increasing size. Really, as one can see from Figure 3, the golden (2x2)-matrix contains one number φ^1 and one number φ^{-1}; the (2^2x2^2)-matrix contains one number φ^2, one number φ^{-2} and two numbers φ^0; the (2^3x2^3)-matrix contains one number φ^3, one number φ^{-3}, three numbers φ, three numbers φ^{-1}, etc. At their appropriate arrangement, which is shown in Figure 5, Pascal's triangle is formed.

Figure 5. The Pascal's triangle for quantities of iterative kinds of numbers in the Kronecker family of the golden matrices from Figure 3. The brackets contain iterative kinds of numbers in the matrix of corresponding size

Matrix size	Pascal's triangle
2^1x2^1	$1(\varphi^1)$ $1(\varphi^{-1})$
2^2x2^2	$1(\varphi^2)$ $2(\varphi^0)$ $1(\varphi^2)$
2^3x2^3	$1(\varphi^3)$ $3(\varphi^1)$ $3(\varphi^{-1})$ $1(\varphi^{-3})$
2^4x2^4	$1(\varphi^4)$ $4(\varphi^2)$ $6(\varphi^0)$ $4(\varphi^{-2})$ $1(\varphi^{-4})$
……	………………………………………………

The molecular system of the genetic alphabet is constructed by nature in such manner that other genetic matrices play the role of quint matrices and golden matrices for other parameters (Petoukhov, 2005). For example, the quantities of atoms in molecular rings of pyrimidines and purines: the ring of purine contains 6 atoms and the ring of pyrimidine contains 9 atoms. From the viewpoint of this kind of parameters, C = U = 6, A = G = 9. The ration 9:6 = 3:2 is equal to the quint. Thus the symbolic matrices [A C; U G]$^{(n)}$, [G C; U A]$^{(n)}$, [A U; C G]$^{(n)}$, [G U; A C]$^{(n)}$ become the threefold quint matrixes in the Kronecker power "n" in the case of replacement of their symbolic elements by these numbers 9 and 6. The square root of such numeric matrices is connected with the golden matrices obviously.

A biological organism is the master on the use of a set of parallel information channels. It is enough to remind about many sensory channels by means of which we obtain sensory information simultaneously: visual, acoustical, tactile, etc. It is probable, that many kinds of genetic matrices are used by organism in parallel information channels as well.

The theory of discrete signals processing utilizes the important notions of the energy and of the power of signals (see details above in the background of this chapter). If one interprets any row of the quint genomatrix $P_{MULT}^{(n)}$ = [C A; U G]$^{(n)}$ as a vector-signal, then the energy of such vector-signal is equal to 13^n and its power is equal to $(13/2)^n$. If one interprets any row of the golden genomatrix $\Phi^{(n)}$ = ([C A; U G]$^{(n)}$)$^{0.5}$ as a vector-signal, then the energy of such vector-signal is equal to 3^n and its power is equal to the value $(3/2)^n$, where the quint ratio participates.

The bi-symmetric genomatrices $\Phi^{(n)}$ and $P_{MULT}^{(n)}$ have unexpected group-invariant property, which is connected with multiplications of matrices and which can be named "mosaic-invariant property". We will explain this property through the example of the matrix $P_{MULT}^{(3)}$ from Figure 2. This matrix consists of four numbers: 8, 12, 18 and 27 only with their special disposition. The numbers 8 and 27 are disposed at matrix diagonals separately in the form of a diagonal cross. The numbers 12 are disposed in matrix cells, a set of which produces a special mosaic. Such mosaic can be referred to as a "symbol 69" conditionally (one can note, that the symbols "6" and "9" are famous in "I Ching" as traditional symbols of Yin and Yang correspondingly, but such coincidence can be accidental). The numbers 18 are disposed in matrix cells, a set of which produces a mirror-symmetrical mosaic in comparison with a 69-mosaic of the previous case. Figure 6 demonstrates these two cases by means of the set of dark matrix cells with numbers 12 (left matrix) and with numbers 18 (right matrix).

Figure 6. The mosaic of cells with number 12 (left, the cells marked by dark) and the mosaic of cells with number 18 (right) from the multiplicative matrix $P_{MULT}^{(3)}$ *(Figure 2)*

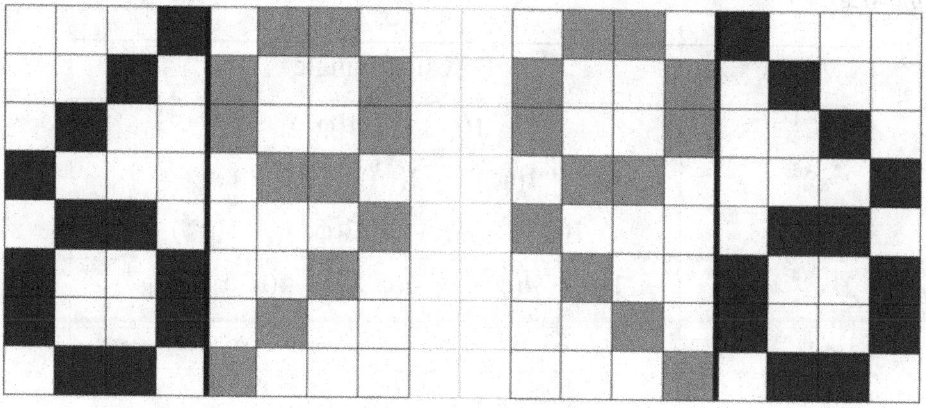

It is known that if an arbitrary octet matrix with four kinds of numbers as its matrix elements is raised to the power of "*n*", the resulting matrix will have usually many more kinds of numbers with very different disposition (up to 64 kinds of numbers for 64 matrix cells). But our bi-symmetrical genetic matrices have the unexpected property of invariance of their numeric mosaic after the operation of raising to the power of "*n*". For example, if the octet matrix $P_{MULT}^{(3)}$ is raised to the power of 2, the resulting octet matrix $(P_{MULT}^{(3)})^2$ will have a new set of four numbers 2197, 2028, 1872 and 1728 (instead of the initial four numbers 27, 18, 12 and 8 correspondingly) with the same disposition inside the octet matrix.

It is essential that this beautiful property of invariance of the numeric mosaic of the genetic matrix is independent of values of numbers. This property is realized for such matrices with the arbitrary set of four numbers a, b, c, d, if they are located in the same manner inside a matrix. Moreover, if we have one matrix *X* with a set of four numbers "*a*", "*b*", "*c*", "*d*" and another matrix *Y* with another set of four numbers "*k*", "*m*", "*p*", "*q*", then the product of these matrices will be the matrix $Z = X*Y$ with a set of new four numbers "*r*", "*g*", "*v*", "*z*" and with the same mosaic of their disposition (Figure 7).

It is obvious that the four symbols (for example, *a, b, c, d*) in such matrices can be not only ordinary numbers, but also arbitrary mathematical objects: complex numbers, matrices, functions of time (for example, it can be that $a=R*cos(wt)$, $b=T*sin(wt)$, ...), etc. In particular, the possibility of the modeling of chronocyclic functions by means of such mosaic-invariance matrices can be useful for the chronocyclic theory of degeneracy of genetic codes, which was described in the previous chapter. Such a mosaic-invariant property of these genetic matrices is the expression of cooperative behavior of its elements, but not the result of the individual behavior of each kind of element. This property is reminiscent some aspects of the cooperative behavior of the elements of biological organisms.

The mathematical analogy exists between the described bi-symmetric (2x2)-genomatrices and the famous matrices of the hyperbolic turn, which are bi-symmetrical also: [sh(x) ch(x); ch(x) sh(x)], where "sh(x)" and "ch(x)" are hyperbolic sine and cosine. This analogy gives us the opportunity to interpret normalized biosymmetric genomatrices in connection with hyperbolic turns, which have the following applications in physics and mathematics:

- Rotation of pseudo- Euclidean space;
- The special theory of relativity;

Figure 7. Multiplication of mosaic-invariant matrices X and Y gives a new matrix Z with the same mosaic of the disposition of its four kinds of numbers. For illustration, cells with numbers "b", " m", "s" in matrices X, Y, Z are marked by dark color

d	c	c	b	c	b	b	a
c	d	b	c	b	c	a	b
c	b	d	c	b	a	c	b
b	c	c	d	a	b	b	c
c	b	b	a	d	c	c	b
b	c	a	b	c	d	b	c
b	a	c	b	c	b	d	c
a	b	b	c	b	c	c	d

*

q	p	p	m	p	m	m	k
p	q	m	p	m	p	k	m
p	m	q	p	m	k	p	m
m	p	p	q	k	m	m	p
p	m	m	k	q	p	p	m
m	p	k	m	p	q	m	p
m	k	p	m	p	m	q	p
k	m	m	p	m	p	p	q

=

z	v	v	g	v	g	g	r
v	z	g	v	g	v	r	g
v	g	z	v	g	r	v	g
g	v	v	z	r	g	g	v
v	g	g	r	z	v	v	g
g	v	r	g	v	z	g	v
g	r	v	g	v	g	z	v
r	g	g	v	g	v	v	z

$X(a, b, c, d)$ $Y(k, m, p, q)$ $Z(r, g, v, z)$

- The geometric theory of logarithms, where properties of logarithms are introduced by hyperbolic turns (Shervatov, 1954).
- Theory of solitons of sine-Gordon equation.

In particular, this coincidence of the genomatrices with the matrices of hyperbolic turns reflects structural connections of the genetic code with the famous psychophysical Weber-Fechner's law. We will return to the bi-symmetric genomatrices in Chapter 8, where their connections with a special kind of hypercomplex number are revealed.

THE GENOMATRICES, MUSICAL HARMONY AND PYTHAGOREAN MUSICAL SCALE

The theme of harmony of living nature is discussed frequently by many authors. The word "harmony" has arisen in Ancient Greece in relation to the Pythagorean musical scale.

In the antique theory of music the word "harmony" has found the modern value - the consent of discordant. Seven musical notes carry names familiar to all: do (C), re (D), mi (E), fa (F), sol (G), la (A), si (B). These seven notes are interrelated among themselves by their frequencies not in an accidental manner, but they form the regular uniform ensemble. Really, it is well-known that the seven notes of the Pythagorean musical scale from appropriate octaves form the regular sequence of the geometric progression on the base of the quint ratio 3:2 between frequencies of the adjacent members of this sequence (Figure 8). The quint 3:2, which is the ratio between frequencies of the third and the second harmonics of an oscillated string, plays the role of the factor of this geometrical progression. The frequency 293 Hz of the note re (D^1) of the first octave stays in the middle of this frequency series. The ratios of the frequencies of all notes to this frequency of the note re (D^1) form the symmetrical series by signs and sizes of their powers of the quint: from the power "-3" up to the power "+3".

The Kronecker family of the genomatrices $P_{MULT}^{(n)} = [3\ 2;\ 2\ 3]^{(n)}$ is connected with the Pythagorean musical scale. Let us consider it more attentively. Each genomatrix of the family $P_{MULT}^{(n)}$ demonstrates the quint (or the perfect fifth) principle of its structure because they have the quint ratio 3:2 at different

Figure 8. The quint (or the fifth) sequence of the 7 notes of the Pythagorean musical scale. The upper row shows the notes. The second row shows their frequencies. The third row shows the ratios between the frequencies of these notes to the frequency 293 Hz of the note re (D^1). The designation of notes is given on Helmholtz system. Values of frequencies are approximated to integers

fa (F)	do (C)	sol (G)	re (D^1)	la (A^1)	mi (E^2)	si (B^2)
87	130	196	293	440	660	990
$(3/2)^{-3}$	**$(3/2)^{-2}$**	**$(3/2)^{-1}$**	**$(3/2)^{0}$**	**$(3/2)^{1}$**	**$(3/2)^{2}$**	**$(3/2)^{3}$**

levels: between numerical sums in top and bottom quadrants, sub-quadrants, sub-sub-quadrants, etc. including quint ratios between neighbor numbers in them. For example, $P_{MULT}^{(3)}$ contains 4 numbers – 27, 18, 12, 8 - with the quint ratio between them: 27/18=18/12=12/8=<u>3/2</u>.

Each quint genomatrix $P_{MULT}^{(n)}$ contains (n+1) kinds of numbers from a geometrical progression, factor of which is equal to the quint 3/2:

$P_{MULT}^{(1)} \Rightarrow 3, 2$
$P_{MULT}^{(2)} \Rightarrow 9, 6, 4$
$P_{MULT}^{(3)}) \Rightarrow 27, 18, 12, 8$
....
$P_{MULT}^{(6)} \Rightarrow 729, 486, 324, 216, 144, 96, 64$
....

Let us write out these kinds of numbers in columns for each genomatrix $P_{MULT}^{(n)}$ to arrive at the "genetic" triangle, which is shown on the left part of the expression:

3	9	27	81	243...	1	3	9	27...
2	6	18	54	162...		2		
	4	12	36	108...			4	
		8	24	72...				8
			16	48...				
				32...				

On the right side in the expression the historically famous numeric triangle by Plato is demonstrated. This triangle was utilized by Ancient Greeks to create the Pythagorean musical scale on the basis of its main proportions. One can see the analogy between the "genetic" triangle and the Plato's triangle.

Moreover, as Professor Jay Kappraff (USA) has informed one of the authors of this book in his private letter, this genetic triangle, which was obtained from the matrices of the genetic code, was known many centuries ago: it is identical to the famous triangle, which was published 2000 years ago by Nichomachus of Gerasa in his famous book "Introduction into arithmetic". Nichomachus belonged to the Pythagorean society, and this triangle was famous for centuries as the bases of the Pythagorean theory of musical harmony and aesthetics. In accordance with this triangle, the Parthenon (Kappraff, 2006) and other great architectural objects were created because architecture was interpreted as the non-movement music, and the music was interpreted as the dynamic architecture. Nichomachus of Gerasa

Figure 9. A presentation of the genomatrix $P_{MULT}^{(3)}$(293/27) in the form of the music-matrix $P_{MUSIC}^{(3)}$ of the frequencies of the musical notes (see Figure 8)*

re (D^1)	sol (G)	sol (G)	do (C)	sol (G)	do (C)	do (C)	fa (F)
sol (G)	re (D^1)	do (C)	sol (G)	do (C)	sol (G)	fa (F)	do (C)
sol (G)	do (C)	re (D^1)	sol (G)	do (C)	fa (F)	sol (G)	do (C)
do (C)	sol (G)	sol (G)	re (D^1)	fa (F)	do (C)	do (C)	sol (G)
sol (G)	do (C)	do (C)	fa (F)	re (D^1)	sol (G)	sol (G)	do (C)
do (C)	sol (g)	fa (F)	do (C)	sol (G)	re (D^1)	do (C)	sol (G)
do (C)	fa (F)	sol (G)	do (C)	sol (G)	do (C)	re (D^1)	sol (G)
fa (F)	do (C)	do (C)	sol (G)	do (C)	sol (G)	sol (G)	re (D^1)

was one of the great persons in the theory of musical harmony and aesthetics. The Cambridge library has the ancient picture, where Nichomachus is shown together with other great persons in this field: Pythagoras, Plato and Boeticus (http://www.jcsparks.com/painted/boethius.html). One can find more details about the triangle by Nichomachus of Gerasa in the publications (Kappraff, 2000, 2002). This unexpected connection of times proves additionally the adequacy of the presented way of the matrix research of genetic systems and the connection of genetic systems with the Pythagorean musical scale, reflected in Nichomachus's triangle.

As we mentioned above, a set of certain kinds of numbers in each genomatrix $P_{MULT}^{(n)} = [3\ 2;\ 2\ 3]^{(n)}$ reproduce fragments of the geometrical progressions with the quint factor. Thus sequences of such kinds of numbers can be compared to quint sequences of musical notes from Figure 8. If one confronts the least number from a quint genomatrix with the musical note "fa" (F), which possesses the least frequency on Figure 8, then all sequences of such kinds of numbers automatically corresponds to the series of the musical notes. For example, the sequence of numbers 8, 12, 18, 27 of $P_{MULT}^{(3)}$ corresponds to the frequency sequence of the notes fa(F) - do(C) - sol(G) - re(D^1). Genomatrix $P_{MULT}^{(6)}$ contains the sequence of 7 numbers, which corresponds to the whole quint sequence of the 7 notes of Figure 8: fa(F) - do(C) - sol(G) - re(D^1) - la (A^1) - mi (E^2) - si (B^2).

For this reason, each genomatrix $P_{MULT}^{(n)}$ can be presented in the form of a matrix $P_{MUSIC}^{(n)}$ of frequencies of notes (or a "music-matrix"). For instance, Figure 9 demonstrates the genomatrix $P_{MULT}^{(3)}$ of the 64 triplets as a music-matrix $P_{MUSIC}^{(3)}$ of frequencies of appropriate four notes (the general factor 293/27 arises for concordance of numeric values of the note frequencies with numbers 8, 12, 18, 27 of the genomatrix $P_{MULT}^{(3)}$). Figure 10 shows the note staff with the notes, the sequence of which corresponds to the sequences of the notes in the music-matrix on Figure 9.

The four numbers 8=2*2*2, 12=2*2*3, 18=2*3*3, 27=3*3*3, which are presented in the genomatrix $P_{MULT}^{(3)}$ on Figure 2, characterize those four kinds of triplets, which differ by their numbers of hydrogen bonds of nitrogenous bases. For instance, number 18=2*3*3 belongs to those triplets, which have one nitrogenous base with 2 hydrogen bond and two bases with 3 hydrogen bonds (the mathematics of genomatrices testifies products of numbers of hydrogen bonds should be taken into account here but not their sums; it has precedents and the justification in information theories, in particular, in the theory of parallel channels of coding and processing the information). Different sequences of these four numbers, for example 12-8-27-12-8-18-18-..., determine appropriate successions of the musical ratios 1:1, $(3:2)^{\pm1}$, $(3:2)^{\pm2}$, $(3:2)^{\pm3}$ (in this example, 3:2 - $(3:2)^3$ – $(2:3)^2$ – (2:3) – $(3:2)^2$ - 1:1 -...). It is obvious that such succession can be interpreted as a kind of genetic music for triplets, which is connected with

Figure 10. The musical presentation of the genomatrix $P_{MULT}^{(3)}$

their hydrogen bonds. Each gene and each part of DNA and RNA have their own genetic "melody of hydrogen bonds" which can be played by means of musical tools.

But the described musical sequence is not the single one in the molecule DNA at all. DNA can be considered as a set of joint sequences, which are very different in their physical-chemical sense: a sequence of nitrogenous bases; a sequence of hydrogen bonds of complementary pairs of these bases; a sequence of triplets; a sequence of rings of nitrogenous bases; a sequence of ensembles of protons in rings of nitrogenous bases, etc. One can note the phenomenological fact that many of these sequences are constructed on such ratios between quantitative characteristics of their neighboring members, which are typical for the Pythagorean musical scale. Correspondingly each of these sequences of ratios can be interpreted as a special kind of genetic musical melody. The whole set of such sequences in DNA can be considered as a polyphonic (coordinated) music ensemble. An investigation of this music ensemble seems to be an important scientific task.

Let us demonstrate a few additional examples of sequences with the musical ratios in DNA. A sequence of triplets in DNA has another kind of genetic music also which is connected with the quantity of protons in molecular rings of nitrogenous bases (Figure 11). The pyrimidines C and T have 40 protons in their rings; the purines A and G have 60 protons in their rings. (Each complementary pair has 100 protons in their rings precisely). The ratio 60:40 is equal to the quint 3:2. Let us present each triplet by the product of the proton numbers 40 and 60 in its rings (as we did above for numbers 2 and 3 of the hydrogen bonds of triplets). Then any triplet has one of four proton numbers: 64000=40*40*40; 96000=40*40*60; 144000=40*60*60; 216000=60*60*60. This proton set of the four numbers is dif-

Figure 11. On top: Complementary pairs of four nitrogenous bases in DNA: A - T and C - G. By a dotted line are specified hydrogen bonds in these pairs. Black circles are atoms of carbon, small white circles - hydrogen, circles with the letter N - nitrogen, and circles with the letter O – oxygen. At bottom: the numerical representations of a sequence of complementary pairs of the bases in DNA as a sequence of numbers of hydrogen bonds in the given pairs (the average row made up on basis of numbers 2 and 3) and as a numerical sequence of protons of molecules rings of these nitrogenous bases

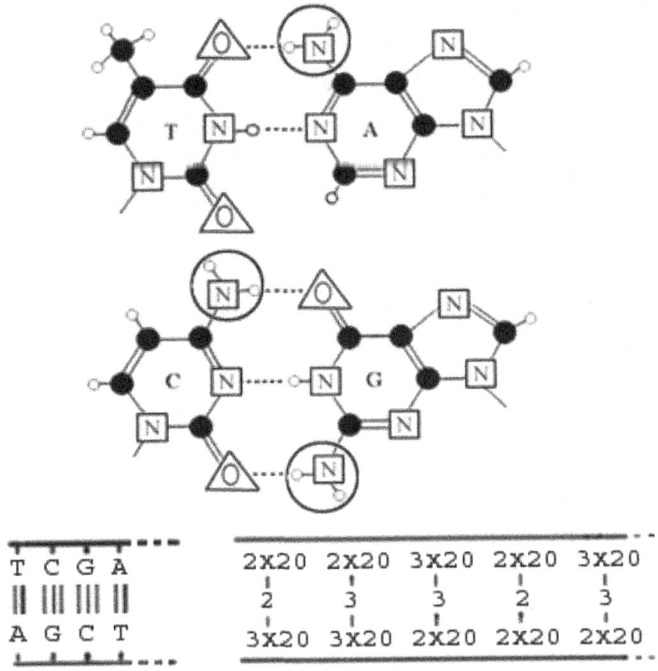

ferent from the considered set of four numbers 8, 12, 18, 27 of hydrogen bonds in triplets by the factor 8000 only. In other words, a ratio between any two numbers from this proton set has a quint character again and is equal to one of the values $(3:2)^k$, where $k = \pm1, 2, 3$. One can note that a sequence of triplets of one DNA-filament has two different sequences with the same typical ratios: one sequence for triplet characteristics of its hydrogen bonds and another sequence for triplet characteristic of protons in triplet rings. These two sequences differ each from other by dispositions of these ratios along DNA-filament, generally speaking (Figure 11). So, any triplet sequence bears on itself two different genetic melodies on these two parameters.

Sequential dispositions of musical ratios for these two parameters of triplets (and of nitrogenous bases also) are different on two filaments of DNA, but they are connected in regular manner due to a fact of complementary pairs of bases. Figuratively speaking, two filaments of DNA bear complementary kinds of genetic music on these parameters.

It should be added about an atomic parameter of nitrogenous bases: the quantity of non-hydrogen atoms in molecular rings of the pyrimidines C and T is equal to 6 and the quantity of non-hydrogen atoms in molecular rings of the purines A and G is equal to 9. Their quint ratio 9:6=3:2 can be considered as a basis for "atomic" genetic music of the nitrogenous bases and triplets along DNA. But these kinds of sequences of ratios are identical to sequences of ratios in the case considered above about 40 and 60 protons in rings of the pyrimidines and the purines. For this reason these sequences have nothing new from musical viewpoint though they can have an important meaning in the ensemble of genetic music because they are organized on the higher – atomic - level.

A sequence of numbers of 2 and 3 of hydrogen bonds between complementary nitrogenous bases along DNA (for instance, 3-2-2-3-2-3-...) determines a sequence of ratios between its neighboring - subsequent and previous - members (in the considered example, 2:3 - 2:2 - 3:2 - 2:3 -....). This simple sequence contains ratios 1:1, 3/2 and 2/3 only. From a viewpoint of musical analogy, this sequence determines a special kind of very simple genetic music.

Quantities of molecular rings in the pyrimidines and the purines are characterized by the octave ratio 2:1. This fact gives an additional possibility to consider sequences of nitrogenous bases and triplets in DNA as genetic melodies. But sequences of ratios in these cases contain the octave ratios only and are not so interesting from musical viewpoint though they can play an important role in the whole ensemble of genetic music.

Total quantities of protons in both pairs of nitrogenous bases A-T and C-G are the same and are equal to 136. On this numeric parameter, a sequence of nitrogenous bases has constant ratios 1:1 along DNA.

The full list of different kinds of such genetic music at different parameters and levels of genetic system permits one to reproduce a musical polyphonic party for each gene and for other parts of the genetic system. These musical sequences were created by nature itself. Each gene and each protein have their own genetic music (or briefly "genomusic"). The natural music of a gene can be reproduced in acoustical diapason not for aesthetic pleasure but for medical therapy, for theoretical needs, etc. This natural genomusic and its compositions can be connected to deep physiological archetypes, which were introduced into science by the creator of analytic psychology Carl Jung. From the viewpoint of musical harmony in structures of molecular-genetic system, outstanding composers are researchers of harmony in the organization of living substance. According to the famous expression by G. Leibnitz, music is the mysterious arithmetic of the soul, which calculates itself without understanding this action.

It is well-known, that some kinds of music stimulate growth of plants, cure people, etc. "American Music Therapy Association" unites more than 5000 members; 2700 musicians are certificated as professional musical therapists there. One should emphasize that "melodies" of such genetic music are not formed by any person in a forcible way, but they are defined by natural sequences of parameters in chain genetic molecules. They are named conditionally as "natural genetic music" to distinguish them from variants of "genetic music", sometimes offered by other authors on the basis of obviously forcible approaches without a sufficient support on molecular features of genetic sequences. The claim is that some authors propose their own "genetic music" on the basis of an arbitrary correspondence of the genetic letters or triplets to musical notes without sufficient attention to the musical correspondence of ratios of natural numeric parameters of adjacent genetic elements. One can find more details about natural genomusic with some examples in the book (Petoukhov, 2008).

All physiological systems should be coordinated structurally with the genetic code for their genetic transfer to next generations and for a survival in a course of biological evolution. For this reason we collect examples of harmonious ratios (first of all, the quint 3:2) in structures and functions on different levels of biological systems including the supra-molecular level. For example, the quint ratio 3:2 exists between:

- durations of phases of the activity and the rest in human cardio-cycles (0.6 sec and 0.4 sec correspondingly);

- plasmatic and globular volumes of blood (60% and 40%);
- albumens and globulins of blood (60% and 40%);
- 60S and 40S sub-particles in the composition of ribosomes (from http://vivovoco.rsl.ru/VV/JOURNAL/NATURE/08_03/KISSELEV.HTM).

In conclusion to this paragraph let us consider a well-known algorithm of the construction of the Pythagorean musical scale from a geometrical progression, which factor is equal to the quint. This algorithm, which is useful for the theme of the next paragraph, creates the sequence of the notes do-re-mi-fa-sol-la-si-do on the interval of frequencies {1, 2} of one octave, the lowermost note "do" which has the conditional frequency 1 and the lowermost note of the next octave has the conditional frequency 2. This algorithm contains the following steps:

1. Taking the first seven members of such geometrical progression with the quint factor 3/2, which begins from the inverse value of the quint: $(3/2)^{-1}$, $(3/2)^0$, $(3/2)^1$, $(3/2)^2$, $(3/2)^3$, $(3/2)^4$, $(3/2)^5$;
2. Returning into the octave interval {1, 2} for those members of this sequence, values of which overstep the limits of this interval; this returning is made for these values by means of their multiplication or division with the number 2. As a result of this operation, the new sequence is appeared (this sequence can be named "the geometrical progression with the returning into the octave "): $2*(3/2)^{-1}$, $(3/2)^0$, $(3/2)^1$, $(3/2)^2/2$, $(3/2)^3/2$, $(3/2)^4/4$, $(3/2)^5/4$;
3. The permutation of these seven members in accordance with their increasing values from 1 up 2 (the number 2 is included in this sequence as the end of the octave): $(3/2)^0$, $(3/2)^2/2$, $(3/2)^4/4$, $2*(3/2)^{-1}$, $(3/2)^1$, $(3/2)^3/2$, $(3/2)^5/4$, 2.

In this last sequence, a ratio of the greater number to the adjacent smaller number refers to as the interval factor. Two kinds of interval factors exist in this sequence only: 9/8, which is named the tone-interval T, and 256/243, which is named the semitone-interval S. One can check that the sequence of interval factors in this case is T-T-S-T-T-T-S. These five tone-intervals and two semitone-intervals cover the octave precisely: $(9/8)^5 * (256/243)^2 = 2$.

It is known that the name "semitone-interval" in the Pythagorean musical scale is utilized by convention only because the semitone-interval 256/243= 1.0545... is not equal to the half of the tone-interval, that is the square root from the tone-interval: $(9/8)^{0.5} = 1.0607$... .The scale of the golden wurf, which is described in the next paragraph, possesses the analogical peculiarities: its semitone-interval differs from the half of its tone-interval.

If one takes not 7, but 6 or 8 members in the initial quint geometrical progression (see the first step of the algorithm), then the same Pythagorean algorithm does not give a binary sequence of interval factors T and S because three kinds of interval factor arise.

The similar algorithm will be used in the next paragraph to construct new mathematical scale on the base of described data about the genetic code and its genomatrices.

A SCALE OF THE GOLDEN WURF, MUSIC AND FIBONACCI NUMBERS

Many theorists of music paid attention to the connection of the structure of many musical compositions of prominent composers with the golden section $\varphi = (1+5^{0.5})/2 = 1.618...$. The results of matrix genetics, which were described above, reveal a new direction of thoughts about a relation between the golden section and music because structures of a genetic code are connected with the golden section.

Similarly to a quint genomatrix $P_{MULT}^{(n)}$, which contains a sequence of $(n+1)$-kinds of numbers from a geometrical progression with the quint factor 3/2, a corresponding golden genomatrix $\Phi^{(n)}$ contains a sequence of $(n+1)$-kinds of numbers from a geometric progression, the factor of which is equal to $\varphi^2 =$ 2.618….:

$$\Phi^{(1)} \Rightarrow \varphi^1, \varphi^{-1}$$
$$\Phi^{(2)} \Rightarrow \varphi^2, \varphi^0, \varphi^{-2}$$
$$\Phi^{(3)} \Rightarrow \varphi^3, \varphi^1, \varphi^{-1}, \varphi^{-3} \quad (4.2)$$

The previous paragraph demonstrated that the Kronecker family of the quint genomatrices is connected with the Pythagorean musical scale. Now we turn to the Kronecker family of the quint genomatrices and to the geometrical progressions with the factor φ^2. Is it possible to apply the described Pythagorean algorithm to such geometrical progressions with factor φ^2 to arrive at a new musical (or mathematical) scale, where only two interval factors exist (as its tone-interval and its semitone-interval) by analogy with the Pythagorean musical scale? Investigation of this question seems to be important because such a new scale or scales can be essential for a theory of musical harmony and for the creation of musical compositions with increased physiological activity.

After research of this question the beautiful positive result is obtained: yes, it is possible every time, when we take one of Fibonacci numbers 2, 3, 5, 8, 13 (see the Figure 1) as the first member of such a geometrical progression (the situation becomes more difficult for the greater Fibonacci numbers 21, 34,…). Mathematical scales, which are formed in these cases, possess such quantities of their tone-intervals and semitone intervals, which are equal to Fibonacci numbers as well. Moreover values of these tone-intervals and semitone-intervals are expressed by means of Fibonacci numbers also.

Such interrelated Fibonacci-stage scale, each of which has interval factors of two kinds only, are named "the scales of the golden wurf" or "wurf-scales" briefly. Let us consider the example of the 8-stage scale of the golden wurf. We should construct a new mathematical scale of frequencies, which fills up the octave {1, 2}, by means of the Pythagorean algorithm with the irrational factor $\boldsymbol{\varphi}^2$ of a geometrical progression instead of the quint ratio 3/2. As a result we should arrive at such a scale, which possesses two kinds of interval factors (a tone-interval and a semitone-interval) only by analogy with the Pythagorean musical scale. One can note that the factor $\boldsymbol{\varphi}^2 = 2.618…$ exceeds the considered interval of the octave {1, 2}. Therefore it is comfortable to use from the very beginning the twice smaller factor $\varphi^2/2 = p = 1.309…$, the value of which belongs to this octave interval. It is easy to check that the final sequence (4.3) of the wurf-scale does not depend on whether we use the factor φ^2 or the factor $\varphi^2/2$, which are equivalent to each other in the given problem. This factor $p = \varphi^2/2$ has been known in the field of investigations of biological symmetries and invariants for a long time under the name of the golden wurf (Petoukhov, 1981, 1989). We will discuss the golden wurf later.

Now let us construct the 8-stage scale of the golden wurf by means of the analogue of the described Pythagorean algorithm, using the factor $p = \varphi^2/2$ in the initial geometric progression (instead of the quint factor 3/2). All three steps of the Pythagorean algorithm are reproduced:

1. Taking the first eight (!) members of such a geometrical progression with the factor $p = \varphi^2/2$, which begins from the inverse value of this factor: $p^{-1}, p^0, p^1, p^2, p^3, p^4, p^5, p^6$;

Figure 12. Sequences of interval factors in the 7-stage Pythagorean scale of C major (the upper row) and in the 8-scale of the golden wurf. Tone-intervals are marked by T, semitone-intervals are marked by S

T	T	S	T		T	T	S
T	T	S	T	S	T	T	S

2. Returning into the octave interval {1, 2} for those members of this sequence, values of which overstep the limits of this interval; this returning is made for these values by means of their multiplication or division with the number 2. As a result of this operation, the new sequence is obtained (this sequence can be named "the geometrical progression with return to the octave"). 2^*p^{-1}, p^0, p^1, p^2, $p^3/2$, $p^4/2$, $p^5/2$, $p^6/4$;
3. The permutation of these seven members in accordance with their increasing values from 1 up to 2 (the number 2 is included in this sequence as the end of the octave):

$$1, p^3/2, p^6/4, p^1, p^4/2, 2^*p^{-1}, p^2, p^5/2, 2 \qquad (4.3)$$

This final sequence (4.3) satisfies the initial condition concerning the existence of two kinds of interval factors only. Really, it is easy to check that all ratios of adjacent members of this sequence are equal to two values only, which play the role of the interval factors. For this sequence (4.3) the tone-interval is $T = p^3/2 = 1.1215...$ and the semitone-interval is $S = 4^*p^{-5} = 1.0407...$. The sequence of these interval factors is T-T-S-T-S-T-T-S. This sequence fills all the octave in accuracy: $(p^3/2)^5 * (4^*p^{-5})^3 = 2$. The quantities of various interval factors are equal to Fibonacci numbers here. Really, the 3 semitone-intervals, 5 tone-intervals and 8 interval factors all exist here. It is interesting, that if we take non-Fibonacci number (for example, 4, 6 or 9) of the first members of the initial geometric progression on the first step of the Pythagorean's algorithm, final sequences arise which have more than two kinds of interval factors.

Let us compare the classical 7-stage Pythagorean musical scale with the obtained 8-stage scale of the golden wurf. The Figure 12 shows the minimal difference between the sequences (musical scales) of the tone-intervals and semitone-intervals inside the octave for both scales. The initial and final parts of both sequences coincide completely, and only one additional semitone-interval arises in the middle part of the octave. This additional semitone-interval exists because the factor "p" is less than the quint factor.

Using the sequence (4.3) of the intervals, one can construct the sequence of tones (musical notes), which is named the "wurf-scale of *C* major" by analogy with Pythagorean scale of *C* major (Figure 13). A choice of frequencies for these tones of the first octave is made in such way that this scale contains the frequency 440 Hz, which corresponds to note "la" in the Pythagorean scale and in *equal temperament* scale and which is used traditionally for tuning in musical instruments. Figure 14 compares the Pythagorean 7-steps scale *C* major and 8-stage scale of the golden wurf for the first octave. Taking into account a minimal difference between both scales, the majority of the notes of the wurf-scale are named by analogy with the appropriate notes of the Pythagorean scale but with the letter "m" in the end (for instance, "rem" instead "re"). The additional fifth note is named "pim".

This scale of the golden wurf, which was constructed in connection with parameters of the genetic code, possesses many analogies with the Pythagorean genetic code by their internal symmetries and proportions. Its main difference from the Pythagorean scale is connected with irrational values of its

Figure 13. The upper row demonstrates the frequencies of the tones in the7-stage Pythagorean scale of C major in the first octave. The bottom row demonstrates the frequencies of the tones in 8-stage scale of the golden wurf of C major in the similar octave. Numbers mean frequencies in Hz. The names of the notes are given

260.7	293.3	330.0	347.6		391.1	440	495.0	521.5
DO$_1$	RE	MI	FA		SOL	LA	SI	DO$_2$
256.8	288.0	323.0	336.1	376.8	392.3	440	493.5	513.6
DOM$_1$	REM	MIM	FAM	PIM	SOLM	LAM	SIM	DOM$_2$

interval factors. This wurf-scale could not be constructed by Pythagoreans who did not know irrational numbers. Irrational factors are used also in the modern equal-temperament scale. According to some data, Ancient Chinese knew about the equal-temperament scale, but neglected it preferring the Pythagorean scale, in which they saw cosmic and biological importance.

The history of attempts at creation of new musical scales includes names of many prominent scientists: J. Kepler, R. Descartes, G. Leibnitz, L. Euler, etc. But these authors had no possibility to use the data about the genetic code in their attempts. In our opinion, the data about the genetic code allow one to create new musical scales with positive physiological potentials. The constructed 8-stage scale of the golden wurf is investigated now in the Moscow State Conservatory by the group of specialists, which is headed by the dean of its Composer Department A. Koblyakov, from the viewpoint of its musical meaning.

The Fibonacci-stage scales are connected with many interesting mathematical and musical materials: the musical generalization of classical Fibonacci's problem, the series of anti-Fibonacci numbers, recurrent algorithms, etc. Many of these materials together with tables of frequencies of musical notes for various Fibonacci-stage scales are published in the book (Petoukhov, 2008). One should note that our attempt to create the mathematical scale of the golden section, where the factor of the geometrical progression is equal to the golden section (but not to the golden wurf), has led to the scale, which differs from the Pythagorean musical scale cardinally and which was not so interesting from the musical viewpoint. Furthermore such scale of the golden section has no evident connection with Fibonacci numbers in its interval factors.

In concluding this paragraph we discuss briefly the golden wurf $p = \varphi^2/2$, which has arisen in biological morphology initially. The wurf or the double ratio is known for a long time in the field of highest geometries as the main invariant of projective geometry. (It is interesting that the finite projective-geometric plane is connected with Hadamard matrices (Sachkov, 2004), which are related to the genetic code as

Figure 14. The helical structure of the human ear cochlea, which is uncoiled into a straight line, with the projective geometry proportion of the golden wurf

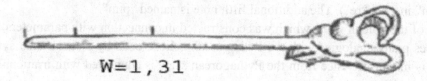

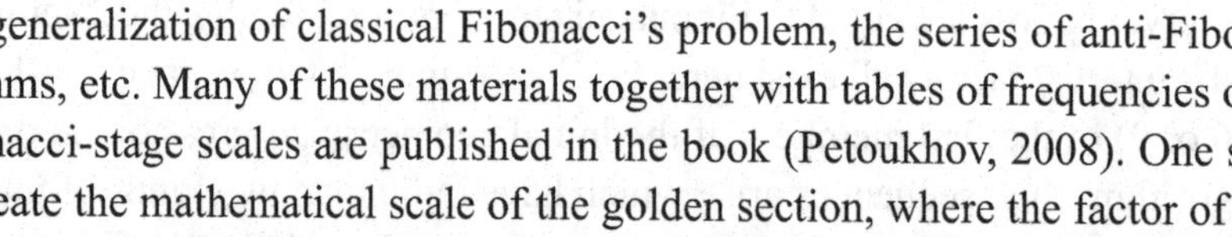

described in Chapter 6). The translation of the notion "wurf" from German language means "throw". The golden wurf was introduced in works (Petoukhov, 1981, 1989), which were devoted to non-Euclidean biological symmetries. The golden wurf has a status of ontogenetic and phylogenetic invariant of aggregated proportions of three-component kinematical blocks of human and animal bodies. The value of the golden wurf concerns acoustic perception also: the human ear cochlea consists of three patterns (three coils of a helix), the ratios of whose lengths form a geometrical progression with the golden section as a factor (see Figure 14). The double ratio of these three lengths is equal to the golden wurf: p = $\varphi^2/2$ = 1.309... (Petoukhov, 1989).

ON HARMONY OF A SCALE OF PROTONS IN THE SET OF AMINO-ACIDS

Can musical principles of organization exist not only in DNA but in other molecular and supramolecular structures of the genetic system? Some facts are revealed the positive answer to this question. Let us consider a few of them related to amino acids and their connections in proteins.

Amino acids are connected in a protein chain by peptide bonds, where the quint ratio 3:2 exists: in a peptide bond its double bond is disposed on 60% in a region of the group C-O and on 40% in a region of C-N (Shults, Schirmer, 1979, Chapter 2). This phenomenological fact was explained by Nobel Prize winner L. Pauling in his resonance theory, which is related to vibration principles.

Now let us consider the set of 20 amino acids of the genetic system, which has the following sequence of quantities of protons (the names of acids are shown in brackets):

40 (Gly), **48** (Ala), **56** (Ser), **62** (Pro), **64** (Cys, Thr, Val), **70** (Asn, Asp),

72 (Ile, Leu), **78** (Gln, Glu), **80** (Lys, Met), **82** (His), **88** (Phe), **94** (Arg),

96 (Tyr), **108** (Trp). (4.4)

It is known that the basic principle of musical scales of all people in all centuries was a principle of octave. The described proton sequence is disposed inside the octave interval from 48 to 96 mainly. One can analyze a disposition of all numbers of this proton set relative to this octave interval 48-96, where number 48 is a tonic. 12 kinds of proton numbers lay inside this interval: 48, 56, 62, 64, 70, 72, 78, 80, 82, 88, 94, 96. Classical construction of the Pythagorean musical scale uses the division of the octave interval by consonant ratios, foremost, by the quint 3:2 and the quart 4:3. In the case of our proton octave 48-96, the quint from the tonic 48 is equal to 72 = 48*3/2 and the quart is equal to 64=48*4/3. Both of these numbers belong to the analyzed proton set. Additionally, a consonant ratio 5:3, named a major sixth, gives one more number 80 = 48*5/3 from this proton set.

Two proton numbers 40 and 108 lay outside the considered interval from both its ends. But the number 40 is equal to 48*5/6; in other words, the number 40 has the consonant ratio 5:6 (its classical name "a minor third") relative to the tonic 48. The number 108 is equal to 96*9/8; in other words, the number 108 has the classical ratio of the whole-tone 9:8 relative to the octave end 96. Number 40 has an octave double - number 80, which belongs to the proton sequence (4.4) as well.

So, we have the numeric proton sequence:

40 – 48 – 64 – 72 – 80 – 96 – 108 (4.5)

members of which are connected by musical ratios. Four numbers 8, 12, 18, 27, which were considered above in the genomatrix $P_{MULT}^{(3)} = (3\ 2;\ 2\ 3]^{(3)}$ as numbers of hydrogen (or one-proton) bonds, have their octave doubles (or twins) in this sequence (4.5): 64 = **8***8, 72 = **18***4, 96 = **12***8, 108 = **27***4. It shows a certain coordination of proton characteristics of genetic components on different levels of the genetic system. It should be emphasized for comparison that the sequence of molecular masses of 20 amino acids has not such a musical scale and is not interesting from the musical viewpoint.

If one takes the number 48 conditionally as the equivalent of the musical note "do(c^1)", then the proton sequence (4.5) is the equivalent of the sequence of the notes on Figure 15.

From a position of the theory of musical harmony, the proton sequence (4.4) has one essential defect: the analyzed octave interval {48-96} does not contain the octave double (or the twin) of the greatest number 108, which is number 54 (though this sequence contains the octave double of the least number 40, which is number 80). Why is an amino acid with 54 protons absent? Perhaps, it was eliminated in the course of biological evolution because of additional reasons? (For example, the rings of each complementary pairs of nitrogenous bases have exactly 54 protons in their 9 atoms of carbon, and it can be one possible reason to avoid a repetition of this proton number in amino acids?).

Or one can find additional 54-proton factor, which is essential for the set of amino acids and which operates with them? It is the open question now, which should be investigated in the future (by the way, the number 54 is equal to the sum of the famous Pythagorean set 1, 2, 3, 2^2, 2^3, 3^2, 3^3). But if one supposes that the proton sequence (4.4) is added by this number 54, the sequence (4.4) gets the very symmetrical form (Figure 16). Really, the analyzed octave {48-96} has 6 equal parts, the boundaries of which are determined by numbers divisible by 8 (upper row of numbers). Each such part has a length, which is equal to 8 and which is divided by quart 3:4 in two subparts with their lengths 6 and 2. The borders between such adjacent subparts correspond to the proton values in the lower row on Figure 16. Perhaps, the theory of atomic memory (Brewer, & Hanh, 1984), which is related to protons and spin echo, can be used for an understanding of such peculiarities of the protons sequence (4.4).

Figure 15. The presentation of the proton sequence (4.5) in the musical form

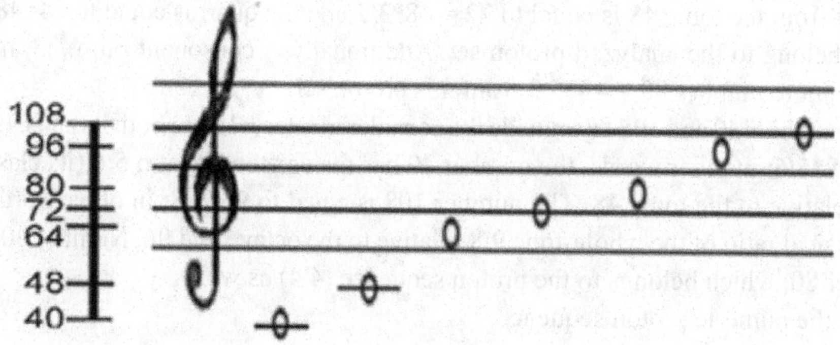

FUTURE TRENDS AND CONCLUSION

The proposed additional approach is the effective scientific instrument to analyze multi-component and multi-parametric ensembles of the molecular-genetic systems by means of numeric genomatrices. It reveals new facts about hidden interrelations among genetic elements and allows a comparison of them with famous facts and theories from other fields of science and culture. Methods of symmetries are not only useful in this approach, but they are needed systematically here to study relations of symmetry among various sets and subsets of the genetic systems. This study leads to important knowledge about internal regular structures of molecular-genetic systems and to new genetic patterns as well. The set of described and future results of investigations of numeric genetic matrices is the significant part of matrix genetics on the whole. This scientific direction permits one to apply effective ideas and methods from other modern sciences for problems in this molecular-genetic field. Taking into account all these data, one can recommend this approach, methods and patterns for intensive application in molecular genetics and in theoretical biology.

The discovery of the connection of the genetic code with the golden section shows the molecular-genetic base of many known facts about physiological and aesthetic meanings of the golden section. Specifically the described facts give new materials for the question about architectural canons, where the golden section is used for a long time; for example, the famous modulor by Sh. Le Corbusier (1948, 1953) is based on the golden section. The mathematical scale of the golden wurf, which was constructed in matrix genetics, can be utilized for architectural proportions (in the role of wurf-modulor or the modulor of the golden wurf).

The new – "matrix-genetic" – definition of the golden section is proposed, and leads to new theoretical investigations about the possible role of the golden section in nature and culture.

The facts described in this chapter about relations of the genetic systems with musical harmony are essential additionally for the problem of genetic bases of aesthetics and inborn feeling of harmony. According to the words of the famous physicist Richard Feynman about feeling of musical harmony, "*we may question whether we [stressed] are any better off than Pythagoras in understanding why [stressed] only certain sounds are pleasant to our ear. The general theory of aesthetics is probably no further advanced now than in the time of Pythagoras*" (Feynman, Leighton, & Sands, 1963, Chapter 50).

A cultural direction of "genetic art" (or briefly "genoart") can be developed additionally due to these data of matrix genetics. The genoart has many patterns, which are revealed by matrix genetics, and can be used to create new works of art, of designs and architectural and musical compositions. For example, the quint genomatrices can be presented in a form of color mosaics if matrix numbers are replaced by colors. It is possible to see regular complication of color mosaics along the family of the genomatrices with an increase of their Kronecker powers.

Figure 16. The proton sequence (4.4) of the amino acids with the additional number 54

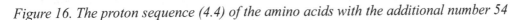

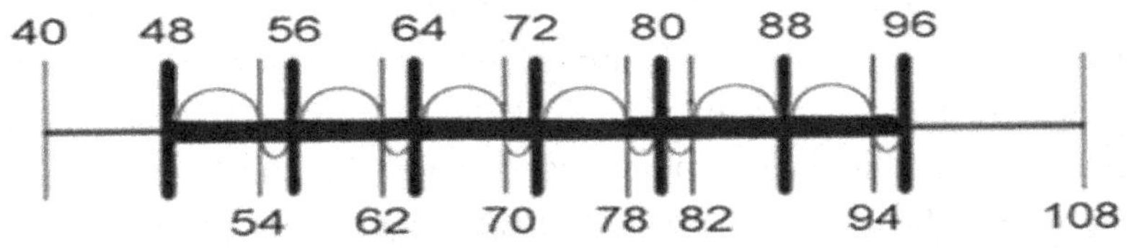

There is no doubt that application of numeric genetic matrices for investigations of the various ensembles of parameters of the genetic system can give many unexpected and useful results in the future as well. This direction of theoretical researches will be developed in parallel with developing matrix application in many other branches of science. The initial results, methods and patterns, which are described in this chapter, will serve as the basis for many new investigations in bioinformatics and theoretical biology. In particular Chapter 8 of this book describes one of the important ways to further mathematical analysis in this direction.

The matrix-genetic approach to phenomena of the golden section in genetic systems, physiology and aesthetics can be developed in many theoretical ways and can give new interesting mathematical models.

According to the described materials, each gene, each DNA, each protein can be characterized by its own musical ensemble. Generally speaking, this genetic music can be reproduced artificially for many practical applications in different fields: medicine, biotechnology, ergonomics, sports, etc. Such genetic musical melodies can be reproduced in sounds, colors ("color music"), electrical stimulus, and impulses of laser beams, etc. for different needs. Musical therapy and other branches of therapy can utilize these new forms of physical influences. Whether such "natural genetic music" (or compositions on its basis) possesses a special physiological effectiveness for the treatment of people and animals, stimulation of growth of plants and microorganisms, and so forth? For example, is it possible to treat patients with diabetes by means of sessions of such musical melodies, which correspond to the quint sequences of the gene of insulin? Future experiments can give the answer only. It seems that a creation of a computer bank of genetic music of various genes and proteins is useful for theoretical and practical needs. One can add here that the creator of analytic psychology Carl Jung, studying archetypes of human consciousness, has created the medical method of amplification. This method is based on an active intercourse of his patients with these archetypes including famous tables of Ancient Chinese "I Ching", which are connected with the genetic matrices (see Chapter 11).

Many composers declared a mysterious connection of music with the golden section early. In our opinion, this connection has the genetic base. The described facts are related to a problem of genetic bases of aesthetics and an inborn feeling of harmony.

Investigations of numeric genetic matrices are the effective scientific instrument to analyze multi-component and multi-parametric ensembles of the molecular-genetic systems. The obtained results give a new vision of connections of genetic systems with famous mathematical objects and theories from other branches of science and culture. Owing to the results of matrix genetics new opportunities arise to demonstrate the close connection between science and culture. The famous ideas about the harmony of biological organisms obtain new essential additions including materials about the golden section and the harmony of the Pythagorean musical scale.

The obtained results show that the system of hydrogen bonds of the complementary nitrogenous bases of the genetic code is not an accidental system, but it is the significant part of the harmonic molecular-genetic system, which is connected with principles of musical harmony and the golden section.

In our opinion, music is not only the tool for a call of emotions and pleasures, but it is also one of the principles of the organization and language of living substance. From the viewpoint of musical harmony in structures of molecular-genetic system, one can think that outstanding composers are researchers of harmony in living substance and in their own organisms. Investigation of musical harmony in genetic molecules and in adjacent systems ("musical bioinformatics") is the new interesting branch in biology. It is useful for education in fields of genetics, bioinformatics, theory of musical harmony, etc.

REFERENCES

Bellman, R. (1960). *Introduction to matrix analysis*. New York, Toronto, London: McGraw-Hill.

Berger, L. G. (2001). *Epistemology of art*. Moscow: Isskusstvo (in Russian).

Brewer, R. G., & Hanh, E. L. (1984). Atomic memory. *Scientific American, 251*(6), 50–57.

Bruhn, S. (1992). Symmetry and irreversibility in the musical language(s) of the 20th century. *Symmetry: Culture and Science, 3*(2), 187–200.

Bruhn, S. (1996). Symmetry in music and poetry [special issue]. *Symmetry: Culture and Science, 7*(2), 113–224.

Cook, T.A. (1914). *The curves of life*. London: Constable and Co.

Darvas, G. (2007). *Symmetry*. Basel: Birkhäuser book.

Feynman, R., Leighton, R., & Sands, M. (1963) *The Feynman lectures*. New York: Pergamon Press.

Goldman, J. (1992). *Healing sounds*. Vermont: Healing Arts Press.

Jean, R. V. (2006). *Phyllotaxis. A systemic study in plant morphogenesis*. Cambridge: Cambridge University Press.

Jenni, H. (1972). *Kymatic*. Dusseldorf: Springer-Verlag.

Kappraff, J. (1990). *Connections, the geometric bridge between art and science*. New York: McGraw Hill.

Kappraff, J. (1992). The relationship between mathematics and mysticism of the golden mean through history. In I. Hargittai (Ed.), *Fivefold symmetry* (pp. 33-66). Singapore: World Scientific.

Kappraff, J. (2000). The arithmetic of Nichomachus of Gerasa and its applications to systems of proportions. *Nexus Network Journal, 2*(4). Retrieved on October 3, 2000, from http://www.nexusjournal.com/Kappraff.html

Kappraff, J. (2002). *Beyond measure: Essays in nature, myth, and number*. Singapore: World Scientific.

Kappraff, J. (2006). Anne Bulckens' analysis of the proportions of the Parthenon. *Symmetry: Culture and Science, 17*(1-2), 91–96.

Le Corbusier, S. (1948). *Modulor*. Boulogne: Collection Ascoral.

Le Corbusier, S. (1953). *Der Modulor*. Stuttgart: DVA.

Lendvai, E. (1993). *Symmetries of music, an introduction to semantics of music*. Kecskemét: Kodály Institute.

Needham, J. (1962). *Science and civilization in China*. Cambridge: Cambridge University Press.

Petoukhov, S. V. (1981). *Biomechanics, bionics, and symmetry*. Moscow: Nauka.

Petoukhov, S. V. (1989). Non-Euclidean geometries and algorithms of living bodies. In I. Hargittai (Ed.), *Computers & Mathematics with Applications, 17*(4-6), 505-534. Oxford: Pergamon Press.

Petoukhov, S. V. (2001). Genetic codes 1: Binary sub-alphabets, bi-symmetric matrices, and golden section; Genetic codes 2: Numeric rules of degeneracy and the chronocyclic theory. *Symmetry: Culture and Science, 12*(1), 255–306.

Petoukhov, S. V. (2003-2004). Attributive conception of genetic code, its bi-periodic tables, and problem of unification bases of biological languages. *Symmetry: Culture and Science, 14-15*(part 1), 281–307.

Petoukhov, S. V. (2005b). Hadamard matrices and quint matrices in matrix presentations of molecular genetic systems. *Symmetry: Culture and Science, 16*(3), 247–266.

Petoukhov, S. V. (2008). *Matrix genetics, algebras of the genetic code, noise-immunity*. Moscow: RCD (in Russian).

Ponnamperuma, C. (1972). *The origin of life*. New York: E. P. Dutton.

Sachkov, V. N. (2004). *Introduction to combinatory methods of discrete mathematics*. Moscow: Binom (in Russian).

Schrodinger, E. (1955). *What is life? The physical aspect of the living cell.* Cambridge: University Press.

Shervatov, V. G. (1954). *The hyperbolic functions.* Moscow: GITTL (in Russian).

Shnoll, S. E. (1989). *Physical-chemical factors of biological evolution*. Moscow: Nauka (in Russian).

Shubnikov, A. V., & Koptsik, V. A. (1974). *Symmetry in science and art*. New York: Plenum Press.

Shults, G. E., & Schirmer, R. H. (1979). *Principles of protein structure*. Berlin: Springer-Verlag.

Trahtman, A. M. (1972). *Introduction in generalized spectral theory of signals*. Moscow: Sovetskoie Radio (in Russian).

Tusa, E. (1994). Lambdoma-"I Ging"-genetic code. *Symmetry: Culture and Science, 5*(3), 305–310.

Voloshinov, A. V. (2000). *Mathematics and arts*. Moscow: Prosveschenie (in Russian).

Weinberger, N. M. (2004). Music and brain. *Scientific American, 291*(5), 88–95.

Chapter 5
Genetic Code and Stochastic Matrices

ABSTRACT

In this chapter, we first use the Gray code representation of the genetic code C = 00, U = 10, G = 11, and A = 01 (C pairs with G, A pairs with U) to generate a sequence of genetic code-based matrices. In connection with these code-based matrices, we use the Hamming distance to generate a sequence of numerical matrices. We then further investigate the properties of the numerical matrices and show that they are doubly stochastic and symmetric. We determine the frequency distributions of the Hamming distances, building blocks of the matrices, decomposition and iterations of matrices. We present an explicit decomposition formula for the genetic code-based matrix in terms of permutation matrices. Furthermore, we establish a relation between the genetic code and a stochastic matrix based on hydrogen bonds of DNA. Using fundamental properties of the stochastic matrices, we determine explicitly the decomposition formula of genetic code-based biperiodic table. By iterating the stochastic matrix, we demonstrate the symmetrical relations between the entries of the matrix and DNA molar concentration accumulation. The evolution matrices based on genetic code were derived by using hydrogen bonds-based symmetric stochastic (2x2)-matrices as primary building blocks. The fractal structure of the genetic code and stochastic matrices were illustrated in the process of matrix decomposition, iteration and expansion in corresponding to the fractal structure of the biperiodic table introduced by Petoukhov (2001a, 2001b, 2005).

INTRODUCTION AND BACKGROUND

The universal genetic code may be viewed as the mapping of nucleic acids into polypeptides that is employed in every organism, organelle and virus with some minor variations. A mathematical view of genetic code is a map

DOI: 10.4018/978-1-60566-124-7.ch005

Figure 1. Creating a 3-bit Gray code from a 2-bit Gray code

A Gray Code for 2 Bits:	**00**	**01**	**11**	**10**				
the 2-bit code with "0" prefixes:	000	001	011	010				
the 2-bit code in reverse order:					10	11	01	00
the reversed code with "1" prefixes:					110	111	101	100
A Gray code for 3 bits:	**000**	**001**	**011**	**010**	**110**	**111**	**101**	**100**

$$g: C \rightarrow A \tag{1}$$

where $C = \{(x_1x_2x_3): x_i \in R = \{A, C, G, U\}\}$ denotes the set of codons and A = {Ala, Arg, Asp, …, Val, UAA, UAG, UGA} denotes the set of amino acids and termination codons. Genetic determinism, which presents the belief that we are controlled by our genes and that no other factor is significant, is now all-pervasive. This viewpoint is emphasized by the statement: "*life is a partnership between genes and mathematics*" (Stewart, 1999, p. xi).

We recall some basic definitions of a stochastic matrix. A square matrix of $P = (p_{ij})$ is a stochastic matrix if all entries of the matrix are nonnegative and the sum of the elements in each row (or column) is unity or a constant. If the sum of the elements in each row and column is unity or the same, the matrix is called doubly stochastic. The term "stochastic matrix" goes back at least to Romanovsky (1931). It plays a large role in the theory of discrete Markov chains. Stochastic matrices and doubly stochastic matrices have many remarkable properties. For example the Birkhoff–von Neumann Theorem says that every doubly stochastic matrix is a convex combination of permutation matrices of the same order and the permutation matrices are the extreme points of the set of doubly stochastic matrices. The properties of stochastic matrices are mainly spectral theoretic and are motivated by Markov chains. Doubly stochastic matrices have additional combinatorial structure.

The so called Gray code is one of the most famous in the theory of signal processing. The Gray code was used in a telegraph demonstrated by French engineer É. Baudot in 1878. The codes were first patented by F. Gray in 1953. The Gray code is a binary code in which consecutive decimal numbers are represented by binary expressions that differ in the state of one, and only one, bit. Gray codes have been extensively studied in other contexts. For example, Gray codes have been used in converting analog information to digital form. Here we review briefly how to construct a Gray code for each positive integer n. One way to construct a Gray code for n bits is to take a Gray code for (n-1) bits with each code prefixed by 0 (for the first half of the code) and append the (n-1) Gray code reversed with each code prefixed by 1 (for the second half). This is called a "binary-reflected Gray code". Figure 1 is an example of creating a 3-bit Gray code from a 2-bit Gray code.

A Gray code representation of the genetic code was proposed in the work (Swanson, 1984). A representation of the genetic code as a six-dimensional Boolean hypercube was proposed in (Jimenéz-Montaño, Mora-Basáñez, & Pöschel, 1994). In (Štambuk, 2000), universal metric properties of the genetic code

Figure 2. The two-level representation of the nitrogenous bases of the genetic code, which is corresponded to the binary 2-bit Gray code

$$C= \begin{array}{|c|}\hline 0 \\ 0 \\ \hline\end{array}\ ; A = \begin{array}{|c|}\hline 0 \\ 1 \\ \hline\end{array}\ ; G = \begin{array}{|c|}\hline 1 \\ 1 \\ \hline\end{array}\ ; U = \begin{array}{|c|}\hline 1 \\ 0 \\ \hline\end{array}$$

were defined by means of the nucleotide base representation on the square with vertices U or T = 0 0, C = 0 1, G = 1 0 and A = 1 1. It was shown that this notation defines the Cantor set and Smale horseshoe map representation of the genetic code. The "Biperiodic table of the genetic code" [C A; U G][(3)] (Figure 3 in Chapter 1), which has demonstrated an important symmetrical structure and has led to many discoveries, was introduced in (Petoukhov, 2001a, 2001b, 2005). This chapter describes stochastic characteristics of the biperiodic table on the basis of their original investigations and considerations in the works (He, 2001, 2003a, 2003b; He, Petoukhov, & Ricci, 2004).

One should recall information about the Hamming distance as well. The Hamming distance D is defined for strings of the same length. For two strings A and B, D(A,B) is the number of places in which the two string differ, i.e., have different characters. More formally, the distance between two strings A and B is $D(A,B) = \Sigma| A_i - B_i |$, sum of the numbers of places strings A and B differ. For example, the string A= 0101 and string B= 0110 has a Hamming distance $D(A,B) = 2$ whereas string A = "Butter" and string B = "ladder" has a Hamming distance $D(A,B) = 4$. This distance is applicable to encoded information, and is a particularly simple metric of comparison.

GENETIC CODE, HAMMING DISTANCE AND STOCHASTIC MATRICES

In this chapter we use the Gray code representation of the genetic code in a special form of a two-level (or double-decker) construction (Figure 2) to generate a sequence of genetic code-based matrices.

This binary representation is correlated to binary symbols of C, A, G, U on Figure 3 in Chapter 1. The reason for such a two-level construction is related to the tabular form of presentation of the genetic code on said figure. In connection with these code-based matrices, we use the Hamming distance to generate a sequence of numerical matrices. We investigate the properties of the numerical matrices and show that they are doubly stochastic and symmetric. We determine the frequency distributions of the Hamming distances, building blocks of the matrices, decomposition and iterations of matrices. We shall present an explicit decomposition formula for the genetic code-based matrix in terms of permutation matrices, which provide a hypercube representation of the genetic code.

We next list the sequences s_n of the Gray codes denoted by G_n in Figure 3.

It's easy to see that every n-bit string appears somewhere in the sequence; adjacent sequences $s_i s_{i+1}$ differ in exactly one bit, $i = 1, 2, \ldots, 2^n - 1$; the last sequence $s_2{}^n$ and the first sequence s_1 differ in exactly one bit in each of the cases on Figure 3. This proves that the n-cube has a Hamiltonian cycle for every positive integer n ^{3} 2, for example, $s_1 s_2, \ldots, s_2{}^n, s_1$ is a Hamiltonian cycle. There is a natural way to relate the genetic codons to Gray code by means of utilizing the Gray code representation of each nitrogenous bases on Figure 2. Examples of such representation of some codons in the two-level form are shown on Figure 4.

Figure 3. Gray code sequences s_n for n-digit cases

G_n	Gray Code Sequence s_n	Number of G_n
G_1	0 1	$2^1 = 2$
G_2	00 01 11 10	$2^2 = 4$
G_3	000 001 011 010 110 111 101 100	$2^3 = 8$
.	…	.
.	…	.
G_n	000…0 000…1 000…11 …010…0 111…0 111…1 111…00 …100…0	2^n

This new approach (He, 2001, 2003a, 2003b) presents each genetic multiplet as a two-level combination (or symbiosis) of two examples of the relevant Gray code: binary numbers of one example of the Gray code is utilized for the upper bit strings of the symbol of the multiplet, and the second example is utilized for its lower bit string. One can see on Figure 4 that a replacement of each of the symbols "1" and "0" by the opposite symbols "0" and "1" correspondingly leads from the Gray code representation of the codon CUG to the Gray code representation of its anticodon. This algorithm holds true for all pairs of "codon-anticodon" of the genetic code. Notice that the upper and lower bit strings of both the codon and anti-codon differ in a single bit. The Gray code arises in genetics as a means of minimizing the mismatches between the digits encoding adjacent bases and therefore the degree of mutation or differences between nearby chromosome segments. The requirement in an encoding scheme is that changing one bit in the segment of the chromosome should cause that segment to map to an element which is adjacent to the pre-mutated element.

Next we formalize our algorithm to generate the Hamming distance-based matrices corresponding to genetic code-based matrices. Let n be the length of strings (binary strings or DNA/RNA strings). We present our constructions for $n = 1$, 2, and 3. The general result for any positive integer n will be summarized following our discussions.

For $n = 1$, the Gray code $\boldsymbol{G}_1 = \{0, 1\}$. We arrange the G_1 in a 2-dimensional table (row/column) and form the table entry by stacking the column code on the top of row code as below. Denote this matrix by $\boldsymbol{H}_{21}$. This is a (2x2)-matrix generated by G_1 (Figure 5, on the left side).

The corresponding genetic code-based matrix with a single base is denoted by $\boldsymbol{C}_{21}$ (Figure 5, in the middle). We next compute the Hamming distance of each entry of the matrix H_{21}. The resulting matrix is denoted by $\boldsymbol{D}_{21}$ (Figure 5, on the right side). This matrix $\boldsymbol{D}_{21}$ has Hamming distances 0's and 1's. The frequencies of the 0's and 1's are 2 and 2, respectively. The total sum of entries of the matrix is 2. The

Figure 4. Examples of the Gray code representation of the codon CUG and of its anticodon GAC

$$\text{CUG} = \boxed{\begin{matrix}011\\001\end{matrix}}\ ;\ \text{GAC} = \boxed{\begin{matrix}100\\110\end{matrix}}$$

Figure 5. On the left side: the table G_1 with the binary two-level numeration of cells of the (2x2)-matrix H_{21}. In the middle: the corresponding genetic code-based matrix C_{21} with the genetic bases A. C, G, U. On the right side: the matrix D_{21}, entries of which are Hamming distances of the bases A. C, G, U in their considered Gray code representation

G_1	**0**	**1**
0	0 0	1 0
1	0 1	1 1

;

C	U
A	G

;

0	1
1	0

common row/column sum is 1. The Hamming distance between any two horizontal and vertical neighboring entries is 1.

For $n = 2$, the Gray code $\boldsymbol{G_2} = \{00, 01, 11, 10\}$. We arrange the G_2 in a 2-dimensional table and form the table entry by stacking the column code on top of the row code (Figure 6, on the left side). Denote this matrix by $\boldsymbol{H_{42}}$. This is a 4x4 matrix generated by G_2. One should emphasize that sequences of numbers of columns and of rows of $\boldsymbol{H_{42}}$ are given here in accordance with the Gray code sequence and they differ from the usual sequence of binary numeration which was used in Chapter 1 on Figure 3.

The corresponding genetic code-based matrix is denoted by $\boldsymbol{C_{42}}$ (Figure 6, in the middle). We next compute the Hamming distance of each entry of the matrix H_{42}. The resulting matrix is denoted by $\boldsymbol{D_{42}}$ (Figure 6, on the right side).

We note that the matrix $\boldsymbol{D_{21}}$ is centrally embedded inside $\boldsymbol{D_{42}}$ and the matrix $\boldsymbol{D_{42}}$ has two (2x2)-matrices as building blocks denoted by B_{21} and B_{22}. In view of this, the matrix $\boldsymbol{D_{42}}$ may be written as the block matrix $[B_{21}B_{22}; B_{22}B_{21}]$.

The frequencies of matrix building blocks B_{21} and B_{22} are 2 and 2, respectively. This matrix $\boldsymbol{D_{42}}$ has Hamming distances 0's, 1's, and 2's. The frequencies of the 0's, 1's and 2's are 4, 8, and 4, respectively. The total sum of entries in the matrix is 16. The common row/column sum is 4. The Hamming distance between any two horizontal and vertical neighboring entries is 1.

Figure 6. On the left side: the table G_2 with the binary two-level numeration of cells of the (4x4)-matrix H_{42}. In the middle: the corresponding genetic code-based matrix C_{42} with the 16 genetic duplets. On the right side: the matrix D_{42}, entries of which are Hamming distances of the genetic duplets in their considered Gray code representation

G_2	**00**	**01**	**11**	**10**
00	00 00	01 00	11 00	10 00
01	00 01	01 01	11 01	10 01
11	00 11	01 11	11 11	10 11
10	00 10	01 10	11 10	10 10

;

CC	CU	UU	UC
CA	CG	UG	UA
AA	AG	GG	GA
AC	AU	GU	GC

;

0	**1**	**2**	**1**
1	**0**	**1**	**2**
2	**1**	**0**	**1**
1	**2**	**1**	**0**

Figure 7. The table G_3 with the binary two-level numeration of cells of the (8x8)-matrix H_{83}

G_3	000	001	011	010	110	111	101	100
000	000 000	001 000	011 000	010 000	110 000	111 000	101 000	100 000
001	000 001	001 001	011 001	010 001	110 001	111 001	101 001	100 001
011	000 011	001 011	011 011	010 011	110 011	111 011	101 011	100 011
010	000 010	001 010	011 010	010 010	110 010	111 010	101 010	100 010
110	000 110	001 110	011 110	010 110	110 110	111 110	101 110	100 110
111	000 111	001 111	011 111	010 111	110 111	111 111	101 111	100 111
101	000 101	001 101	011 101	010 101	110 101	111 101	101 101	100 101
100	000 100	001 100	011 100	010 100	110 100	111 100	101 100	100 100

For $n = 3$, the Gray code $\boldsymbol{G}_3 = \{000, 001, 011, 010, 110, 111, 101, 100\}$. The matrix $\boldsymbol{H}_{83}$ is a (8x8)-matrix on Figure 7.

The corresponding genetic code-based matrix is denoted by $\boldsymbol{C}_{83}$ (Figure 8). In this matrix we mark by dark (white) colors each of those cells which contains a "black" ("white") triplet by analogy with Figure 2 in Chapter 2.

The genetic matrix $\boldsymbol{C}_{83}$ differs from the genetic matrix [C A; U G]$^{(3)}$ which was considered in other chapters of the book (see Figure 3 in Chapter 1). It proposes an original variant of matrix presentation of the genetic code. The relevant Hamming distance-based matrix $\boldsymbol{D}_{83}$ is shown on Figure 9.

This matrix $\boldsymbol{D}_{83}$ has Hamming distances 0's, 1's, 2's and 3's. The frequencies of the 0's, 1's, 2's and 3's are 8, 24, 24, and 8, respectively. The total sum of the matrix $\boldsymbol{D}_{83}$ is 96. The common row/column sum is 12. The Hamming distance between any two horizontal and vertical neighboring entries is 1. We also note that the matrices $\boldsymbol{D}_{21}$ and $\boldsymbol{D}_{42}$ are centrally embedded inside $\boldsymbol{D}_{83}$ and the matrix $\boldsymbol{D}_{83}$ has three 2x2 matrices building blocks B_{21}, B_{22} and B_{23}. It is obvious from Figure 9 that the matrix $\boldsymbol{D}_{83}$ may be also written in the following form of a block matrix (Figure 10).

The frequencies of matrix building blocks B_{21}, B_{22} and B_{23} are 4, 8, and 4, respectively. The distribution of the codons at separate magnitudes of the Hamming distance is shown on Figure 11 together with frequencies of meeting of these magnitudes of Hamming distances in the matrix $\boldsymbol{D}_{83}$.

Let us consider generalization of such matrices. In a general case of the matrices $\boldsymbol{C}_{21}$, $\boldsymbol{C}_{42}$, $\boldsymbol{C}_{83}$, … and of the matrices $\boldsymbol{D}_{21}$, $\boldsymbol{D}_{42}$, $\boldsymbol{D}_{83}$, …we will use the general symbols $C_{2^n n}$ and $D_{2^n n}$ correspondingly. Here $2^n n$ is the lower index in both cases. For general positive integer n, we have the following results (He, 2003a).

Let n be the length of binary or DNA/RNA strings and G_n be the n-bit Gray code. Then

Figure 8. The genetic code-based matrix C_{83} with the 64 genetic triplets. Black (white) codons are disposed in dark (white) cells

CCC	CCU	CUU	CUC	UUC	UUU	UCU	UCC
CCA	CCG	CUG	CUA	UUA	UUG	UCG	UCA
CAA	CAG	CGG	CGA	UGA	UGG	UAG	UAA
CAC	CAU	CGU	CGC	UGC	UGU	UAU	UAC
AAC	AAU	AGU	AGC	GGC	GGU	GAU	GAC
AAA	AAG	AGG	AGA	GGA	GGG	GAG	GAA
ACA	ACG	AUG	AUA	GUA	GUG	GCG	GCA
ACC	ACU	AUU	AUC	GUC	GUU	GCU	GCC

Figure 9. The relevant Hamming distance-based matrix $\mathbf{D}_{83}$

0	1	2	1	2	3	2	1
1	0	1	2	3	2	1	2
2	1	0	1	2	1	2	3
1	2	1	0	1	2	3	2
2	3	2	1	0	1	2	1
3	2	1	2	1	0	1	2
2	1	2	3	2	1	0	1
1	2	3	2	1	2	1	0

Figure 10. The presentation of the matrix $\mathbf{D}_{83}$ *as a block matrix*

B_{21}	B_{22}	B_{23}	B_{22}
B_{22}	B_{21}	B_{22}	B_{23}
B_{23}	B_{22}	B_{21}	B_{22}
B_{22}	B_{23}	B_{22}	B_{21}

Figure 11. The distributions of the codons at separate magnitudes 0, 1, 2, 3 of the Hamming distance in the matrix $\mathbf{D}_{83}$*. Frequencies of meeting these magnitudes are shown in the third column*

Distance	Codons	Frequency
0	CCC CCG CGG CGC GGC CGG GCG GCC	8
1	ACC ACG AGC AGG CAC CAG CCA CGA CCU CGU CUC CUG GAC GAG GCA GGA GCU GGU GUC GUG UCC UGC UCG UGG	24
2	AAC AAG ACA ACU AGA AGU AUC AUG CAA CAU CUA CUUGAA GAU GUA GUU UAC UAG UCA UCU UGA UGU UUC UUG	24
3	AAA AAU AUA AUU UAA UAU UUA UUU	8

1. The genetic code-based matrix $C_{2\ n}^{\ n}$ is a (2^nx2^n)-matrix with RNA bases of length n. Each two neighboring entries of the genetic code both from vertical and horizontal direction differs exactly one base.
2. The Hamming distance-based matrix $D_{2\ n}^{\ n}$ is also a (2^nx2^n)-matrix with Hamming distances of 0, 1, 2, …, n. The common row/column sum of the matrix $D_{2\ n}^{\ n}$ equals $n2^{n-1}$ and the total summation of the entries of matrix $D_{2\ n}^{\ n}$ is $n2^{2n-1}$.
3. The matrix $D_{2\ n}^{\ n}$ is a doubly stochastic and symmetric.
4. The frequency distributions denoted by f_{nk} ($n = 2, 3, \ldots, k = 1, 2, \ldots$) of Hamming distances of 0, 1, 2, …, n is shown below for $n = 1, 2, 3, 4$, and 5 on Figure 12.

The same table can be presented in another form (Figure 13).

The general relationships of the frequencies are determined by a recurrence formula:

$$f_{21} = 2, f_{22} = 2,$$

$$f_{nk} = 2\,(f_{(n-1)(k-1)} + f_{(n-1)k})$$

The frequency distribution of the Hamming distances is the **Pascal triangle** with a multiple of 2^n. The solution to this recurrence relation is

Figure 12. Frequency distribution of Hamming distances for different numbers n

n	Hamming Distances	Frequency Distributions	Frequency notation
1	0 1	2 2	$f_{21}\, f_{22}$
2	0 1 2	4 8 4	$f_{31}\, f_{32}\, f_{33}$
3	0 1 2 3	8 24 24 8	$f_{41}\, f_{42}\, f_{43}\, f_{44}$
4	0 1 2 3 4	16 64 96 64 16	$f_{51}\, f_{52}\, f_{53}\, f_{54}\, f_{55}$
5	0 1 2 3 4 5	32 160 320 320 160 32	$f_{61}\, f_{62}\, f_{63}\, f_{64}\, f_{65}\, f_{66}$

Figure 13. Another form of presentation of frequency distribution of Hamming distances for different numbers n

n	Hamming Distances	Frequency Distributions	Frequency notation
1	0 1	2* 1 1	$f_{21} f_{22}$
2	0 1 2	4* 1 2 1	$f_{31} f_{32} f_{33}$
3	0 1 2 3	8* 1 3 3 1	$f_{41} f_{42} f_{43} f_{44}$
4	0 1 2 3 4	16* 1 4 6 4 1	$f_{51} f_{52} f_{53} f_{54} f_{55}$
5	0 1 2 3 4 5	32* 1 5 10 10 5 1	$f_{61} f_{62} f_{63} f_{64} f_{65} f_{66}$

$f_{nk} = 2^n C(n,k)$, $k = 1, 2, \ldots, n$.

5. The matrix $D_{2\,n}^{\,n}$ consists of (n-1) (2x2)-matrix building blocks $B_{21}, B_{22}, \ldots, B_{2(n-1)}$. The previous matrix $D_{2\,(n-1)}^{\,n-1}$ is centrally embedded inside the next matrix $D_{2\,n}^{\,n}$. The frequencies of matrix building blocks $B_{21}, B_{22}, \ldots, B_{2(n-1)}$ are $f_{(n-1)1}, f_{(n-1)2}, \ldots, f_{(n-1)(n-1)}$, respectively.

Next we illustrate the stochastic and hypercube structure of the genetic code based matrix $C_{2\,n}^{\,n}$ via the structure of matrix of $D_{2\,n}^{\,n}$.

As we have noted, the matrix $D_{2\,n}^{\,n}$ is a symmetric and doubly stochastic matrix. For its simplicity, we consider the case when $n = 3$, i.e. $D_{83,}$ the entry of the matrix is a RNA codon. Here we list some basic properties of the matrix $D_{83.}$

- The matrix D_{83} is symmetric since $D_{83} = D_{83}^{\,T}$ (the transpose of a matrix).
- The matrix D_{83} is singular since Det (D_{83}) = 0 (determinant of a matrix).
- The eigenvalues of D_{83} is $\{l_1, l_2, \ldots l_8\}$={-4, -4, -4, 0, 0, 0, 0, 12}.
- The eigenvectors are {0, -1, -1, 0, 1, 0, 0, 1}, {0, 1, 0, -1, -1, 0, 1, 0}, {-1, -1, 0, 0, 1, 1, 0, 0}, {0, -1, 1, 0, -1, 0, 0, 0},{1, -2, 1, 0, -1, 0, 1, 0}, {1, -1, 0, 0, -1, 1, 0, 0},{-1, 1, -1, 1, 0, 0, 0, 0}, {1, 1, 1, 1, 1, 1, 1, 1}. Furthermore these 8 vectors are linearly independent. They form a basis for a vector space of dimension of 8.
- Trace of matrix D_{83} = sum of eigenvalues = 0 + 0 + 0 + 0 - 4 - 4 - 4 + 12 =0.

Since the matrix D_{83} is doubly stochastic, the matrix D_{83} can be decomposed as a convex combination of finitely many permutation matrices (Bapat, & Raghavan, 1997); that is,

$$D_{83} = a_1 P_1 + a_2 P_2 + \ldots + a_8 P_8,$$

where $P_1, P_2, \ldots, P_8$ are permutation matrices and $0 \le a_1, a_2, \ldots, a_8 \le 12$, $a_1 + a_2 + \ldots + a_8 = 12$. A permutation matrix can be obtained from an identity matrix by permuting its rows and columns. Explicitly we have the following result.

The matrix $D_{83} = 0\, P_1 + 1\, (P_2 + P_3 + P_4) + 2\, (P_5 + P_6 + P_7) + 3\, P_8$, where P_1 = Table 1.

The corresponding codons (or vertices/nodes of a graph) of this matrix P_1 are {CCC, CCG, CGG, CGC, GGC, GGG, GCG, GCC}.

P_2 = Table 2.

The corresponding codons (or vertices/nodes of a graph) of P_2 are {CCA, CCU, CGA, CGU, GGA, GGU, GCU, GCA}.

P_3 = Table 3.

The corresponding codons (or vertices/nodes of a graph) of this matrix P_3 are {CAC, CAG, CUG, CUC, GUC, GUG, GAG, GAC}.

P_4 = Table 4.

The corresponding codons (or vertices/nodes of a graph) of this matrix P_4 are {ACC, ACG, AGG, AGC, UGC, UGG, UCG, UCC}.

P_5 = Table 5.

The corresponding codons (or vertices/nodes of a graph) of this matrix P_5 are {ACA, ACU, AUG, AGA, UGA, UGU, UCU, UCA}.

P_6 = Table 6.

The corresponding codons (or vertices/nodes of a graph) of matrix P_6 are {CAA, CAU, CUU, CUA, GUA, GUU, GAU, GAA}.

P_7 = Table 7.

The corresponding codons (or vertices/nodes of a graph) of this matrix P_7 are {AAC, AAG, AGU, AUC, UCC, UGG, AUG, UAC}.

P_8 = Table 8.

The corresponding codons (or vertices/nodes of a graph) of the 8th matrix P_8 are {AAA AAU, AUU, AUA, UUA, UUU, UAU, UAA}.

GENETIC CODE, ATTRIBUTIVE MAPPING AND STOCHASTIC MATRICES

Chapter 1 has demonstrated already the three binary sub-alphabets of the genetic alphabet which allow creating the described tabular and matrix form of presentation of ensembles of molecular elements of the genetic code (Figures 2 and 3 from Chapter 1). These sub-alphabets are based on the three kinds of binary-oppositional attributes of the nitrogenous bases A, C, G, U/T. From the viewpoint of the first kind of the binary-oppositional attributes (pyrimidines-purine), the following pairs of equivalent genetic letters exist: C = U and A = G (here "=" is the symbol of equivalence). From the viewpoint of the second kind of the binary-oppositional attributes (amino-mutating and non-amino-mutating), the following pairs of

Table 1. P_1 =

1	0	0	0	0	0	0	0
0	1	0	0	0	0	0	0
0	0	1	0	0	0	0	0
0	0	0	1	0	0	0	0
0	0	0	0	1	0	0	0
0	0	0	0	0	1	0	0
0	0	0	0	0	0	1	0
0	0	0	0	0	0	0	1

Table 2. P_2 =

0	1	0	0	0	0	0	0
1	0	0	0	0	0	0	0
0	0	0	1	0	0	0	0
0	0	1	0	0	0	0	0
0	0	0	0	0	1	0	0
0	0	0	0	1	0	0	0
0	0	0	0	0	0	0	1
0	0	0	0	0	0	1	0

equivalent genetic letters exist: C = A and U = G. From the viewpoint of the third kind of the binary-oppositional attributes (2 and 3 hydrogen bonds), the following pairs of equivalent genetic letters exist: C = G and A = U.

The works (He, 2001, 2003) contain an analysis of three mapping relations on the basis of these attributes for generating new interesting matrices and for studying their properties and symmetries.

Here we further investigate the symmetrical structures of the genetic matrix $[C\ A;\ U\ G]^{(3)}$ from the viewpoint of the third kind of the binary oppositional attributes. The complementary letters C and G have 3 hydrogen bonds (C = G = 3) and the complementary letters A and U have 2 hydrogen bonds (A = U = 2). Let us replace each multiplet in the genetic matrix $[C\ A;\ U\ G]^{(3)}$ by the sum of these numbers of its hydrogen

Table 3. $P_3 =$

0	0	0	1	0	0	0	0
0	0	1	0	0	0	0	0
0	1	0	0	0	0	0	0
1	0	0	0	0	0	0	0
0	0	0	0	0	0	0	1
0	0	0	0	0	0	1	0
0	0	0	0	0	1	0	0
0	0	0	0	1	0	0	0

Table 4. $P_4 =$

0	0	0	0	0	0	0	1
0	0	0	0	0	0	1	0
0	0	0	0	0	1	0	0
0	0	0	0	1	0	0	0
0	0	0	1	0	0	0	0
0	0	1	0	0	0	0	0
0	1	0	0	0	0	0	0
1	0	0	0	0	0	0	0

Table 5. $P_5 =$

0	0	0	0	0	0	1	0
0	0	0	0	0	0	0	1
0	0	0	0	1	0	0	0
0	0	0	0	0	1	0	0
0	0	1	0	0	0	0	0
0	0	0	1	0	0	0	0
1	0	0	0	0	0	0	0
0	1	0	0	0	0	0	0

Table 6. $P_6 =$

0	0	1	0	0	0	0	0
0	0	0	1	0	0	0	0
1	0	0	0	0	0	0	0
0	1	0	0	0	0	0	0
0	0	0	0	0	0	1	0
0	0	0	0	0	0	0	1
0	0	0	0	1	0	0	0
0	0	0	0	0	1	0	0

Table 7. $P_7 =$

0	0	0	0	1	0	0	0
0	0	0	0	0	1	0	0
0	0	0	0	0	0	1	0
0	0	0	0	0	0	0	1
1	0	0	0	0	0	0	0
0	1	0	0	0	0	0	0
0	0	1	0	0	0	0	0
0	0	0	1	0	0	0	0

Table 8. $P_8 =$

0	0	0	0	0	1	0	0
0	0	0	0	1	0	0	0
0	0	0	0	0	0	0	1
0	0	0	0	0	0	1	0
0	1	0	0	0	0	0	0
1	0	0	0	0	0	0	0
0	0	0	1	0	0	0	0
0	0	1	0	0	0	0	0

bonds. For example, the triplet CAU will be replaced by number 7 (= 3+2+2). (Figure 2 in Chapter 4 presented another variant, when symbolic multiplets were replaced by product of numbers of their hydrogen bonds). In this "additive" case, we get the following numeric matrix, denoted by *G(i,j)* (Figure 14).

One can easily see that the matrix *G(i,j)* has common row sum and common column sum of 60. It implies that the matrix *G(i, j)* is a doubly stochastic (8x8)-matrix. In next paragraph, we explore the properties of this matrix and its relationship with DNA and protein sequences.

Here we list some basic properties of the matrix *G(i,j)*.

- The matrix *G(i,j)* is symmetric since $G(i,j) = G(i,j)^T$.
- The matrix *G(i,j)* is singular since Det (*G(i,j)*) = 0.
- The eigenvalues of *G(i,j)* is $\{l_1, l_2, \ldots l_8\}$={0, 0, 0, 0, 4, 4, 4, 60}.
- The eigenvectors are {2, -1, -1, 0, -1, 0, 0, 1}, {1, 0, -1, 0, -1, 0, 1, 0}, {1, -1, 0, 0, -1, 1, 0, 0}, {1, -1, -1, 1, 0, 0, 0, 0},{-1, 0, 0, 1, -1, 0, 0, 1}, {0, -1, 0,-1, 1, 1, 0, 1, 0},{0, 0, -1, -1, 1, 1, 0, 0}, {1, 1, 1, 1, 1, 1, 1, 1}. Furthermore these 8 vectors are linearly independent. They form a basis for a vector space of dimension of 8.
- Trace of matrix *G(i,j)* = Sum of eigenvalues = 0 + 0 + 0 + 0 + 4 + 4 + 4 + 60 =72.

Since the matrix *G(i,j)* is doubly stochastic, the matrix *G(i,j)* can be decomposed as a convex combination of finitely many permutation matrices (Bapat, & Raghavan, 1997); that is,

$$G(i,j) = a_1P_1 + a_2P_2 + \ldots + a_mP_m,$$

where $P_1, P_2, \ldots, P_m$ are permutation matrices and $0 \le a_1, a_2, \ldots, a_m \le 60$, $a_1 + a_2 + \ldots + a_m = 60$. A permutation matrix can be obtained from an identity matrix by permuting its rows and columns. Explicitly we have

$$G(i,j) = 9\,P_1 + 8\,(P_2 + P_3 + P_4) + 7\,(P_5 + P_6 + P_7) + 6\,P_8,\ \text{where}$$

P_1 = Table 9.
P_2 = Table 10.
P_3 = Table 11.

Figure 14. The transformation of the genomatrix [C A; U G]$^{(3)}$ *into the numeric genomatrix* **G(i,j)**, *each entry of which is equal to the sum of hydrogen bonds of the relevant codon*

G(i,j) =

9	8	8	7	8	7	7	6
8	9	7	8	7	8	6	7
8	7	9	8	7	6	8	7
7	8	8	9	6	7	7	8
8	7	7	6	9	8	8	7
7	8	6	7	8	9	7	8
7	6	8	7	8	7	9	8
6	7	7	8	7	8	8	9

P_4 = Table 12.
P_5 = Table 13.
P_6 = Table 14.
P_7 = Table 15.
P_8 = Table 16.

Each permutation matrix is also doubly stochastic and symmetric. Each matrix can be viewed as a vertex of genetic cube illustrated in (Petoukhov, 2001). One may note that this genetic cube can be iter-

Table 9. P_1 =

1	0	0	0	0	0	0	0
0	1	0	0	0	0	0	0
0	0	1	0	0	0	0	0
0	0	0	1	0	0	0	0
0	0	0	0	1	0	0	0
0	0	0	0	0	1	0	0
0	0	0	0	0	0	1	0
0	0	0	0	0	0	0	1

Table 10. P_2 =

0	1	0	0	0	0	0	0
1	0	0	0	0	0	0	0
0	0	0	1	0	0	0	0
0	0	1	0	0	0	0	0
0	0	0	0	0	1	0	0
0	0	0	0	1	0	0	0
0	0	0	0	0	0	0	1
0	0	0	0	0	0	1	0

Table 11. P_3 =

0	0	1	0	0	0	0	0
0	0	0	1	0	0	0	0
1	0	0	0	0	0	0	0
0	0	0	0	0	0	1	0
0	1	0	0	0	0	0	0
0	0	0	0	0	0	0	1
0	0	0	0	1	0	0	0
0	0	0	0	0	1	0	0

Table 12. P_4 =

0	0	0	0	1	0	0	0
0	0	0	0	0	1	0	0
0	0	0	0	0	0	1	0
0	0	0	0	0	0	0	1
1	0	0	0	0	0	0	0
0	1	0	0	0	0	0	0
0	0	1	0	0	0	0	0
0	0	0	1	0	0	0	0

Table 13. P_5 =

0	0	0	1	0	0	0	0
0	0	1	0	0	0	0	0
0	1	0	0	0	0	0	0
1	0	0	0	0	0	0	0
0	0	0	0	0	0	0	1
0	0	0	0	0	0	1	0
0	0	0	0	0	1	0	0
0	0	0	0	1	0	0	0

Table 14. P_6 =

0	0	0	0	0	1	0	0
0	0	0	0	1	0	0	0
0	0	0	0	0	0	0	1
0	0	0	0	0	0	1	0
0	1	0	0	0	0	0	0
1	0	0	0	0	0	0	0
0	0	0	1	0	0	0	0
0	0	1	0	0	0	0	0

Table 15. $P_7 =$

0	0	0	0	0	0	1	0
0	0	0	0	0	0	0	1
0	0	0	0	1	0	0	0
0	0	0	0	0	1	0	0
0	0	1	0	0	0	0	0
0	0	0	1	0	0	0	0
1	0	0	0	0	0	0	0
0	1	0	0	0	0	0	0

Table 16. $P_8 =$

0	0	0	0	0	0	0	1
0	0	0	0	0	0	1	0
0	0	0	0	0	1	0	0
0	0	0	0	1	0	0	0
0	0	0	1	0	0	0	0
0	0	1	0	0	0	0	0
0	1	0	0	0	0	0	0
1	0	0	0	0	0	0	0

ated by taking the power of the matrix $G(i,j)$. A fractal structure of the genetic cube will emerge from one generation to another.

GENETIC CODE, POWER OF MATRICES AND STOCHASTIC MATRICES

We next recall a well-known result on the power of matrix. If A is the adjacency matrix of a simple graph, the ij-th entry of A^m is equal to the number of paths of length m from vertex i to vertex j, m=1, 2, 3,.... To apply this result to the matrix D_{83}, we conclude that the number of paths of length m is equal to the entries of m-th power of an adjacency matrix D_{83} corresponding to a simple graph with codons as vertices.

For $m = 1, 2, 3,\ldots$, we denote $D_{83}^{\,m}$ the m-th power of matrix $G(i,j)$. It's easy to see that the matrices $D_{83}^{\,1}, D_{83}^{\,2}, \ldots, D_{83}^{\,m}$ are doubly stochastic, their eigenvalues are $\{ (l_1)^m, (l_2)^m, \ldots (l_8)^m \} = \{0, 0, 0, 0, (-4)^m, (-4)^m, (-4)^m, 12^m\}$ with the same eigenvectors of D_{83}.

Here we illustrate the powers of matrix D_{83} when $m = 2$, and 3, respectively.

$(D_{83})^2$ = Table 17.

The next iteration is the 3rd power of matrix D_{83}. The resulting matrix is

$(D_{83})^3$ = Table 18.

As the power m increases, the number of paths increases rapidly. This kind of hypercomplex number is considered in Chapter 9 under the name "hyperbolic matrions". One can extend this result into the general case of matrix $D_{2^n n}$. If the length of DNA/RNA sequences is n, then all possible Hamming distances among the entries of the matrix $D_{2^n n}$ are 0, 1, 2, ..., n. The dimension of this matrix is 2^n by 2^n. Each entry of the matrix is a chain of DNA/RNA bases of length n. The iterations of the matrices provide a way of knowing the number of paths traveling from one entry to another within the matrix.

Chemical analysis of the molar content of the bases (generally called the base composition) adenine, thymine (uracil), guanine, and cytosine in DNA molecules isolated from many organisms provided the important known fact that [A] = [U] and [G] = [C], in which [ ] denotes molar concentration, from which followed the corollary [A+G] = [U+C] or [purines] = [pyrimindines]. These chemical properties are well linked with the iterations of the genetic matrix. For n=1, 2, 3,..., we denote $G(i,j)^n$ the n-th power of matrix $G(i,j)$. It's easy to see that the matrices $G(i,j)^1, G(i,j)^2, \ldots, G(i,j)^n$ are doubly stochastic, their eigenvalues are $\{ (l_1)^n, (l_2)^n, \ldots (l_8)^n \} = \{0, 0, 0, 0, 4^n, 4^n, 4^n, 60^n\}$ with the same eigenvectors of $G(i,j)$.

Table 17. $(D_{83})^2 =$

24	20	16	20	16	12	16	20
20	24	20	16	12	16	20	16
16	20	24	20	16	20	16	12
20	16	20	24	20	16	12	16
16	12	16	20	24	20	16	20
12	16	20	16	20	24	20	16
16	20	16	12	16	20	24	20
20	16	12	16	20	16	20	24

Table 18. $(D_{83})^3 =$

192	208	224	208	240	224	224	208
208	192	208	224	224	240	208	224
224	208	192	208	224	208	240	224
208	224	208	192	208	224	224	240
240	224	224	208	192	208	224	208
224	240	208	224	208	192	208	224
224	208	240	224	224	208	192	208
208	224	224	240	208	224	208	192

The iteration of the matrix $G(i,j)$ gives us an indicator for molar concentration accumulations. We illustrate this process of iteration by computing $G(i,j)^2$ and $G(i,j)^3$ respectively.

$G(i,j)^2$ = Table 19.

It's easy to see that the sum of the corresponding entries of the first row (or column) and of the last row (or column) has common sum of 900. This also applies to 2nd row (or column) with 7th row (or column), 3rd row (or column) with the 6th row (or column) and 4th row (or column) with the 5th row (or column). These properties are corresponding to the molar concentration accumulations under multiplication and addition. For example the entry at the first row and the first column 456 is a result of accumulation of the following codons:

456 = CCC*CCC+CCA*CCU+CAC*CUC+CAA*UCC+ACC*CUU+ACA*UCU+AAC*UUC +AAA*UUU =

= (3+3+3)*(3+3+3)+(3+3+2)*(3+3+2)+(3+2+3)*(3+2+3)+(3+2+2)*(2+3+3) +(2+3+3)*(3+2+2)+(2+3+2)*(2+3+2)+(2+2+3)*(2+2+3)+(2+2+2)*(2+2+2)

The entry at the 8th row and the first column 444 is a result of accumulation of the following codons:

444= UUU*CCC+UUG*CCU+UGU*CUC+UGG*UCC+GUU*CUU+GUG*UCU+GGU*UUC +GGG*UUU =

Table 19. $G(i,j)^2 =$

456	452	452	448	452	448	448	444
452	456	448	452	448	452	444	448
452	448	456	452	448	444	452	448
448	452	452	456	444	448	448	452
452	448	448	444	456	452	452	448
448	452	444	448	448	456	448	452
448	444	452	448	452	448	456	452
444	448	448	452	448	452	452	456

=(2+2+2)*(3+3+3)+(2+2+3)*(3+3+2)+(2+3+2)*(3+2+3)+(2+3+3)*(2+3+3)+(3+2+2)*(3+2+2)+(3+2+3)*(2+3+2)+(3+3+2)*(2+2+3)+(3+3+3)*(2+2+2)

The sum to these two entries equals 456+444=900, which leads to common row (column) sum of 4*900 = 3600. These common sums were governed by the fact that [A] = [U], [C] = [G], and [A+ G] = [C + U].

The next iteration is the 3rd power of matrix $G(i,j)$. The resulting matrix is $G(i, j)^3$ = Table 20.

In this case, we have the common row (or column) sum of 216000 = 4*5400. The value 5400 = 60*900 = 4*15*900 was derived from previous sum accumulation. This matrix iteration shows us the process of molar accumulation and demonstrates various symmetrical structure embedded in the molar concentration.

Next we illustrate a model of genetic code evolution based on the Kronecker family of the genomatrices [C A; U G]$^{(n)}$. A fractal character of hierarchic structure of this family was described in (Petoukhov,

Table 20. $G(i, j)^3$ =

27024	27008	27008	26992	27008	26992	26992	26976
27008	27024	26992	27008	26992	27008	26976	26992
27008	26992	27024	27008	26992	26976	27008	26992
26992	27008	27008	27024	26976	26992	26992	27008
27008	26992	26992	26976	27024	27008	27008	26992
26992	27008	26976	26992	27008	27024	26992	27008
26992	26976	27008	26992	27008	26992	27024	27008
26976	26992	26992	27008	26992	27008	27008	27024

Figure 15. Fractal structure of the genetic code or a possible model of its three-stages evolution (from the table with one 'initial" nitrogen bases to octet table with 64 triplets) by means of standardizing quaternary partition of each table cell at a transition to the next table of this sequence

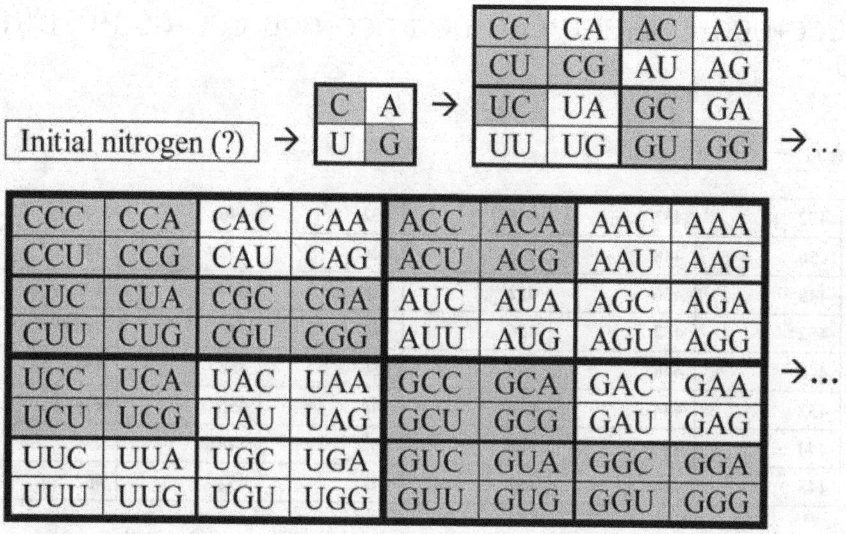

2001). This fractal character is connected with properties of the Kronecker algorithm of construction of this family: [C A; U G]⊗[C A; U G]⊗[C A; U G]⊗..., where ⊗ is the symbol of Kronecker multiplication. For example the (8x8)-genomatrix $[C\ A;\ U\ G]^{(3)}$ is divided into four (4x4)-quadrants with certain dispositions of letters C, A, U, G on the first positions of their multiplets; each of these (4x4)-quadrants is divided into four (2x2)-sub-quadrants with similar dispositions of letters C, A, U, G on the first positions of their multiplets; each of these (2x2)-sub-quadrants is divided into four cells C, A, U, G. It permits one to represent building algorithms of this table $[C\ A;\ U\ G]^{(n)}$. in a form of three generations of 4-ary or quaternary divisions (partitions or "reproductions") of its cells ensembles (Figure 15). One can suppose that evolution of the genetic code may be organized in a similar fractal way with quaternary partitions of elements at each of three stages (Petoukhov, 2001).

In the (16x16)-matrix $[C\ A;\ U\ G]^{(4)}$, each entry has four bases and it can form two possible codons in a linear chain from left to right. For example, four bases in the 4-plet CCCA may form two codons CCC and CCA in a sequential order. The next (32x32)-matrix $[C\ A;\ U\ G]^{(5)}$ has 1024 entries. In this case, each entry has five bases and forms three possible codons in a sequential order. For example the entry CCCAC may form three codons CCC, CCA, CAC. The next standardizing quaternary partition leads to a (64x64)-matrix $[C\ A;\ U\ G]^{(4)}$ with 4096 entries. Each entry has six bases and it may form 4 possible codons. For example, the codons CAC, ACC, CCC and CCA are formed from the entry CACCCA. We use Tables 21, 22 and 23 to illustrate this process by summarizing the matrix dimensions, number of bases with possible codons and total number of entries of such matrices.

In general, we can form Table 24 to summarize this process for $k = 1, 2, 3, \ldots$.

Figure 16 shows the corresponding sequence of matrices, where the first matrix contains numbers 2 and 3 of hydrogen bonds of the genetic bases C, A, U, G and where each next matrix is generated by means of standardizing quaternary partition.

Table 21. Phase I: First phase of evolution in three stages

Dimensions = 2^0x2^0 No. of Entries =1 No. of Bases=0	Dimensions = 2^1x2^1 No. of Entries =4 No. of Bases =1	Dimensions = 2^2x2^2 No. of Entries =16 No. of Bases =1	Dimensions = 2^3x2^3 No. of Codons =64 No. of Bases =1 No. of Possible Codons=1

Table 22. Phase II: Second phase of evolution in three stages

Dimensions = 2^3x2^3 No. of Entries =64 No. of Bases =3 No. of Possible Codons=1	Dimensions = 2^4x2^4 No. of Entries =256 No. of Bases =4 No. of Possible Codons=2	Dimensions = 2^5x2^5 No. of Entries =1024 No. of Bases =5 No. of Possible Codons=3	Dimensions = 2^6x2^6 No. of Entries =4096 No. of Bases =6 No. of Possible Codons=4

Table 23. Phase III: Third phase of evolution in three stages

Dimensions = 2^6x2^6 No. of Entries =4096 No. of Bases =6 No. of Possible Codons=4	Dimensions = 2^7x2^7 No. of Entries =16384 No. of Bases =7 No. of Possible Codons=5	Dimensions = 2^8x2^8 No. of Entries =65536 No. of Bases =8 No. of Possible Codons=6	Dimensions = 2^9x2^9 No. of Entries =262144 No. of Bases =9 No. of Possible Codons=7

Table 24.

Dimensions = 2^{3k}x2^{3k} No. of Entries =4^{3k} No. of Bases =3*k* No. of Possible Codons=3*k*-2	Dimensions = 2^{3k+1}x2^{3k+1} No. of Entries =4^{3k+1} No. of Bases =3*k*+1 No. of Possible Codons=3*k*-1	Dimensions = 2^{3k+2}x2^{3k+2} No. of Entries =4^{3k+2} No. of Bases =3*k*+2 No. of Possible Codons=3*k*	Dimensions = $2^{3(k+1)}$x$2^{3(k+1)}$ No. of Entries =$4^{3(k+1)}$ No. of Bases =3(*k*+1) No. of Possible Codons=3*k*+1

Figure 16. The beginning of the sequence of numeric matrices on the base of the algorithm of standardizing quaternary partition

0

→

3	2
2	3

→

6	5	5	4
5	6	4	5
5	4	6	5
4	5	5	6

→

9	8	8	7	8	7	7	6
8	9	7	8	7	8	6	7
8	7	9	8	7	6	8	7
	8	8	9	6	7	7	8
8	7	7	6	9	8	8	7
7	8	6	7	8	9	7	8
7	6	8	7	8	7	9	8
6	7	7	8	7	8	8	9

→...

Figure 17. Hydrogen-bonds triangle

N = Length of DNA bases	$S(X_n)$ = Sums of Hydrogen bonds
0 (initial)	0
1	2 3
2	4 5 6
3 (codons)	**6 7 8 9**
4	8 9 10 11 12
5	10 11 12 13 14 15
6 (dipeptide)	**12 13 14 15 16 17 18**
7	14 15 16 17 18 19 20 21
8	16 17 18 19 20 21 22 23 24
9 (tripeptide)	**18 19 20 21 22 23 24 25 26 27**
10	20 21 22 23 24 25 26 27 28 29 30
.	... ...
.	H-Bonds Triangle

For n = 0,1 2, 3, …, let X_n denotes a DNA sequence of length n and $S(X_n)$ =sums of hydrogen bonds of DNA bases. Then $S(X_n) = 2n + k$, k = 0, 1, 2, …, n. Furthermore, $S(X_n) = S(X_{n-1})$ + (2 or 3) for n =1, 2, 3, …

Construction of the sums $S(X_n)$ of hydrogen bonds may be illustrated by a Hydrogen-bonds triangle (Figure 17).

This leads us to find building blocks of genetic code based stochastic matrices. The first building block matrix is a (2x2)-matrix B_2 (Table 25).

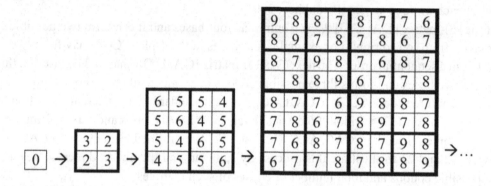

Table 25.

3	2
2	3

Figure 18.

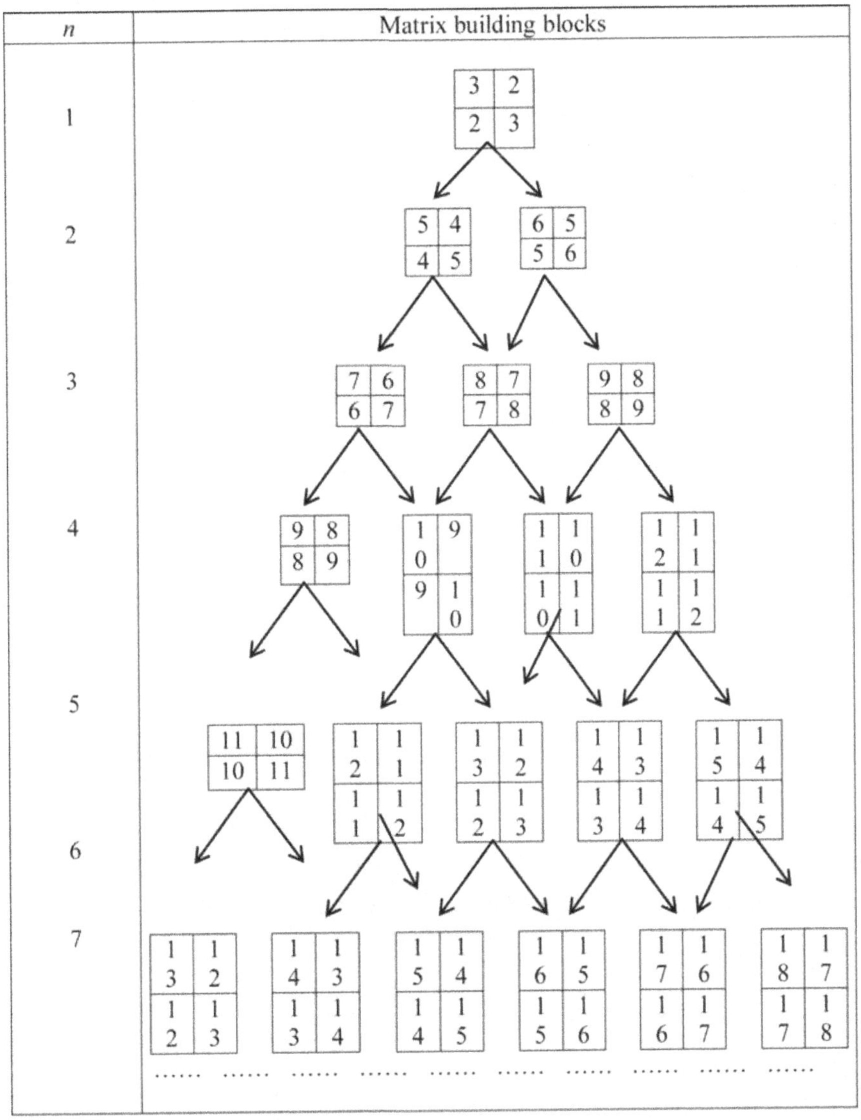

As the length of RNA sequences increases, the matrix building blocks grow and the frequency of the occurrence of the building block in evolution matrix increases as well. We illustrate this process by constructing these building blocks (Figure 18).

Frequency of building blocks in the stochastic matrices may be illustrated by a triangle scheme (Figure 19).

Figure 19.

Matrix	Building Blocks	Frequency of Building Blocks
E_2	B_{11}	1
E_4	B_{21} B_{22}	2 2
E_8 (codons)	**B_{31} B_{32} B_{33}**	**4 8 4**
E_{16}	B_{41} B_{42} B_{43} B_{44}	8 16 16 8
E_{32}	B_{51} B_{52} B_{53} B_{54} B_{55}	16 32 64 32 16
E_{64} (dipeptide)	**B_{61} B_{62} B_{63} B_{64} B_{65} B_{66}**	**32 64 128 128 64 32**
E_{128}	B_{71} B_{72} B_{73} B_{74} B_{75} B_{76} B_{77}	64 128 256 512 256 128 64
E_{512}	B_{81} B_{82} B_{83} B_{84} B_{85} B_{86} B_{87} B_{88}	128 256 512 1024 1024 512 256 128
E_{1024} (tripeptide)	**B_{91} B_{92} B_{93} B_{94} B_{95} B_{96} B_{97} B_{98} B_{99}**	**256 512 1024 2048 4096 2048 1024 512 256**
......	... ...	... ...

Table 26.

n+2	n+1
n+1	n+2

Frequency of building blocks in the stochastic matrices may be illustrated by a triangle scheme.

These (2x2)-matrices are of the form the following matrix B_{nk} (n = 1, 2, 3, …, k = 1, 2, …, n) (Table 26).

The row (or column) sum equals to (2n+3). The determinant of the matrix B_{nk} also equals to (2n+3). The eigenvalues of matrix B_{nk} equal to 1 or (2n+3). Furthermore, we have

$$B_{nk} + [2] = B_{(n+1)k}$$

$$B_{nk} + [3] = B_{(n+1)(k+1)}$$

$$B_{(n+1)k} + [1] = B_{(n+1)(k+1)}$$

Where **[1]**, **[2]**, **[3]** are 2 x 2 matrices with values of 1, 2, 3, respectively in all four entries of each matrix. We use the following diagram to illustrate those relation between building blocks. They form a triangular type of structure (Figure 20).

Each matrix is doubly stochastic and symmetric. It can be expressed as a convex combination of symmetric permutation matrices. It forms a polyhedron with each permutation matrix as vertex.

Figure 20.

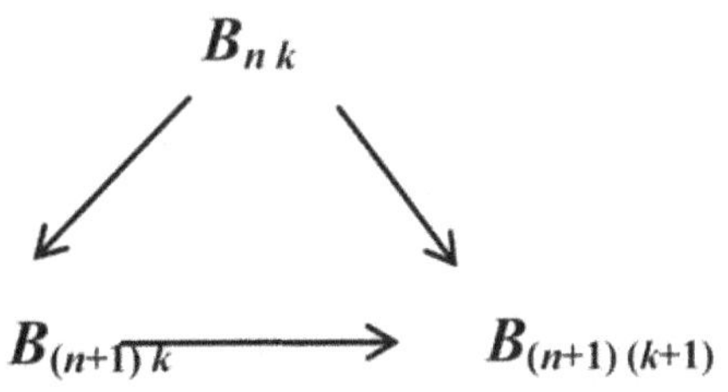

Table 27.

$2n + k$	$2n + k$-1
$2n + k$-1	$2n + k$

Table 28.

<table>
<tr><th>Length of bases n</th><th>B_{n1}</th><th>B_{n2}</th><th>...</th><th colspan="2">middle term(s)</th><th>...</th><th>$B_{n(n-1)}$</th><th>B_{nn}</th></tr>
<tr><td>n is odd</td><td>$2^{(n-1)}$</td><td>2^n</td><td>...</td><td colspan="2">$2^{(n+[(n+1)/2]-2)}$</td><td>...</td><td>2^n</td><td>$2^{(n-1)}$</td></tr>
<tr><td>n is even</td><td>$2^{(n-1)}$</td><td>2^n</td><td>...</td><td>$2^{(n+[n/2]-2)}$</td><td>$2^{(n+[n/2]-2)}$</td><td>...</td><td>2^n</td><td>$2^{(n-1)}$</td></tr>
</table>

B_{nk} ($k = 1, 2, \ldots, n$) is a sequence of (2x2)-symmetric matrices which are building blocks of the evolution matrix $E_2^{\,n}$. The matrix B_{nk} has an explicit form shown in Table 27.

The determinant of B_{nk} equals $4n$+$2k$-1, which is the same as its row/column sum. The matrix B_{nk} has two different eigenvalues 1, $4n$+$2k$-1. It has a pair of orthogonal eigenvectors {-1, 1} and {1, 1}. The frequency of these building blocks in the evolution matrix is listed in Table 28.

Next we demonstrate a process of formulation of evolution matrices $E_2^{\,n}$ from these proper numbers of building blocks in each stage. We will generate the evolution matrices E_2, E_4, E_8 (codons), E_{16}, E_{32}, E_{64} (dipeptides), E_{128}, E_{512}, E_{1024} (tripeptides).... It is easy to note that the building process connects together the previous step and the present step. Each resulting evolution matrix is a doubly stochastic and symmetric matrix. They can be decomposed into a convex combination of permutation matrices with corresponding matrix dimension. Each permutation matrix is vertex of polyhedron (in analogues with polypeptide). Since the sequence of RNA has a length of n, we call this sequence a n-sided polypeptide (similar to n-sided polygons). Note that any n-sided polygon consists of (n-2) triangles. Any n-sided linear polypeptide may be decomposed into (n-2) codons (triplets). The next level of building blocks of proteins could be tripeptides-a chain of three amino acids.

This building process of evolution matrices may be illustrated by Figure 21 from E_2, E_4, E_8 (codons), to $E_{16.,}$ respectively.

We summarize our results here. Let n = length of RNA sequences, B_{nj} ($j = 1, 2, \ldots n$) be (2x2)-matrix building blocks of the evolution matrices $E_{D(Xn)}$ and $D(X_n)$ = the dimension of evolution matrix E. Then

Figure 21.

B_{11}

B_{22}	B_{21}
B_{21}	B_{22}

B_{33}	B_{32}	B_{32}	B_{31}
B_{32}	B_{33}	B_{31}	B_{32}
B_{32}	B_{31}	B_{33}	B_{32}
B_{31}	B_{32}	B_{32}	B_{33}

B_{44}	B_{43}	B_{43}	B_{42}	B_{43}	B_{42}	B_{42}	B_{41}
B_{43}	B_{44}	B_{42}	B_{43}	B_{42}	B_{43}	B_{41}	B_{42}
B_{43}	B_{42}	B_{44}	B_{43}	B_{42}	B_{41}	B_{43}	B_{42}
B_{42}	B_{43}	B_{43}	B_{44}	B_{41}	B_{42}	B_{42}	B_{43}
B_{43}	B_{42}	B_{42}	B_{41}	B_{44}	B_{43}	B_{43}	B_{42}
B_{42}	B_{43}	B_{41}	B_{42}	B_{43}	B_{44}	B_{42}	B_{43}
B_{42}	B_{41}	B_{43}	B_{42}	B_{43}	B_{42}	B_{44}	B_{43}
B_{41}	B_{42}	B_{42}	B_{43}	B_{42}	B_{43}	B_{43}	B_{44}

each building block is a stochastic symmetric (2x2)-matrix, the number of building blocks $| B_{nj} | = n$ and $D(X_n) = 2^n$.

FUTURE TRENDS AND CONCLUSION

The first part of this chapter showed a close relation between the genetic code and the doubly stochastic matrix by using Hamming distance via the Gray code correspondence. The Hamming distance is applicable to encoded information, and is a particularly simple metric of comparison for error detections. The second part of the chapter showed a close relation between the genetic code and doubly stochastic matrix by using genetic attribute based mapping based on hydrogen bonds. Similar studies can be applied to other attributive mappings based on other chemical properties of DNA bases.

The matrices are storages of digital data. The matrices appear in various dimensions with different shapes. Stochastic matrices motivated by the language of probability show up repeatedly in nature. Biological evolution can be interpreted as a process of deployment and duplication of the certain forms of ordering. Having advanced in the understanding of structurally functional features of base systems of genetic coding, mankind extracts simultaneously an opportunity to advance in different areas of biology, which are built in consent with these base systems. The considered stochastic matrices seem to be connected with mechanisms of order production in inheritable biological systems. It is hoped that

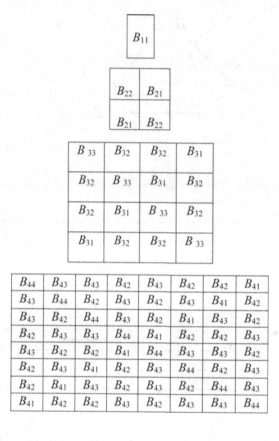

Figure 22. The Hadamard matrix, which corresponds to the matrix C_{83} from Figure 8 by means of the U-algorithm. Cells with the entry "+1" ("-1") are denoted by black (white) color

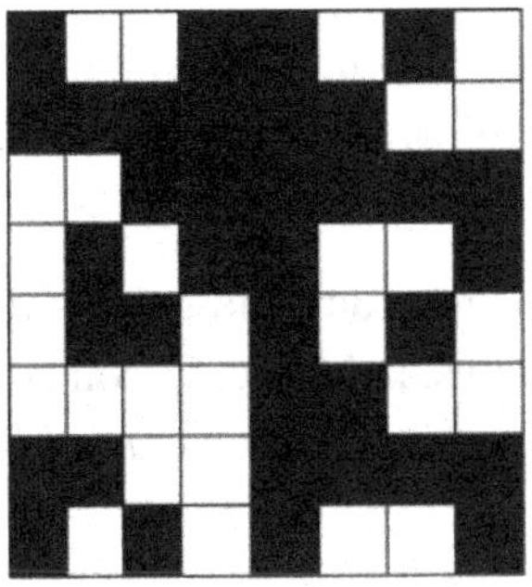

relationships among genetic code, Hamming distance, and stochastic matrices will help us explore the structure of the genetic code.

One of the interesting directions of future investigations is connected with relations between the matrices, which are described in this chapter, and famous Hadamard matrices, which are considered in the next Chapter 6 together with a special U-algorithm of transformation of some genetic matrices into relevant Hadamard matrices. One can check easily that the mosaic (8x8)-matrix C_{83} (Figure 8), which was constructed in this chapter on the basis of the Gray code numeration of columns and rows, is transformed by the same U-algorithm into one of Hadamard (8x8)-matrices (Figure 22).

REFERENCES

Bapat, R. B., & Raghavan, T. E. S. (1997). *Nonnegative matrices and applications*. Cambridge: Cambridge University Press.

He, M. (2001). Double helical sequences and doubly stochastic matrices. In S. Petoukhov (Ed.), *Symmetry in genetic information* (pp. 307-330). Budapest: International Symmetry Foundation

He, M. (2003a). Symmetry in structure of genetic code. In *Proceedings of the Third All-Russian Interdisciplinary Scientific Conference "Ethics and the Science of Future. Unity in Diversity,"* February 12-14, Moscow.

He, M. (2003b). Genetic code, attributive mappings and stochastic matrices. *Bulletin of Mathematical Biology*, *66*(5), 965–973. doi:10.1016/j.bulm.2003.10.002

He, M., Petoukhov, S. V., & Ricci, P. E. (2004). Genetic code, Hamming distance, and stochastic matrices. *Bulletin of Mathematical Biology*, *66*(5), 965–973. doi:10.1016/j.bulm.2003.10.002

Jimenéz-Montaño, M. A., Mora-Basáñez, C. R., & Pöschel, T. (1994). On the hypercube structure of the genetic code. In A. Hwa, Lim & Cantor, C. A. (Eds.), *Proc. 3. Int. Conf. on Bioinformatics and Genome Research* (p. 445). World Scientific.

Petoukhov, S. V. (1999). Genetic code and the ancient Chinese book of changes. *Symmetry: Culture and Science*, *10*(3-4), 211–226.

Petoukhov, S. V. (2001a). *The bi-periodic table of genetic code and number of protons*. Foreword of K. V. Frolov, Moscow, 258 (in Russian).

Petoukhov, S. V. (2001b). Genetic codes 1: Binary sub-alphabets, bi-symmetric matrices, and golden section; Genetic codes 2: Numeric rules of degeneracy and the chronocyclic theory. *Symmetry: Culture and Science*, *12*(1), 255–306.

Petoukhov, S. V. (2005a). The rules of degeneracy and segregations in genetic codes. The chronocyclic conception and parallels with Mendel's laws. In M. He, G. Narasimhan & S. Petoukhov (Eds.), *Advances in Bioinformatics and its Applications, Series in Mathematical Biology and Medicine, 8*, (pp. 512-532). Singapore: World Scientific.

Romanovsky, V. (1931). Sur les zeros des matrices stocastiques. *C. R. Acad. Sci. Paris*, *192*, 266–269.

Štambuk, N. (2000). Universal metric properties of the genetic code. *Croatica Chemica Acta*, *73*(4), 1123–1139.

Stewart, I. (1999). *Life's other secret: The new mathematics of the living world.* New York: Penguin.

Swanson, R. (1984). A unifying concept for the amino acid code. *Bulletin of Mathematical Biology*, *46*(2), 187–203.

Chapter 6
The Genetic Code, Hadamard Matrices, Noise Immunity, and Quantum Computers

ABSTRACT

This chapter continues an analysis of the degeneracy of the vertebrate mitochondrial genetic code in the matrix form of its presentation, which possesses the symmetrical black-and-white mosaic. Taking into account a symmetry breakdown in molecular compositions of the four letters of the genetic alphabet, the connection of this matrix form of the genetic code with a Hadamard (8x8)-matrix is discovered. Hadamard matrices are one of the most famous and the most important kinds of matrices in the theory of discrete signals processing and in spectral analysis. The special U-algorithm of transformation of the symbolic genetic matrix $[C\ A;\ U\ G]^{(3)}$ into the appropriate Hadamard matrix is demonstrated. This algorithm is based on the molecular parameters of the letters A, C, G, U/T of the genetic alphabet. In addition, the analogical relations is shown between Hadamard matrices and other symmetrical forms of genetic matrices, which are produced from the symmetrical genomatrix $[C\ A;\ U\ G]^{(3)}$ by permutations of positions inside triplets. Many new questions arise due to the described fact of the connection of the genetic matrices with Hadamard matrices. Some of them are discussed here, including questions about an importance of amino-group NH_2 in molecular-genetic systems, and about possible relations with the theory of quantum computers, where Hadamard gates are utilized. A new possible answer is proposed to the fundamental question concerning reasons for the existence of four letters in the genetic alphabet. Some thoughts about cyclic codes and a principle of molecular economy in genetic informatics are presented as well.

INTRODUCTION AND BACKGROUND

We continue to investigate connections of the genetic matrices with matrix formalisms of the theory of discrete signals processing. One of the most famous and the most important kinds of matrices in this

DOI: 10.4018/978-1-60566-124-7.ch006

Figure 1. The family of Hadamard matrices $H(2^k)$ based on the Kronecker product

$$H(2) = \begin{bmatrix} 1 & 1 \\ -1 & 1 \end{bmatrix}; \quad H(4) = \begin{bmatrix} 1 & 1 & 1 & 1 \\ -1 & 1 & -1 & 1 \\ -1 & -1 & 1 & 1 \\ 1 & -1 & -1 & 1 \end{bmatrix}$$

$$H(2^k) = \begin{bmatrix} H(2^{k-1}) & H(2^{k-1}) \\ -H(2^{k-1}) & H(2^{k-1}) \end{bmatrix}$$

theory are the so called Hadamard matrices. These matrices are used also in many other fields due to their advantageous properties: in error-correcting codes such as the Reed-Muller code; in spectral analysis; in multi-channels spectrometers with Hadamard transformations; in quantum computers with Hadamard gates (or logical operators), in quantum mechanics as unitary operators, etc.

Does any natural connection exist between the genetic matrices, which were described in previous chapters, and Hadamard matrices? This question should be investigated especially because a possible positive answer to it may lead to many significant consequences and new thoughts about structures of the genetic code. This chapter demonstrates the existence of such a connection and analyzes some questions related to it.

A huge number of scientific publications are devoted to Hadamard matrices. These matrices give effective opportunities for information processing.

By definition a Hadamard matrix of dimension "n" is the (nxn)-matrix $H(n)$ with elements "+1" and "-1". It satisfies the condition $H(n)*H(n)^T = n*I_n$, where $H(n)^T$ is the transposed matrix and I_n is the (nxn)-identity matrix. The Hadamard matrices of dimension 2^k are given by the recursive formula $H(2^k) = H(2)^{(k)} = H(2)\otimes H(2^{k-1})$ for $2 \leq k \in N$, where $\otimes$ denotes the Kronecker (or tensor) product, (k) means the Kronecker exponentiation, k and N are integers, $H(2)$ is demonstrated in Figure 1.

Rows of a Hadamard matrix are mutually orthogonal. It means that every two different rows in a Hadamard matrix represent two perpendicular vectors, a scalar product of which is equal to 0. The element "-1" can be disposed in any of four positions in the Hadamard matrix $H(2)$. Such matrices are used in many fields due to their advantageous properties: in error-correcting codes such as the Reed-Muller code; in spectral analysis and multi-channel spectrometers with Hadamard transformations; in quantum computers with Hadamard gates, etc. It was discovered unexpectedly that Hadamard matrices reflect essential peculiarities of molecular genetic systems (Petoukhov, 2005, 2006, 2008a-d).

Normalized Hadamard (2x2)-matrices are matrices of rotation on 45^0 or 135^0 depending on an arrangement of signs of its individual elements. A Kronecker product of two Hadamard matrices is a Hadamard matrix as well. A permutation of any columns or rows of a Hadamard matrix leads to a new Hadamard matrix.

Hadamard matrices and their Kronecker powers are used widely in spectral methods of analysis and processing of discrete signals and in quantum computers. A transform of a vector $\bar{a}$ by means of a Hadamard matrix H gives the vector $\bar{u} = H*\bar{a}$, which is named Hadamard spectrum. A greater analogy between Hadamard transforms and Fourier transforms exists (Ahmed & Rao, 1975). In particular the fast Hadamard transform exists in parallel with the fast Fourier transform. The whole class of multichannel "spectrometers with Hadamard transforms" is known (Tolmachev, 1976), where the principle of tape masks (or chain masks) is used, and it reminds one of the principles of a chain construction of genetic texts in DNA. Hadamard matrices are used widely in the theory of coding (for example, they are connected with Reed-Muller error correcting codes and with Hadamard codes (Peterson & Weldon, 1972; Solovieva, 2006), the theory of compression of signals and images, a realization of Boolean functions by means of spectral methods, the theory of planning of multiple-factor experiments and in many other branches of mathematics.

Biological organisms are sets of biochemical molecules. The Hadamard matrices in analytical chemistry have been introduced (Pan, 2007). This work pays a special attention to applications of Hadamard matrices to enhance signal-to-noise ratio. It is explained in a simple example of weighing. The basic idea is connected to weighing of the objects in groups but not separately for a determination of their individual weights more accurately. For example, in a case of four objects, we can weigh them by two different ways. By the first way we can weigh each of them individually by means of a single pan spring balance which is well calibrated to give us correct values Ψ_1, Ψ_2, Ψ_3, Ψ_4 for these four objects 1, 2, 3, 4 correspondingly with a small random error "e". By the second way we can weigh all four objects in groups by means of a two-pan balance to receive their general weights η_1, η_2, η_3, η_4 in the next four weighing with appropriate random errors e_1, e_2, e_3, e_4:

$$\eta_1 = \Psi_1 + \Psi_2 + \Psi_3 + \Psi_4 + e_1$$

$$\eta_2 = \Psi_1 - \Psi_2 + \Psi_3 - \Psi_4 + e_2$$

$$\eta_3 = \Psi_1 + \Psi_2 - \Psi_3 - \Psi_4 + e_3$$

$$\eta_4 = \Psi_1 - \Psi_2 - \Psi_3 + \Psi_4 + e_4$$

Here the measurement with a negative value means that the object is placed on the opposite pan of the balance. From these equations one can easily calculate values Ψ_1, Ψ_2, Ψ_3, Ψ_4. This final result will be much more accurate than in the previous case of weighing of each object individually (see details in (Pan, 2007)). The disposition of signs "+" and "-" in this system of the four equations is identical to their disposition in the relevant Hadamard (4*4)-matrix. In such way applications of Hadamard transforms enhance the signal-to-noise ratio.

Rows of Hadamard matrices are named Walsh functions or Hadamard functions. Walsh functions can be represented in terms of product of Rademacher functions $r_n(t) = \text{sign}(\sin 2^n \pi t)$, $n = 1,2,3,\ldots$, which accept the two values "+1" and "-1" only (here "sign" is the function of a sign on argument). Sets of numerated Walsh functions (or Hadamard functions), when they are united in square matrices, form systems depending on features of such union. Figure 2 shows two examples of systems of such functions, which are used widely in the theory of digital signals processing.

They are connected with (8x8)-matrices by Hadamard and with the Walsh-Hadamard transform, which is the most famous among non-sinusoidal orthogonal transforms and which can be calculated by means of mathematical operations of addition and subtraction only (see more detail in (Ahmed & Rao, 1975; Trahtman & Trahtman, 1975; Yarlagadda, & Hershey, 1997). Hereinafter we will use the simplified designations of matrix elements on illustrations of Hadamard matrices: the symbol "+" or the black color of a matrix cell means the element "+1"; the symbol "-" or the white color of a matrix cell means the element "-1". The theory of discrete signals pays special attention to quantities of changes of signs

Figure 2. Examples of the two systems of Walsh functions (or Hadamard functions), which are used frequently in the theory of digital signals processing. On the left side: the Walsh-Hadamard system. On the right side: the Walsh-Paley system. Quantities of changes of signs "+" and "-" are shown for each row and each column

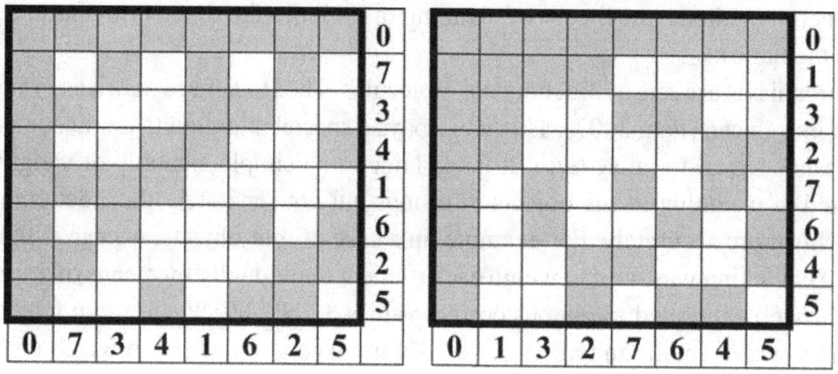

"+" and "-" along each row and each column in Hadamard matrices. These quantities are connected with important notion of "sequency" as a generalization of notion of "frequency" (Ahmed & Rao, p.85). Figure 2 shows these quantities for each row and each column in presented matrix examples.

Normalized Hadamard matrices are unitary operators. They serve as one of the important instruments to create quantum computers, which utilize so called Hadamard gates (as evolution of the closed quantum system is unitary) (Nielsen & Chuang, 2001).

THE GENETIC CODE AND HADAMARD MATRICES

The molecular composition of the letters A, C, G, U/T of the genetic alphabet is characterized by one special disturbance of symmetry: the three nitrogenous bases A, C, G have one amides (amino-group) NH_2, but the fourth basis U/T has not it (Figure 1 of Chapter 1).

From the viewpoint of existence of the amino-group NH_2, the letters A, C, G are identical to each other and the fourth letter U/T is opposite to them. This fact of existence or absence of the amino-group NH_2 in certain genetic letters can be reflected in the alphabetic genomatrix P = [C A; U G] by symbols "+1" instead of the letters A, C, G and by the symbol "-1" instead of the letter U. In this case this genomatrix is transformed into the Hadamard genomatrix $P_{H(2)}$ = H(2) = [1 1; -1 1]. All other variants of the Kronecker families of the alphabetic genomatrices, which were considered in Chapter 2 on Figure 11, become the Kronecker families of Hadamard matrices by such a way as well (Figure 3 demonstrates examples).

The detection of natural realization of Hadamard matrices (and systems of orthogonal functions, which are connected with them) on the basis of parameters of the molecular-genetic system, which serves to transfer discrete genetic information, show that all known advantages of Hadamard matrices can be utilized in bioinformatics. Taking into account a possible important role of Hadamard matrices in the genetic signals processing, one can consider genetic sequences as lattice functions, for which a substantial class of discrete logical operations exists. This class contains logical addition, logical subtraction, logical product, logical shift, logical convolution and logical differentiation. All these operations can be applied to the analysis of problems of genetic information processing.

Figure 3. Examples of transforms of the Kronecker families of the alphabetical genomatrices from Figure 11 of Chapter 2 into appropriate Kronecker families of Hadamard matrices

$$[C\ A;\ U\ G]^{(n)} = \begin{vmatrix} C & A \\ U & G \end{vmatrix}^{(n)} \Rightarrow (H^{CAUG})^{(n)} = \begin{vmatrix} 1 & 1 \\ -1 & 1 \end{vmatrix}^{(n)}$$

$$[U\ G;\ C\ A]^{(n)} = \begin{vmatrix} U & G \\ C & A \end{vmatrix}^{(n)} \Rightarrow (H^{UGCA})^{(n)} = \begin{vmatrix} -1 & 1 \\ 1 & 1 \end{vmatrix}^{(n)}$$

$$[G\ U;\ A\ C]^{(n)} = \begin{vmatrix} G & U \\ A & C \end{vmatrix}^{(n)} \Rightarrow (H^{GUAC})^{(n)} = \begin{vmatrix} 1 & -1 \\ 1 & 1 \end{vmatrix}^{(n)}$$

$$[A\ C;\ G\ U]^{(n)} = \begin{vmatrix} C & A \\ G & U \end{vmatrix}^{(n)} \Rightarrow (H^{ACGU})^{(n)} = \begin{vmatrix} 1 & 1 \\ 1 & -1 \end{vmatrix}^{(n)}$$

One should note that the attribute of absence of the amides (amino-group) NH_2 picks out those nitrogenous bases T and U, which differ from other genetic letters A, C, G by one specific property else. These letters T and U replace each other at transition from DNA to RNA for an unknown reason. These double differences of the first sub-set of genetic alphabet A, C, G with the second subset U/T lead to the identical transformations of the alphabetic genomatrices into Hadamard matrices.

ABOUT THE IMPORTANCE OF AMINO-GROUPS NH_2

The importance of compounds of nitrogen for molecular genetics is reflected even in names: "amino acids" (the organic acids containing amino-groups); "the nitrogenous bases"; "the N-end" of nucleotide circuit, with which synthesis of proteins begins always, etc. All proteins are polyamides. The lack of proteins of food leads to a number of heavy infringements of a nitrogenous exchange. The amino-group of amino acids bears a base function to provide recognition of an amino acid by enzyme (Shapeville & Haenni, 1974).

Beginning with the works (Schuster & Schramm, 1958; Gierer &Mundry, 1958), it is known that action of the nitrous acid NHO_2 on RNA leads to amino-mutation of RNA. More precisely this action deletes amino-group at the nitrogenous bases A and C and leads finally to a replacement of the nitrogenous bases A and C by the bases G and U correspondingly: A→G, C→U. In a certain sense, the bases A and G (C and U) can be interpreted as the two states of the same letter. One can note that objects with such "trigger" properties are used to construct computers. These amino-mutations A→G, C→U are utilized traditionally to demonstrate molecular mechanisms of an origin of genetic mutations. The nitrogenous acid exists only in the diluted water solutions, which are similar to solutions in biological organisms.

The work (Wittmann, 1961) has demonstrated a degeneracy of the genetic code by means of the following method. The author has grouped all 64 triplets into 8 octets, each of which begins with maximal amino-mutation triplets, which are transformed step by step into more and more stable triplets, non-mutating under the action of nitrous acid NHO_2. These octets by Wittmann, which take an important place in the history of the discovery of the genetic code (Ycas, 1969), coincide with the columns of the genomatrix [C A; U G]$^{(3)}$ on Figure 2 of Chapter 2. Taking all these facts into account, one should pay great attention to amino-groups in future development of knowledge about genetic code systems. One can raise Hadamard (2x2)-matrices from Figure 3 into the third Kronecker power to receive Hadamard (8x8)-matrices. One such octet matrix is shown in Figure 2 on the left side. But mosaics of the Hadamard matrices, which are obtained by this method, differ from mosaics of the octet genomatrices of triplets, which were presented in Chapter 2. Really the mosaics of the genomatrices of triplets contain 32 black cells and 32 white cells (Figures 2, 4, and 5 of Chapter 2), whereas Hadamard (8x8)-matrices contain 28 cells of one color and 36 cells of another color.

The question arises as to whether any simple algorithmic connection exists, which is connected to specifics of the genetic code, between the mosaic of the genomatrix [C A; U G]$^{(3)}$ = P^{CAUG}_{123} (see Figure 2 of Chapter 2) and the black-and-white mosaic of some matrix from the set of Hadamard (8x8)-matrices?

The answer to this question is positive: such algorithmic connection exists. It is mated with the fundamental and enigmatic features of the genetic code: firstly, the mutual replacement of the letters U and T in RNA and DNA and, secondly, the difference of these letters from other letters A, C, G by the absence of amids (amino-groups) in them. Really, let us replace black (white) cells of the genomatrix [C A; U G]$^{(3)}$ in Figure 2 of Chapter 2 by the number "+1" ("-1"). As a result we obtain the matrix B_{123} from Figure 7 of Chapter 2. After this we invert the signs in cells of this matrix B_{123} every time when the particular letter U occupies the first or the third positions of a triplet. We name this algorithm of inverting the "U-algorithm" conditionally. For example, by this U-algorithm the cells with the triplets UCA and GAU change their sign one time, and the cell with the triplet UAU changes its sign twice (it means that this cell does not change its sign at all). As a result of such algorithmic changes of signs, the mosaic genomatrix [C A; U G]$^{(3)}$ from Figure 2 of Chapter 2 becomes one of the Hadamard matrices (see the first matrix in Figure 4).One can suppose that the described "genetic" U-algorithm (of inverting the signs every time when the particular letter U or T appear in an odd position of triplets) is connected with the biological mechanism of mutual replacement of the letters U and T at transition from RNA to DNA and vice versa.

The five genomatrices P^{CAUG}_{231}, P^{CAUG}_{312}, P^{CAUG}_{132}, P^{CAUG}_{213}, P^{CAUG}_{321} from Chapter 2 (Figures 4 and 5 of Chapter 2), which are produced from the matrix P^{CAUG}_{123} = [C A; U G]$^{(3)}$ by positional permutations inside triplets, are transformed into other Hadamard matrices by the analogical algorithm. It is obvious because, as we mentioned above, a transform of the genomatrix by positional permutation inside triplets is identical to its transformation by the appropriate permutation of its columns and rows; but permutations of columns or rows in Hadamard matrices give new Hadamard matrices always.

Figure 4 shows six Hadamard matrices, which correspond to the six mentioned genomatrices. One can check that any of these octet matrices satisfies the definition of Hadamard matrices: if the matrix is multiplied on transposed matrix, the result is the unitary matrix with the factor 8. Each (4x4)-quadrant and each (2x2)-sub-quadrant of these Hadamard (8x8)-matrices is a Hadamard matrix as well. In other words, "Hadamard fractals" are presented in the genomatrices, the mosaic of which reflects the specific character of degeneracy of the genetic code. The total quantity of Hadamard matrices of different sizes in these six Hadamard (8x8)-matrices is equal to 126.

One should note a special feature of the genetic Hadamard matrices on Figure 4: a quantity of changes of signs "+" and "-" is equal to 14 for each of halves of these matrices (we say about upper, lower, left and right halves). It can be named conditionally as "a rule of halves of a lunar month" (this numeric coincidence with the halves of the quantity of 28 days in a lunar month is accidental till that

Figure 4. The six balanced Hadamard matrices, which are produced from the six mentioned genomatrices by means of the U-algorithm. The black cells correspond to elements "+1" and the white cells correspond to elements "-1". Numbers of changes of signs "+" and "-" (or changes of colors) are shown for each row and each column

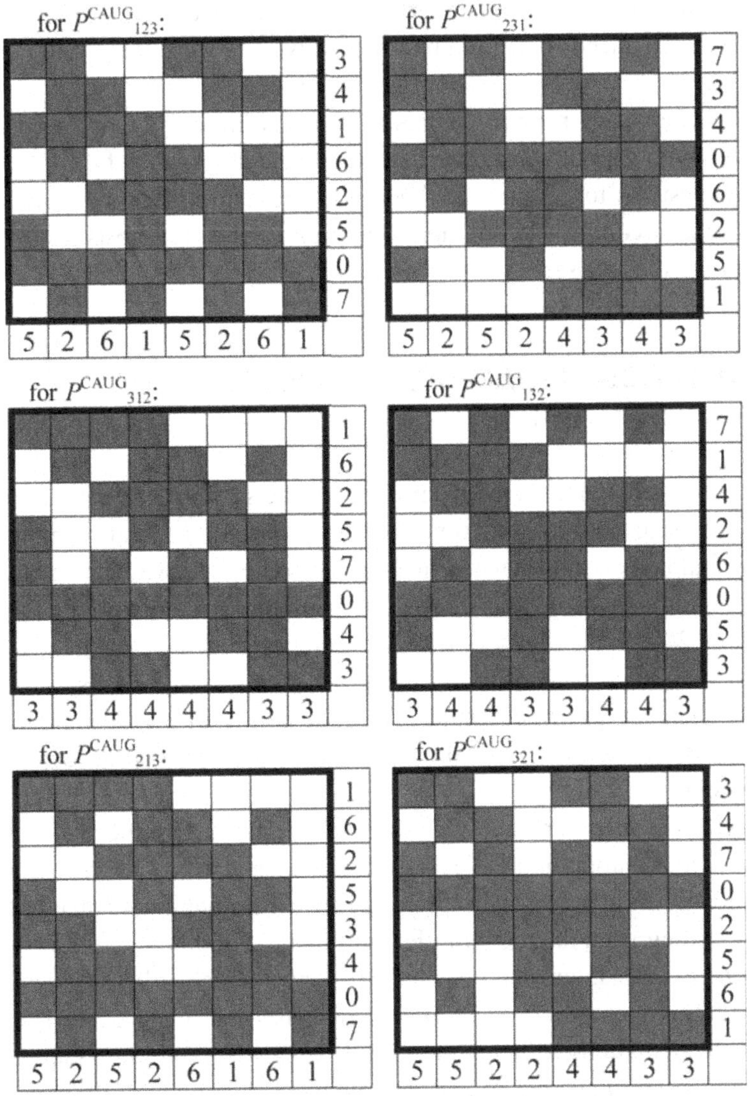

moment when somebody proves a contrary statement). Such "symmetrical" feature is a typical feature for many genetic Hadamard matrices which are presented in Chapters 6, 7, 11 of our book. We name Hadamard matrices with such feature as "balanced" Hadamard matrices. One can check that each of (4x4)-quadrants of these Hadamard (8x8)-matrices is a balanced Hadamard matrix as well. This feature distinguishes the described genetic Hadamard matrices from some types of Hadamard matrices which are used in technical applications widely. For example the Hadamard matrices with the Walsh-Paley system (Figure 2, on the right side) or with the system {wal(w,x)} (Trahtman & Trahtman, 1975, p. 47) have not such feature. The nature has chosen by some reasons the genetic code, which is connected with balanced Hadamard matrices.

Other octet genomatrices, which were considered in Chapter 2, are transformed into appropriate Hadamard matrices by means of the analogical U-algorithm. Some of these matrices are demonstrated in Figure 5 with an indication of those genomatrices, from which they are produced. All of the genetic matrices on Figure 5 are balanced Hadamard matrices as well because they fit to the "rule of halves of a lunar month".

All such kinds of Hadamard matrices represent various basic systems of orthogonal functions, which are coordinated with structural peculiarities of molecular systems of the genetic code. They can be utilized in genetic systems for spectral methods of genetic information processing with the use of noise-immunity coding, of compression of signals and of other useful possibilities, which Hadamard matrices and Walsh functions possess.

GENETIC INFORMATICS, HADAMARD MATRICES AND QUANTUM COMPUTERS

Investigations of structural-functional analogues between the system of genetic coding and computers have been conducted in science for a long time. In the last years a general opinion has arisen that the future of computer technology is connected with quantum computers, which possess fantastic possibilities in comparison to traditional computers due to new principles in their workings (Nielsen, & Chuang, 2001; Valiev, & Kokin, 2001). The theory of quantum computers, which are the new type of computers in principal, is developed intensively. The following question arises. Is the system of genetic coding closer to classical or quantum type of computers from the point of view of computer analogies? From the point of view of analogues with classical or quantum computers is it necessary to investigate the molecular-genetic system?

Genetic molecules exist in accordance with principles of quantum mechanics (from the point of view of classical mechanics, atoms and molecules cannot exist at all). Therefore it is natural to believe in the presence of a relationship between molecular-genetic informatics and quantum computers and to comprehend genetic coding from the point of view of this type of computers. Let us recall the famous data about the advantages of quantum computers, the theory of which utilizes Hadamard matrices (more precisely, Hadamard gates) intensively (see for example (Nielsen, & Chuang, 2001)). Classical computers possess restrictions in calculations on many practically interesting and important classical algorithms, when speed is about increasing number of data and exponential growth of time of calculation. One of famous examples is the question about decomposition (factorization) of number N on prime factors. Classical theory of calculations works with such algorithms of calculations, where number of steps grows as a polynomial of a small power of the size of the entrance data (for example, a polynomial of the second or third power). But in the mentioned question of factorization, the best such algorithms lead to an exponential growth in the number of steps at increasing size of entrance data. For this reason, the time of calculations becomes huge. For instance, in 1994, 129-unit number was factorized on 1600 workstations distributed worldwide. The time of the factorization was equal to 8 months. The estimated time, which the same 1600 workstations require for the factorization of 1 250-unit number, is one million years. Accordingly, factorization of 1000-unit number requires 10^{25} years, that is much more than the age of the Universe. This abstract problem of factorization of great numbers has a direct relation to

Figure 5. The 12 balanced Hadamard matrices, which are produced from the indicated 12 genomatrices of triplets by means of the U-algorithm. Black cells correspond to elements "+1" and white cells correspond to elements "-1"

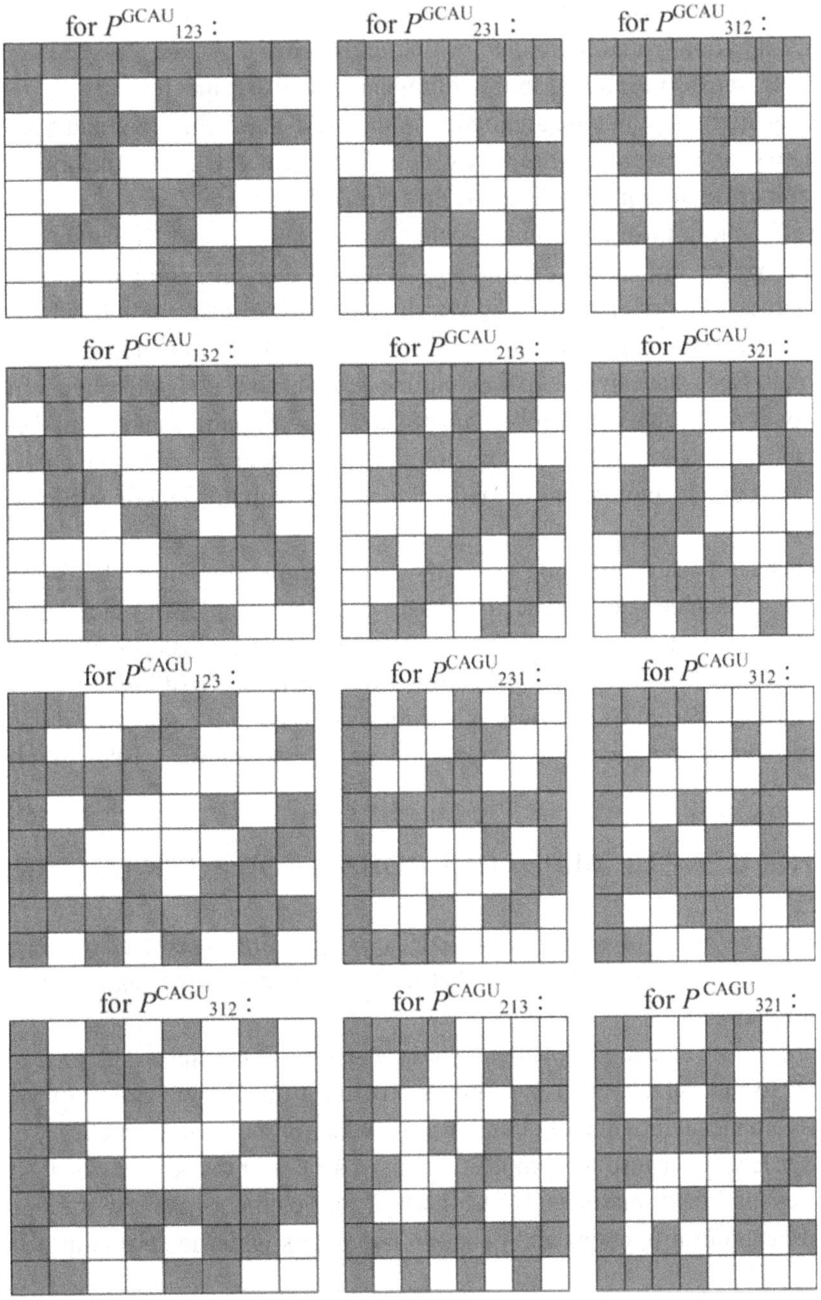

systems of cryptography with the open key, which are utilized in bank systems widely. One should note for comparison that algorithms in quantum computers calculate such factorization of a 1000-unit number by means of a few millions steps only. Classical computers do not allow one to model chemical reactions and systems, where many quantum effects should be taken into account with necessary completeness. It is important for science that quantum computers will allow one to calculate a structure and functioning of quantum systems, including molecules of proteins and DNA.

Classical computer networks consist of wires and a set of logic gates (a set of transistors). Standard electric signals are transferred through wires, and logic gates transform information signals passing through them. A single non-trivial logic gate, which transforms 1 byte of classical information, is the NOT-gate. This gate transforms signals in the following way: $0 \rightarrow 1$ and $1 \rightarrow 0$.

In the case of quantum computers, analogues of the NOT-gate look like matrixes of special kind. Exists of Hadamard gates among them are very useful. The Hadamard gate is the normalized Hadamard matrix, the determinant of which is equal to the unit (see Figure 6).

The amazing efficiency of work of quantum computers is connected with quantum parallelism, which is a fundamental property of quantum calculations. It allows quantum computers to calculate function $f(x)$ for various values x simultaneously. The Walsh-Hadamard transformation, which is a Kronecker product of Hadamard operators, is especially useful there. This Walsh-Hadamard transformation makes a superposition of all basic states with equal amplitude and it is extremely effective for the construction of superposition of 2^n states that use number "n" of gates only. Bioinformatics and the theory of quantum computers can enrich each other by means of analysis of heuristic analogies between them. Data described in our book about connections of Hadamard matrix with the genetic system can promote this mutual enrichment.

As modern science opens an amazing efficiency of quantum computers, the following question is natural. Does living substance s not use the advantages of their principles of functioning in its self-organizing? Possibly, it does, and many new connections between living substance and quantum computers will be revealed by science in the future. One can note that the problem of understanding the biological phenomena from the viewpoint of quantum mechanics and quantum computers draws the increasing attention of theorists in the last years (see for example the books (Penrose, 1989, 2004)).

WHY DOES THE GENETIC ALPHABET CONSIST OF FOUR LETTERS?

Genetic molecules are objects of quantum mechanics, where normalized Hadamard matrices play an important role as unitary operators (it is known that an evolution of a closed quantum system is described by unitary transformations).

Why has nature chosen a genetic alphabet which consists of four letters? The following new answer of matrix genetics to this fundamental question is possible from the viewpoint of the importance of principles of quantum mechanics and of quantum computers for molecular genetics. The genetic alphabet consists of four letters because the simplest unitary matrices in two-dimensional space, first of all, Hadamard matrices (and Pauli matrices, etc.) consist of four elements exactly. It seems very probable that principles of quantum mechanics and quantum computers underlie structural peculiarities of the genetic code.

Figure 6.

$$H = 2^{-0.5} * \begin{vmatrix} 1 & 1 \\ 1 & -1 \end{vmatrix}$$

One can also note that Hadamard matrices arise not only in connection with the mentioned pair of binary-oppositional attributes of "existence or absence of amino-group NH_2" in genetic letters. For example, other variants of Hadamard genomatrices correspond to binary-oppositional attributes of "existence or absence of atoms of oxygen" in the genetic letters: the letters C, G, U/T contain atoms of oxygen, but the letter A does not. In DNA, Hadamard matrices arise in connection with binary-oppositional attributes of "existence or absence of five atoms of carbon": each of the letters A, G, T contains five atoms of carbon in its molecular construction, but the letter C contains four atoms of carbon only (see Figure 1 of Chapter 1). In our opinion, the significance of matrices, which correspond to these kinds of attributes, is less in comparison with the considered case of attributes of amino-group NH_2.

CYCLIC SHIFTS, CYCLIC CODES AND THE PRINCIPLE OF MOLECULAR ECONOMY IN GENETIC INFORMATICS

It was mentioned in Chapter 1, when analogues between matrices of diadic shifts and genomatrices were considered, that matrices of diadic shifts possess a block character and is connected with cyclic shifts of their blocks. More precisely, the identity of quadrants along each diagonal in such matrices allows us considering them as block (2x2)-matrices, in which both rows are mutually connected by a transformation of cyclic shift. The tessellation of a plane with the mosaic of the genomatrix on Figure 3 of Chapter 2 has the character of the tessellation on the base of cyclic shifts of black and white modular units. In addition, cyclic shifts of positions inside triplets (see Chapter 2) allow one to consider hidden regularities in the structure of degeneracy of the genetic code. One can see also that the disposition of a series of elements in the genomatrices of hydrogen bonds (Figure 2 of Chapter 4) correspond to the disposition of the series of elements in the matrices of diadic shifts (Figure 5 of Chapter 1). And what is known about cyclic shifts in the theory of noise-immunity codes in general?

This theory includes an important family of so called "cyclic codes" (Arshinov, & Sadovskiy, 1983; Peterson, & Weldon, 1972). Their name is connected with the fact that these codes are based on the cyclic shifts. Some authors consider cyclic codes as the most valuable achievement of the theory of coding because they allow very compact descriptions, easy algorithms of coding and decoding, a simplicity in their realization (Arshinov, & Sadovskiy, 1983). These codes are related to matrices of cyclic shifts, where rows-vectors differ by their shift or by a cyclic permutation of components.

Some interesting investigations in the field of molecular genetics are known already. The authors of which connect principles of constructions of genetic sequences with the idea of cyclic codes in some sense (Lassez, 1976; Arques, Michel, 1996, 1997; Frey, Michel, 2003, 2006; Stambuk, 1999, etc). Of course, these authors did not know about the data of matrix genetics, which are described in our book, and did not utilize them.

The question about cyclic codes is not so simple. A big number of cyclic codes exist. They can be applied in various combinations. In particular, the data described above about connections of the genetic matrices with block matrices of diadic shifts allow us to suppose the following. It is important to study genetic sequences in a connection with those codes, which are based on matrices of diadic shifts with their block-shift character, in other words, in a connection with "block-cyclic" codes.

The term "cyclic" is very attractive in applications to biological systems, which are characterized by so many cyclic processes. The famous theory of hyper-cycles reflects the importance of cyclic biochemical processes for biological organisms to some extent (Eigen, 1979, 1988, 1992; Eigen, & Winkler, 1993). The chronocyclic conception described above in Chapter 3 is related to cyclic processes as well.

If rows of the genetic matrices are interpreted as code vectors, then the described algorithmic connection between the rows allows one to think about the following. In system of genetic coding, not all code vectors of cyclic or diadic codes should exist necessarily at each moment of time in the form of a system of parameters of real molecular structures. It is enough if only a part of a set of code vectors exist; other code vectors can be calculated by a molecular computer of an organism by means of corresponding algorithms or they can be synthesized in the molecular forms temporarily (including re-packing molecular components). Such a principle gives general economy of molecular materials. This hypothesis about the molecular-economy principle in the field of genetic informatics should be studied in the future.

FUTURE TRENDS AND CONCLUSION

Spectral methods of decomposition of signals on orthogonal systems of functions have proved themselves for a long time as especially important in the theory of signals and informatics in general. Researchers of genetic informatics attempt to address to them already (see, for example, the works (Kargupta, 2001; Lobzin & Chechetkin, 2000), which pay attention to the importance of spectral methods in this field). But an infinite quantity of orthogonal systems of functions exists. It is difficult for researchers of molecular-genetic systems to make a choice of one of infinite number of possible orthogonal systems as an adequate one for spectral methods in the field of genetic informatics. They should make here rather a volitional choice, risking the waste of many years of work in the case of the failure of such choice. They make this choice usually, proceeding from secondary reasons, which do not have direct relation to genetic systems. For example, they choose the system of orthogonal harmonious functions, which is applied in the classical frequency Fourier-analysis, for the reason, that this system has extensive applications in technical fields.

The results described in our book show the orthogonal systems of functions, which are connected with Hadamard matrices and which possess a special meaning for genetic informatics and its spectral methods. The orthogonal systems of functions connected with Hadamard matrixes are picked out by nature from the infinite set of basic systems for their deep connection with an essence of molecular-genetic coding. A consistent investigation of bioinformatics systems should be done from the viewpoint of the theory of genetic Hadamard matrices and their applications. In particular, the comparative analysis of various genetic sequences on their Hadamard spectrums is interesting. The described results give important help in a choice of research tool from an infinite set of orthogonal systems of functions and from a set of variants of noise-immunity codes.

In the spectral analysis of genetic sequences (for example, their correlation functions), it is meaningful to spend their decomposition on orthogonal vectors-rows of Hadamard genomatrices, instead of on trigonometric functions of the frequency Fourier-analysis. Investigations of Hadamard spectrums in mathematical genetics are perspective and well-founded. Especially since some works are already known as applications of Walsh functions (alongside with other systems of basic functions) to spectral analysis of various aspects of genetic algorithms and sequences (Forrest, & Mitchell, 1991; Geadah & Corinthios 1977; Goldberg, 1989; Lee, & Kaveh, 1986; Shiozaki, 1980; Vose & Wright, 1998; Waterman, 1999). Here we emphasize a possible benefit for bioinformatics to use genetic Hadamard matrices which are connected with a phenomenon of degeneracy of the genetic code and form a special subset of a set of Hadamard matrices. The book (Zalmanzon, 1989, p. 416) contains a review of investigations made by various authors about Walsh orthogonal functions in physiological systems of supra-cellular levels. Hadamard matrices have been used in molecular genetics in a connection with Hadamard conjugation for evolutionary trees and with phenomena of cyclic gene expressions (see Chapter 11). In our view, genetic Hadamard matrices described can be useful for developing these branches of molecular biology.

One can also mention that for application of spectral methods to problems of genetic coding it is important not only to choose an adequate basic system of orthogonal functions, but to determine successfully a numerical form of representation of genetic sequences as well. The matter is that spectral methods in the theory of discrete signals operate with numerical sequences or numerical vectors, but genes appear in the literature usually in a form of symbolic sequences of molecular triplets like AUC-UCG-CCG-… . A great number of ways of transformations of such symbolic genetic sequences into their numerical form exist in principle, for example, by means of replacement of each triplet by the number of its atoms, or by the number of its hydrogen bonds, etc. Which kind from this set of possible forms of numerical representation should be chosen and should be investigated first of all for a deep understanding the genetic code? It is one of the questions, which are studied in our book by means of researches of various variants of such numeric presentation, etc.

The discovery of connections of the genetic matrices with Hadamard matrices leads to many new thoughts and possible investigations using methods of symmetries, of spectral analysis, etc. One can expect that those Walsh-Hadamard functions, which are related to the described genetic Hadamard matrices (Figures 4, 5 of Chapter 6) will be used effectively in the spectral analysis of genetic sequences. It seems that investigations of structural and functional principles of bio-information systems from the viewpoint of quantum computers and of unitary Hadamard operators are very perspective. A comparison of orthogonal systems of Walsh-Hadamard functions in molecular-genetic structures and in genetically inherited macro-physiological systems can give new understanding to an interrelation of various levels in biological organisms. Data about the genetic Hadamard matrices together with data about algebras of the genetic code, which are described in the next chapters, can lead to new understanding of genetic code systems, to new effective algorithms of information processing and, perhaps, to new decisions in the field of quantum computers. In our opinion, interesting data will be obtained about cyclic and diadic codes in the genetic systems in the near future. The proposed hypothesis about the molecular-economy principle can be useful to understand some aspects of an effective organization of the molecular-genetic systems.

The genetic matrix $[C\ A;\ U\ G]^{(3)}$, which possesses the certain black-and-white mosaic of degeneracy of the vertebrate mitochondrial genetic code, is connected with the mosaic Hadamard (8x8)-matrix by means of the special U-algorithm. The genetic matrices, which are produced from the genomatrix $[C\ A;\ U\ G]^{(3)}$ by means of permutations of positions inside triplets, are connected with the appropriate Hadamard matrices as well. These mathematical facts give new important data about connections of structural-functional organization of genetic code systems with many methods and fields, where Hadamard matrices play a significant role.

REFERENCES

Ahmed, N., & Rao, K. (1975). *Orthogonal transforms for digital signal processing*. New York: Springer-Verlag Inc.

Arques, D., & Michel, C. (1996). A complementary circular code in the protein coding genes. *Journal of Theoretical Biology, 182*, 45–56. doi:10.1006/jtbi.1996.0142

Arques, D., & Michel, C. (1997). A circular code in the protein coding genes of mitochondria. *Journal of Theoretical Biology, 189*, 45–58. doi:10.1006/jtbi.1997.0513

Arshinov, M., & Sadovskiy, L. (1983). *Codes and mathematics*. Moscow: Znanie (in Russian).

Eigen, M. (1979). *The hypercycle-a principle of natural self-organization*. Berlin: Springer Verlag.

Eigen, M. (1988). *Perspektiven der wissenschaft*. Stuttgart: Deutsche Verlagsanstalt.

Eigen, M. (1992). *Steps towards life. A perspective of evolution*. Oxford: Oxford University Press.

Eigen, M., & Winkler, R. (1993). *Laws of the game. How the principles of nature govern chance*. Princeton: Princeton University Press.

Forrest, S., & Mitchell, M. (1991). The performance of genetic algorithms on Walsh polynomials: Some anomalous results and their explanation. In R. K. Belew & L. B. Booker (Eds.), *Proceedings of the Fourth International Conference on Genetic Algorithms* (pp.182-189). San Mateo, CA: Morgan Kaufmann.

Frey, G., & Michel, C. (2003). Circular codes in archaeal genomes. *Journal of Theoretical Biology, 223*, 413–431. doi:10.1016/S0022-5193(03)00119-X

Frey, G., & Michel, C. J. (2006). Identification of circular codes in bacterial genomes and their use in a factorization method for retrieving the reading frames of genes. *Computational Biology and Chemistry, 30*, 87–101. doi:10.1016/j.compbiolchem.2005.11.001

Geadah, Y. A., & Corinthios, M. J. (1977). Natural, dyadic, and sequency order algorithms and processors for the Walsh-Hadamard transform. *IEEE Transactions on Computers, C-26*, 435–442. doi:10.1109/TC.1977.1674860

Gierer, A., & Mundry, K. W. (1958). Production of mutants of tobacco mosaic virus by chemical alteration of its ribonucleic acid in vitro. *Nature, 182*, 1457–1458. doi:10.1038/1821457a0

Goldberg, D. E. (1989). Genetic algorithms and Walsh functions: Part 1, a gentle introduction. *Complex systems, 2*(2), 129-152.

Goldberg, D. E. (1989). Genetic algorithms and Walsh functions: Part II, deception and its analysis. *Complex systems, 3*(2), 153-171.

Kargupta, H. (2001). A striking property of genetic code-like transformations. *Complex systems, 11*, 43-50.

Lassez, J.-L. (1976). Circular codes and synchronization. *Intern. J. Comp. & Infor. Sci., 5*, 201–208. doi:10.1007/BF00975632

Lee, M. H., & Kaveh, M. (1986). Fast Hadamard transform based on a simple matrix factorization. *IEEE Transactions on Acoustics, Speech, and Signal Processing, ASSSP-34*(6), 1666–1667. doi:10.1109/TASSP.1986.1164972

Lobzin, V. V., & Chechetkin, V. R. (2000). Order and correlations in genomic DNA sequences. The spectral approaches. *Uspehi phizicheskih nauk, 170*(1), 57-81 (in Russian; an English version is at http://ufn.ru/en/articles/2000/1/c/).

Nielsen, M. A., & Chuang, I. L. (2001). *Quantum computation and quantum information.* Cambridge: Cambridge University Press.

Pan, C. (2007). *Applications of Hadamard transform in analytical chemistry.* Alabama: University of Alabama, Department of Chemistry, Graduate students seminar series (www.bama.ua.edu/~chem/seminars/student_seminars/spring07/s07-papers/pan-sem.pdf)

Penrose, R. (1989). *The emperor's new mind.* Oxford: Oxford University Press.

Penrose, R. (2004). *The road to reality. A complete guide to the laws of the universe.* London: Jonathan Cape.

Peterson, W. W., & Weldon, E. J. (1972). *Error-correcting codes.* Cambridge: MIT Press.

Petoukhov, S. V. (2005b). Hadamard matrices and quint matrices in matrix presentations of molecular genetic systems. *Symmetry: Culture and Science, 16*(3), 247–266.

Petoukhov, S. V. (2006). Bioinformatics: matrix genetics, algebras of the genetic code, and biological harmony. *Symmetry: Culture and Science, 17*(1-4), 251–290.

Petoukhov, S. V. (2008a). *Matrix genetics, algebras of the genetic code, noise-immunity.* Moscow: RCD (in Russian).

Petoukhov, S. V. (2008b). *The degeneracy of the genetic code and Hadamard matrices* (pp. 1-8).

Petoukhov, S. V. (2008c). *Matrix genetics, part 1: Permutations of positions in triplets and symmetries of genetic matrices* (pp. 1-12). Retrieved on March 8, 2008, from http://arXiv:0803.0888

Petoukhov, S. V. (2008d). Matrix genetics, part 2: The degeneracy of the genetic code and the octave algebra with two quasi-real units (the "yin-yang octave algebra") (pp. 1-27). Retrieved on March 8, 2008, from http://arXiv:0803.3330

Retrieved on February 8, 2008, from http://arXiv:0802.3366

Schuster, H., & Schramm, G. (1958). Bestimmung der biologisch wirksamen Einheit in der RNS des TMV auf chemischen Wege. *Zeitschrift fur Naturforschung. Teil B. Anorganische Chemie, Organische Chemie, Biochemie, Biophysik, Biologie, 13b*, 697–704.

Shapeville, F., & Haenni, A. L. (1974). *Biosynthese des proteins.* Paris: Hermann.

Shiozaki, A. (1980). A model of distributed type associated memory with quantized Hadamard transform. *Biological Cybernetics, 38*(1), 19–22. doi:10.1007/BF00337397

Solovieva, F. I. (2006). *Inroduction to the theory of coding.* Novosibirsk: Novosibirsk University (in Russian).

Stambuk, N. (1999). Circular coding properties of gene and protein sequences. *Croatica Chemica Acta, 72*(4), 999–1008.

Tolmachev, Y. A. (1976). *New optic spectrometers*. Leningrad: Leningrad University (in Russian).

Trahtman, A. M., & Trahtman, V. A. (1975). *The foundations of the theory of discrete signals on finite intervals*. Moscow: Sovetskoie Radio (in Russian).

Valiev, K. A., & Kokin, A. A. (2001). *Quantum computers: Hopes and reality*. Moscow: RCD (in Russian).

Vose, M., & Wright, A. (1998). The simple genetic algorithm and the Walsh transform: Part I, theory. *Journal of evolutionary computation, 6*(3), 253-274.

Waterman, M. S. (Ed.). (1999). *Mathematical methods for DNA sequences*. Florida: CRC Press, Inc.

Wittmann, H. G. (1961). Ansatze zur entschlusselung des genetishen codes. *Naturwissenschaften, 48*(24), 55. doi:10.1007/BF00590622

Yarlagadda, R., & Hershey, J. (1997). *Hadamard matrix analysis and synthesis with applications to communications and signal/image processing*. New York: Kluwer Academic Publ.

Ycas, M. (1969). *The biological code*. Amsterdam, London: North-Holland Publishing Company.

Zalmanzon, L. A. (1989). *Fourier, Walsh, and Haar transformations and their application in control, communication, and other systems*. Moscow: Nauka (in Russian).

Section 3
Algebras of Genetic Codes

Section 3 is organized into three chapters. It presents genomatrices and the genetic octet Yin-Yang-algebras, the evolution of the genetic code from the viewpoint of the genetic 8-dimensional Yin-Yang-algebra (or the genetic bipolar algebra), and multidimensional numbers and the genomatrices of hydrogen bonds.

Chapter 7
Genomatrices and the Genetic Octet Yin-Yang-Algebras

ABSTRACT

*Algebraic properties of the genetic code are analyzed. The investigations of the genetic code on the basis of matrix approaches ("matrix genetics") are described. The degeneracy of the vertebrate mitochondrial genetic code is reflected in the black-and-white mosaic of the (8*8)-matrix of 64 triplets, 20 amino acids, and stop-signals. The special algorithm, which is based on features of genetic molecules, exists to transform the mosaic genomatrix into the matrices, which are members of the special 8-dimensional algebra. Main mathematical properties of this genetic algebra and its relations with other algebras are analyzed together with some important consequences from the adequate algebraic models of the genetic code. Elements of new "genovector calculation" and ideas of "genetic mechanics" are discussed. The revealed fact of the relation between the genetic code and these genetic algebras, which define new multi-dimensional numeric systems, is discussed in connection with the famous idea by Pythagoras: "All things are numbers." Simultaneously, these genetic algebras can be utilized as the algebras of genetic operators in biological organisms. The described results are related to the problem of algebraization of bioinformatics. They draw attention to the question: what is life from the viewpoint of algebra?*

INTRODUCTION AND BACKGROUND

Does the genetic system possess its own algebra? Why is it important to study the question about the proper algebra of the genetic code? These questions are analyzed in this chapter first of all. Modern science knows that different natural systems can possess their own individual geometries and their own individual algebras (see, for example, the book (Kline, 1980)). The example of Hamilton, who wasted 10 years in

DOI: 10.4018/978-1-60566-124-7.ch007

his attempts to solve the task of description of transformations of 3D space by means of 3-dimensional algebras without a success, is very demonstrative one. This example implies that if a scientist does not guess right what type of algebras are adequate for the natural system which is investigated by him he can waste many years without any result in analogy with Hamilton. One can add that geometrical and physical-geometrical properties of separate natural systems (including laws of conservations, theories of oscillations and waves, theories of potentials and fields, etc.) can depend on the type of algebras which are adequate for them.

Matrix genetics have important analogues with matrix forms of presentations of hypercomplex numbers. Investigations of these analogues have led to adequate models of the genetic code in forms of multi-dimensional numeric systems, which are connected with appropriate multi-dimensional algebras. Such algebraic models of the genetic code put forward many new ideas and thoughts about interrelations among genetic elements and about relations of structures of the genetic code with many other biological, physical, information and mathematical structures.

Does the genetic system possess its own algebra? Why is it important to study the question about the proper algebra of the genetic code? To get answers on these questions and to understand their importance, the following background is useful.

The notion of "number" is the main notion of mathematics. In accordance with the famous thesis, the complexity of civilization is reflected in the complexity of the numbers which are utilized by the civilization. "*Number is one of the most fundamental concepts not only in mathematics, but also in all natural sciences. Perhaps, it is the more primary concept than such global categories, as time, space, substance or a field.*" (Pavlov, 2004)

After the establishment of real numbers in the history of the development of the notion of "number", complex numbers x_0+i*x_1 have appeared. These 2-dimensional numbers have played the role of the magic tool for development of theories and calculations in problems of heat, light, sounds, fluctuations, elasticity, gravitation, magnetism, electricity, current of liquids, and quantum-mechanical phenomena. It seems that modern atomic stations, airplanes, rockets and many other things would not exist without knowledge of complex numbers because the appropriate physical theories are based on these numbers. C. Gauss, J. Argand and C.Wessel have demonstrated that a plane with its properties fits 2-dimensional complex numbers. W. Hamilton has proved that the properties of our 3-dimensional physical space fit mathematical properties of the special quaternions. Hamilton's quaternions have played the significant role in the history of mathematical natural sciences as well. For example, the classical vector calculation is deduced from the theory of these quaternions. This chapter will show that the genetic code is connected with a special 8-dimensional numeric system, which is defined by the appropriate 8-dimensional algebra.

The notion "algebra", which we use in our book, has two main senses. According to the first sense, which is famous more widely, the algebra is the whole section of mathematics involving mathematical operations with mathematical symbols. According to the second sense, which is utilized in this book, algebra is a mathematical object with certain properties or, better to say, arithmetic of multidimensional numbers.

By definition in the frame of this second sense, algebra A with its dimension "n" over a field P is a set of expressions $x_0*\mathbf{i}_0+x_1*\mathbf{i}_1+x_2*\mathbf{i}_2+\ldots+x_{n-1}*\mathbf{i}_{n-1}$ (where $x_0, x_1,\ldots, x_{n-1}$ belong P; $\mathbf{i}_0, \mathbf{i}_1, \ldots\mathbf{i}_{n-1}$ are basic elements of vectors, which fit such expressions). This set is provided with the operation of multiplication by elements "k" from the field P to determine the formula $k*(x_0*\mathbf{i}_0 + x_1*\mathbf{i}_1 +x_2*\mathbf{i}_2 +\ldots+ x_{n-1}*\mathbf{i}_{n-1}) = k*x_0*\mathbf{i}_0 + k*x_1*\mathbf{i}_1 + k*x_2*\mathbf{i}_2 +\ldots+ k*x_{n-1}*\mathbf{i}_{n-1}$. This set is provided with the following operation of addition

Figure 1. The upper row: complex numbers in their matrix form of presentation and their decomposition on the basic elements "1" and "i", which are shown in their matrix forms of presentation as well. The matrix cells with positive coordinates are marked by dark color and the cell with negative coordinate is marked by white color. The lower row: the multiplication table of the basic elements "1" and "i"

$$z = x_0*\mathbf{1} + x_1*\mathbf{i} = \begin{bmatrix} x_0 & x_1 \\ -x_1 & x_0 \end{bmatrix} = x_0*\begin{bmatrix} 1 & 0 \\ 0 & 1 \end{bmatrix} + x_1*\begin{bmatrix} 0 & 1 \\ -1 & 0 \end{bmatrix}$$

	1	**i**
1	**1**	**i**
i	**i**	**-1**

as well: $(x_0*\mathbf{i}_0+x_1*\mathbf{i}_1+x_2*\mathbf{i}_2+...+x_{n-1}*\mathbf{i}_{n-1}) + (y_0*\mathbf{i}_0+y_1*\mathbf{i}_1+y_2*\mathbf{i}_2+ ...+y_{n-1}*\mathbf{i}_{n-1}) = (x_0+y_0)*\mathbf{i}_0 + (x_1+y_1)*\mathbf{i}_1 + ... + (x_{n-1}+y_{n-1})*\mathbf{i}_{n-1}$. This set is provided with the operation of multiplication between symbols i_r, which is given by a specific multiplication table $\mathbf{i}_r*\mathbf{i}_v = w_{rv,0}*\mathbf{i}_0 + w_{rv,1}*\mathbf{i}_1 +... w_{rv,n-1}*\mathbf{i}_2$. This multiplication table is utilized to find the result of multiplications $(x_0*\mathbf{i}_0+x_1*\mathbf{i}_1+x_2*\mathbf{i}_2+...+x_{n-1}*\mathbf{i}_{n-1})*(y_0*\mathbf{i}_0+y_1*\mathbf{i}_1+y_2*\mathbf{i}_2+ ...+y_{n-1}*\mathbf{i}_{n-1})$. Any algebra is defined completely by its multiplication table, that is, by a certain set of numbers $w_{rv,u}$. These numbers do not subordinate to any conditions, and any such set of numbers defines the particular algebra.

Algebras of complex and hypercomplex numbers $x_0*\mathbf{1}+x_1*\mathbf{i}_1+...+x_k*\mathbf{i}_k$ are well-known. It is also known that complex and hypercomplex numbers have not only linear or vector forms of their presentations, but also matrix forms of their presentation. For example complex numbers $z = x*\mathbf{1}+y*\mathbf{i}$ (where **1** is the real unit and ***i*** is the imaginary unit: $\mathbf{i}^2 = -1$) possess the following matrix form of their presentation (Figure 1). By the way, complex numbers are utilized in computers in this matrix form.

The quaternions by Hamilton $Q = x_0*\mathbf{1} + x_1*\mathbf{i}_1 + x_2*\mathbf{i}_2 + x_3*\mathbf{i}_3$ (where $\mathbf{i}_1^2 = \mathbf{i}_2^2 = \mathbf{i}_3^2 = -1$, $\mathbf{i}_1*\mathbf{i}_2 = -\mathbf{i}_2*\mathbf{i}_1 = \mathbf{i}_3$, $\mathbf{i}_1*\mathbf{i}_3 = -\mathbf{i}_3*\mathbf{i}_1 = -\mathbf{i}_2$, $\mathbf{i}_2*\mathbf{i}_3 = -\mathbf{i}_3*\mathbf{i}_2 = \mathbf{i}_1$), which are utilized widely in physics and mathematics as well, have their matrix form of presentation as well. Figure 2 shows this matrix form and its decomposition on the basic elements $\mathbf{1}, \mathbf{i}_1, \mathbf{i}_2, \mathbf{i}_3$ in their matrix forms of presentation as well. In addition the multiplication table of these basic elements $\mathbf{1}, \mathbf{i}_1, \mathbf{i}_2, \mathbf{i}_3$ is demonstrated.

Is the mosaic genetic matrix $P^{(3)}$ = [C A; U G]$^{(3)}$ (Figure 2 in Chapter 2), which was analyzed in the previous chapters, connected with a matrix form of presentation of any multi-dimensional numeric system? This chapter gives a positive answer to this question.

THE GENETIC OCTET MATRIX AS THE MATRIX FORM OF PRESENTATION OF THE OCTET ALGEBRA

Let us return to the genetic matrix $P^{CAUG}{}_{123}{}^{(3)}$ = [C A; U G]$^{(3)}$ (Figure 2 in Chapter 2), which possesses 32 "black" triplets and 32 "white" triplets disposed in matrix cells of appropriate colors. The black-and-white mosaic of this matrix reflects the specificity of degeneracy of the vertebrate mitochondrial genetic code as was described in Chapter 2. Taking into account the molecular characteristics of the nitrogenous bases A, C, G, U/T of the genetic alphabet, one can reform this genomatrix [C A; U G]$^{(3)}$ into the new matrix YY_8 algorithmically (Figure 3).

Figure 2. The upper row: quaternions by Hamilton in the matrix form of their presentation; cells with positive coordinates are marked by dark color and the cells with negative coordinates are marked by white color. The middle row: the decomposition of quaternions in their matrix form on the basic elements **1**, $\mathbf{i}_1$, $\mathbf{i}_2$, $\mathbf{i}_3$, *which are shown in their matrix forms of presentation as well. The lower row shows the multiplication table of these basic elements*

$$Q = x_0{*}\mathbf{1} + x_1{*}\mathbf{i}_1 + x_2{*}\mathbf{i}_2 + x_3{*}\mathbf{i}_3 = \begin{vmatrix} x_0 & x_1 & x_2 & x_3 \\ -x_1 & x_0 & -x_3 & x_2 \\ -x_2 & x_3 & x_0 & -x_1 \\ -x_3 & -x_2 & x_1 & x_0 \end{vmatrix} =$$

$$= x_0{*}\begin{vmatrix} 1&0&0&0 \\ 0&1&0&0 \\ 0&0&1&0 \\ 0&0&0&1 \end{vmatrix} + x_1{*}\begin{vmatrix} 0&1&0&0 \\ -1&0&0&0 \\ 0&0&0&-1 \\ 0&0&1&0 \end{vmatrix} + x_2{*}\begin{vmatrix} 0&0&1&0 \\ 0&0&0&1 \\ -1&0&0&0 \\ 0&-1&0&0 \end{vmatrix} + x_3{*}\begin{vmatrix} 0&0&0&1 \\ 0&0&-1&0 \\ 0&1&0&0 \\ -1&0&0&0 \end{vmatrix}$$

	1	$\mathbf{i}_1$	$\mathbf{i}_2$	$\mathbf{i}_3$
1	**1**	$\mathbf{i}_1$	$\mathbf{i}_2$	$\mathbf{i}_3$
$\mathbf{i}_1$	$\mathbf{i}_1$	**-1**	$\mathbf{i}_3$	$-\mathbf{i}_2$
$\mathbf{i}_2$	$\mathbf{i}_2$	$-\mathbf{i}_3$	**-1**	$\mathbf{i}_1$
$\mathbf{i}_3$	$\mathbf{i}_3$	$\mathbf{i}_2$	$-\mathbf{i}_1$	**-1**

The cells of the matrix YY_8, which are occupied by components with the sign "+", are marked by dark color. The cells of the matrix YY_8, which are occupied by components with the sign "-", are marked by white color. Such black-and-white mosaic of the matrix YY_8 is identical to the black-and-white mosaic of the genomatrix [C A; U G]$^{(3)}$ (Figure 2 of Chapter 2). The matrix YY_8 has the 8 independent parameters x_0, x_1, x_2, x_3, x_4, x_5, x_6, x_7, which are interpreted as real numbers here. It has been discovered that the matrix YY_8 is the matrix form of presentation of the special 8-dimensional algebra (or the 8-dimensional algebra over the field of real numbers) and of the appropriate 8-dimensional numeric system. Below we shall list the other structural analogies of the genomatrix [C A; U G]$^{(3)}$ with the matrix YY_8, the set of which allows one to consider that this matrix YY_8 and its algebra play the role of the adequate model of the genetic code. But initially we pay attention to the "alphabetic" algorithm of Yin-Yang-digitization of 64 triplets, which produces the matrix YY_8 from the genomatrix [C A; U G]$^{(3)}$. This algorithm has received such an unusual name because of special properties of the matrix YY_8 and its algebra (Petoukhov, 2008a-f).

THE ALPHABETIC ALGORITHM OF THE YIN-YANG-DIGITIZATION OF 64 TRIPLETS

This algorithm is based on utilizing the two following binary-oppositional attributes of the genetic letters A, C, G, U/T: "purine or pyrimidine" and "2 or 3" hydrogen bonds. It uses also the famous thesis of molecular genetics that different positions inside triplets have different code meanings. For example

Figure 3. The matrix YY_8, the black cells of which contain coordinates with the sign „+" and the white cells of which contain coordinates with the sign „-". The numeration of the comuns and the rows is identical to the numeration of the columns and the rows of the matrix [C A; U G]$^{(3)}$ on Figure 3 in chapter 1

$YY_8 =$

	000 (0)	001 (1)	010 (2)	011 (3)	100 (4)	101 (5)	110 (6)	111 (7)
000 (0)	x_0	x_1	$-x_2$	$-x_3$	x_4	x_5	$-x_6$	$-x_7$
001 (1)	x_0	x_1	$-x_2$	$-x_3$	x_4	x_5	$-x_6$	$-x_7$
010 (2)	x_2	x_3	x_0	x_1	$-x_6$	$-x_7$	$-x_4$	$-x_5$
011 (3)	x_2	x_3	x_0	x_1	$-x_6$	$-x_7$	$-x_4$	$-x_5$
100 (4)	x_4	x_5	$-x_6$	$-x_7$	x_0	x_1	$-x_2$	$-x_3$
101 (5)	x_4	x_5	$-x_6$	$-x_7$	x_0	x_1	$-x_2$	$-x_3$
110 (6)	$-x_6$	$-x_7$	$-x_4$	$-x_5$	x_2	x_3	x_0	x_1
111 (7)	$-x_6$	$-x_7$	$-x_4$	$-x_5$	x_2	x_3	x_0	x_1

the article (Konopelchenko, & Rumer, 1975) has described that two first positions of each triplet form "the root of the codon" and that they differ drastically from the third position by their essence and by their special role. In view of this "alphabetic" algorithm, the transformation of the genomatrix [C A; U G]$^{(3)}$ into the matrix YY_8 is not an abstract and arbitrary action at all, but such a transformation can be utilized by bio-computer systems of organisms materially.

The alphabetic algorithm of the Yin-Yang-digitization defines the special scheme of reading each triplet: the first two positions of the triplet are read by genetic systems from the viewpoint of one attribute and the third position of the triplet is read from the viewpoint of another attribute. By this alphabetic algorithm, which allows one to recode the symbolic matrix [C A; U G]$^{(3)}$ into the numeric Yin-Yang-matrix YY_8 (see below), each triplet is read in the following way:

- Two first positions of each triplet are filled out by the symbol "α" instead of the complementary letters C and G on these positions and by the symbol "β" instead of the complementary letters A and U correspondingly;
- The third position of each triplet is filled out by the symbol "γ" instead of the pyrimidine (C or U) on this position and by the symbol "δ" instead of the purine (A or G) correspondingly;
- The triplets, which have the letters C or G in their first position, receive the sign "-" in those cases only for which their second position is occupied by the letter A. The triplets, which have the letters A or U on their first position, receive the sign "+" in those cases only for which their second positions is occupied by the letter C.

For example, the triplet CAG receives the symbol "-αβδ", because its first letter C is symbolized by "α", its second letter A is symbolized by "β", and its third letter G is symbolized by "δ". This triplet possesses the sign "-" because its first position has the letter C and its second position has the letter A. One can see that this algorithm recodes all triplets from the traditional alphabet C, A, U, G into the new alphabet α, β, γ, δ. In the result, each triplet receives one of the following 8 expressions: $\alpha\alpha\gamma = x_0$, $\alpha\alpha\delta = x_1$, $\alpha\beta\gamma = x_2$, $\alpha\beta\delta = x_3$, $\beta\alpha\gamma = x_4$, $\beta\alpha\delta = x_5$, $\beta\beta\gamma = x_6$, $\beta\beta\delta = x_7$. We will suppose that the symbols "α", "β", "γ", "δ" are real numbers. This algorithm transforms the initial symbolic matrix [C A; U G]$^{(3)}$

into the numeric matrix YY_8 with the 8 coordinates $x_0, x_1, x_2, x_3, x_4, x_5, x_6, x_7$. We shall name these matrix components $x_0, x_1, \ldots, x_7$, which are real numbers, as "*YY*-coordinates" (see Figure 4).

Let us pay some attention now to algebraic properties of the matrix YY_8.

THE GENOMATRIX YY_8 AS THE ELEMENT OF THE OCTET YIN-YANG-ALGEBRA

By analogy with decompositions of the matrices of complex numbers and of quaternions by Hamilton (Figure 1 and Figure 2), one can represent the 8-parametric matrix YY_8 (Figure 3) as the sum of the 8 basic matrices, each of which is connected with one of the coordinates $x_0, x_1, x_2, x_3, x_4, x_5, x_6, x_7$ (Figure 5). Let us symbolize any basic matrix, which is related to any of *YY*-coordinates x_0, x_2, x_4, x_6 with even indexes, by the symbol $\mathbf{f}_k$ (where "f" is the first letter of the word "female" and $k = 0, 2, 4, 6$). And let us symbolize any matrix, which is related to any of *YY*-coordinates x_1, x_3, x_5, x_7 with odd indexes, by the symbol $\mathbf{m}_s$ (where "m" is the first letter of the word "male" and $s = 1, 3, 5, 7$). In this case one can present the matrix YY_8 by the expression (1), the matrix form of which is shown on Figure 5.

Figure 4. The result of the algorithmic transformation of 64 triplets into the numeric coordinates x_0, x_1, ..., x_7, which are based on the four symbols "α", "β", "γ", "δ"

	000 (0)	001 (1)	010 (2)	011 (3)	100 (4)	101 (5)	110 (6)	111 (7)
000 (0)	CCC ααγ x_0	CCA ααδ x_1	CAC -αβγ $-x_2$	CAA -αβδ $-x_3$	ACC βαγ x_4	ACA βαδ x_5	AAC -ββγ $-x_6$	AAA -ββδ $-x_7$
001 (1)	CCU ααγ x_0	CCG ααδ x_1	CAU -αβγ $-x_2$	CAG -αβδ $-x_3$	ACU βαγ x_4	ACG βαδ x_5	AAU -ββγ $-x_6$	AAG -ββδ $-x_7$
010 (2)	CUC αβγ x_2	CUA αβδ x_3	CGC ααγ x_0	CGA ααδ x_1	AUC -ββγ $-x_6$	AUA -ββδ $-x_7$	AGC -βαγ $-x_4$	AGA -βαδ $-x_5$
011 (3)	CUU αβγ x_2	CUG αβδ x_3	CGU ααγ x_0	CGG ααδ x_1	AUU -ββγ $-x_6$	AUG -ββδ $-x_7$	AGU -βαγ $-x_4$	AGG -βαδ $-x_5$
100 (4)	UCC βαγ x_4	UCA βαδ x_5	UAC -ββγ $-x_6$	UAA -ββδ $-x_7$	GCC ααγ x_0	GCA ααδ x_1	GAC -αβγ $-x_2$	GAA -αβδ $-x_3$
101 (5)	UCU βαγ x_4	UCG βαδ x_5	UAU -ββγ $-x_6$	UAG -ββδ $-x_7$	GCU ααγ x_0	GCG ααδ x_1	GAU -αβγ $-x_2$	GAG -αβδ $-x_3$
110 (6)	UUC -ββγ $-x_6$	UUA -ββδ $-x_7$	UGC -βαγ $-x_4$	UGA -βαδ $-x_5$	GUC αβγ x_2	GUA αβδ x_3	GGC ααγ x_0	GGA ααδ x_1
111 (7)	UUU -ββγ $-x_6$	UUG -ββδ $-x_7$	UGU -βαγ $-x_4$	UGG -βαδ $-x_5$	GUU αβγ x_2	GUG αβδ x_3	GGU ααγ x_0	GGG ααδ x_1

Figure 5. The presentation of the matrix YY_8 as the sum of the 8 basic matrices. The left column shows the basic matrices, which are related to the coordinates x_0, x_2, x_4, x_6 with the even indexes. The right column shows the basic matrices, which are related to the coordinates x_1, x_3, x_5, x_7 with the odd indexes

$$YY_8 = x_0*\begin{bmatrix}1&0&0&0&0&0&0&0\\1&0&0&0&0&0&0&0\\0&0&1&0&0&0&0&0\\0&0&1&0&0&0&0&0\\0&0&0&0&1&0&0&0\\0&0&0&0&1&0&0&0\\0&0&0&0&0&0&1&0\\0&0&0&0&0&0&1&0\end{bmatrix} + x_1*\begin{bmatrix}0&1&0&0&0&0&0&0\\0&1&0&0&0&0&0&0\\0&0&0&1&0&0&0&0\\0&0&0&1&0&0&0&0\\0&0&0&0&0&1&0&0\\0&0&0&0&0&1&0&0\\0&0&0&0&0&0&0&1\\0&0&0&0&0&0&0&1\end{bmatrix} +$$

$$+ x_2*\begin{bmatrix}0&0&-1&0&0&0&0&0\\0&0&-1&0&0&0&0&0\\1&0&0&0&0&0&0&0\\1&0&0&0&0&0&0&0\\0&0&0&0&0&0&-1&0\\0&0&0&0&0&0&-1&0\\0&0&0&0&1&0&0&0\\0&0&0&0&1&0&0&0\end{bmatrix} + x_3*\begin{bmatrix}0&0&0&-1&0&0&0&0\\0&0&0&-1&0&0&0&0\\0&1&0&0&0&0&0&0\\0&1&0&0&0&0&0&0\\0&0&0&0&0&0&0&-1\\0&0&0&0&0&0&0&-1\\0&0&0&0&0&1&0&0\\0&0&0&0&0&1&0&0\end{bmatrix} +$$

$$+ x_4*\begin{bmatrix}0&0&0&0&1&0&0&0\\0&0&0&0&1&0&0&0\\0&0&0&0&0&0&-1&0\\0&0&0&0&0&0&-1&0\\1&0&0&0&0&0&0&0\\1&0&0&0&0&0&0&0\\0&0&-1&0&0&0&0&0\\0&0&-1&0&0&0&0&0\end{bmatrix} + x_5*\begin{bmatrix}0&0&0&0&0&1&0&0\\0&0&0&0&0&1&0&0\\0&0&0&0&0&0&0&-1\\0&0&0&0&0&0&0&-1\\0&1&0&0&0&0&0&0\\0&1&0&0&0&0&0&0\\0&0&0&-1&0&0&0&0\\0&0&0&-1&0&0&0&0\end{bmatrix} +$$

$$+ x_6*\begin{bmatrix}0&0&0&0&0&0&-1&0\\0&0&0&0&0&0&-1&0\\0&0&0&0&-1&0&0&0\\0&0&0&0&-1&0&0&0\\0&0&-1&0&0&0&0&0\\0&0&-1&0&0&0&0&0\\-1&0&0&0&0&0&0&0\\-1&0&0&0&0&0&0&0\end{bmatrix} + x_7*\begin{bmatrix}0&0&0&0&0&0&0&-1\\0&0&0&0&0&0&0&-1\\0&0&0&0&0&-1&0&0\\0&0&0&0&0&-1&0&0\\0&0&0&-1&0&0&0&0\\0&0&0&-1&0&0&0&0\\0&-1&0&0&0&0&0&0\\0&-1&0&0&0&0&0&0\end{bmatrix}$$

$$YY_8 = x_0*\mathbf{f}_0+x_1*\mathbf{m}_1+x_2*\mathbf{f}_2+x_3*\mathbf{m}_3+x_4*\mathbf{f}_4+x_5*\mathbf{m}_5+x_6*\mathbf{f}_6+x_7*\mathbf{m}_7 \quad (1)$$

The important and unexpected fact is that the set of these 8 basic matrices $\mathbf{f}_0$, $\mathbf{m}_1$, $\mathbf{f}_2$, $\mathbf{m}_3$, $\mathbf{f}_4$, $\mathbf{m}_5$, $\mathbf{f}_6$, $\mathbf{m}_7$ forms the closed set relative to multiplications: a multiplication between any two matrices from this set generates a matrix from this set again. The table on Figure 6 presents the results of multiplications among these 8 matrices. The result of multiplying any two basic elements, which are taken from the left column and the upper row, is shown in the cell on the intersection of its row and column (for example, in accordance with this multiplication table $\mathbf{f}_2*\mathbf{m}_5 = -\mathbf{m}_7$).

We have noted above, that such multiplication tables define appropriate algebras over a field. Correspondingly the multiplication table on Figure 6 defines the genetic 8-dimensional algebra YY_8. Mul-

Figure 6. The multiplication table of the basic matrices $\mathbf{f_0}$, $\mathbf{m_1}$, $\mathbf{f_2}$, $\mathbf{m_3}$, $\mathbf{f_4}$, $\mathbf{m_5}$, $\mathbf{f_6}$, $\mathbf{m_7}$ *of the matrix* YY_8 *from Figure 3 and Figure 5*

	f_0	m_1	f_2	m_3	f_4	m_5	f_6	m_7
f_0	f_0	m_1	f_2	m_3	f_4	m_5	f_6	m_7
m_1	f_0	m_1	f_2	m_3	f_4	m_5	f_6	m_7
f_2	f_2	m_3	$-f_0$	$-m_1$	$-f_6$	$-m_7$	f_4	m_5
m_3	f_2	m_3	$-f_0$	$-m_1$	$-f_6$	$-m_7$	f_4	m_5
f_4	f_4	m_5	f_6	m_7	f_0	m_1	f_2	m_3
m_5	f_4	m_5	f_6	m_7	f_0	m_1	f_2	m_3
f_6	f_6	m_7	$-f_4$	$-m_5$	$-f_2$	$-m_3$	f_0	m_1
m_7	f_6	m_7	$-f_4$	$-m_5$	$-f_2$	$-m_3$	f_0	m_1

tiplication of any two members of the octet algebra YY_8 generates a new member of the same algebra. Concerning to multiplication of such numbers in their matrix forms of presentation, it means that both factors have the identical matrix disposition of their 8 parameters $x_0, x_1, \ldots, x_7$ (in the first factor) and $y_0, y_1, \ldots, y_7$ (in the second factor) and the final matrix has the same matrix disposition of its 8 relevant parameters $z_0, z_1, \ldots, z_7$. This situation is similar to the situation of real numbers (or of complex numbers, or of hypercomplex numbers) when multiplication of any two members of the numeric system generates a new member of the same numerical system. In other words, the expression $YY_8=x_0*\mathbf{f_0}+x_1*\mathbf{m_1}+x_2*\mathbf{f_2}+x_3*\mathbf{m_3}+x_4*\mathbf{f_4}+x_5*\mathbf{m_5}+x_6*\mathbf{f_6}+x_7*\mathbf{m_7}$ is some kind of 8-dimensional numbers ("octet genonumber") (Petoukhov, 2008a, 2008d). We mark this algebra and these octet genonumbers by the same symbol YY_8 conditionally.

Let us give a numeric example of multiplication of two octet genonumbers: $V=3*\mathbf{f_0}+2*\mathbf{m_1}-4*\mathbf{f_2}+1*\mathbf{m_3}-5*\mathbf{f_4}+6*\mathbf{m_5}+8*\mathbf{f_6}-7*\mathbf{m_7}$ and $W=2*\mathbf{f_0}-4*\mathbf{m_1}+5*\mathbf{f_2}+3*\mathbf{m_3}-6*\mathbf{f_4}-8*\mathbf{m_5}-1*\mathbf{f_6}+5*\mathbf{m_7}$. The result of multiplication depends on the order of factors because of the non-symmetrical character of the multiplication table relative to its main diagonal, which means that the algebra YY_8 is non-commutative:

$$V*W= 18*\mathbf{f_0} -14*\mathbf{m_1} +24*\mathbf{f_2}+40*\mathbf{m_3} -30*\mathbf{f_4} -62*\mathbf{m_5} -16*\mathbf{f_6} +0*\mathbf{m_7}$$

$$W*V=128*\mathbf{f_0}-124*\mathbf{m_1}-60*\mathbf{f_2}+88*\mathbf{m_3}+48*\mathbf{f_4}-100*m_5+92*\mathbf{f_6} +40*\mathbf{m_7}$$

These results can be arrived at multiplication of appropriate matrix forms of presentation of the octet genonumbers V and W or by multiplication of linear forms of their presentation using the multiplication table on Figure 6.

One should pay special attention to the cells on the main diagonal of the multiplication table (Figure 6). These cells contain squares of the basic elements. In cases of hypercomplex numbers these diagonal cells contain elements "±1" typically (for example, see multiplication tables of complex numbers and of quaternions by Hamilton on Figure 1 and Figure 2). In our case these diagonal cells contain no real units at all but all diagonal cells are occupied by elements "$\pm\mathbf{f_0}$" and "$\pm\mathbf{m_1}$". Thereby the set of the 8 basic matrices $\mathbf{f_0}$, $\mathbf{m_1}$, $\mathbf{f_2}$, $\mathbf{m_3}$, $\mathbf{f_4}$, $\mathbf{m_5}$, $\mathbf{f_6}$, $\mathbf{m_7}$ is divided into two equal subsets by criterion of their squares. The first subset consists of elements with the even indexes: $\mathbf{f_0}$, $\mathbf{f_2}$, $\mathbf{f_4}$, $\mathbf{f_6}$. The squares of members of this $\mathbf{f_0}$-subset are equal to $\pm\mathbf{f_0}$ always. The second subset consists of elements with the odd indexes: $\mathbf{m_1}$, $\mathbf{m_3}$, $\mathbf{m_5}$, $\mathbf{m_7}$. The squares of members of this $\mathbf{m_1}$-subset are equal to $\pm\mathbf{m_1}$ always.

The basic element $\mathbf{f}_0$ possesses all properties of the real unit in relation to the members of the $\mathbf{f}_0$-subset: $\mathbf{f}_0^2 = \mathbf{f}_0$, $\mathbf{f}_0*\mathbf{f}_2=\mathbf{f}_2*\mathbf{f}_0=\mathbf{f}_2$, $\mathbf{f}_0*\mathbf{f}_4=\mathbf{f}_4*\mathbf{f}_0=\mathbf{f}_4$, $\mathbf{f}_0*\mathbf{f}_6=\mathbf{f}_6*\mathbf{f}_0=\mathbf{f}_6$. But the element $\mathbf{f}_0$ does not possess the commutative property of real unit in relation to the members of the $\mathbf{m}_1$-subset: $\mathbf{f}_0*\mathbf{m}_p \neq \mathbf{m}_p*\mathbf{f}_0$, where $p = 1,3,5,7$. For this reason $\mathbf{f}_0$ is named "quasi-real unit from the $\mathbf{f}_0$-subset".

The basic element $\mathbf{m}_1$ possesses all properties of the real unit in relation to the members of the $\mathbf{m}_1$-subset: $\mathbf{m}_1^2=\mathbf{m}_1$, $\mathbf{m}_1*\mathbf{m}_3=\mathbf{m}_3*\mathbf{m}_1=\mathbf{m}_3$, $\mathbf{m}_1*\mathbf{m}_5=\mathbf{m}_5*\mathbf{m}_1=\mathbf{m}_5$, $\mathbf{m}_1*\mathbf{m}_7=\mathbf{m}_7*\mathbf{m}_1=\mathbf{m}_7$. But the element $\mathbf{m}_1$ does not possess the commutative property of real unit in relation to the members of the $\mathbf{f}_0$-subset: $\mathbf{m}_1*\mathbf{f}_k \neq \mathbf{f}_k*\mathbf{m}_1$, where $k = 0,2,4,6$. For this reason $\mathbf{m}_1$ is named "quasi-real unit from the $\mathbf{m}_1$-subset".

The principle "even-odd" exists in this algebra YY_8. Really all members of the $\mathbf{f}_0$-subset and their coordinates x_0, x_2, x_4, x_6 have even indexes and they are disposed in columns with the even numbers 0, 2, 4, 6 in the matrix YY_8 (Figure 3) and in its multiplication table (Figure 6) as well. These coordinates x_0, x_2, x_4, x_6 correspond to triplets with the pyrimidine suffixes C and U (Figure 4). For this reason the $\mathbf{f}_0$-subset can be called as the "pyrimidine subset".

All members of the $\mathbf{m}_1$-subset and their coordinates x_1, x_3, x_5, x_7 have the odd indexes and they are disposed in columns with the odd numbers 1, 3, 5, 7 in the matrix YY_8 (Figure 3) and in its multiplication table (Figure 6) as well. These coordinates x_1, x_3, x_5, x_7 correspond to triplets with the purine suffixes A and G (Figure 4). For this reason the $\mathbf{m}_1$-subset can be called as the "purine subset".

In accordance with Pythagorean and Ancient-Chinese traditions, all even numbers are named "female" numbers or Yin-numbers, and all odd numbers are named "male" numbers or Yang-numbers. From the viewpoint of this tradition, the elements $\mathbf{f}_0$, $\mathbf{f}_2$, $\mathbf{f}_4$, $\mathbf{f}_6$, x_0, x_2, x_4, x_6 with the even indexes play the role of "female" elements or Yin-elements, and the elements $\mathbf{m}_1$, $\mathbf{m}_3$, $\mathbf{m}_5$, $\mathbf{m}_7$, x_1, x_3, x_5, x_7 with the odd indexes play the role of "male" or Yang-elements. Correspondingly the 8-dimensional algebra YY_8 can be named the octet Yin-Yang-algebra (or the even-odd-algebra, or the bipolar algebra, or the bisex-algebra, or the pyrimidine-purine-algebra for triplets with pyrimidine suffixes and with purine suffixes). Such algebra, which possesses two quasi-real units and no one real unit, gives new effective possibilities to model binary oppositions in biological objects at different levels, including sets of triplets, amino acids, male and female gametal cells, male and female chromosomes, etc.

The octet Yin-Yang-numbers YY_8 (octet genonumbers) differ essentially from classical hypercomplex numbers, which have the real unit in the set of their basic elements. By traditional definition, hypercomplex numbers are the elements of the algebras with the real unit. Complex and hypercomplex numbers were constructed historically as generalizations of real numbers with the obligatory inclusion of the real unit in sets of their basic elements. The octet Yin-Yang-numbers YY_8 have not the real unit in the set of their basic elements at all, but they have two quasi-real units $\mathbf{f}_0$ and $\mathbf{m}_1$. In comparison with hypercomplex numbers, Yin-Yang-numbers are the new category of numbers in the mathematical natural sciences in principle. In our opinion, knowledge of this category of numbers is necessary for deep understanding of biological phenomena, and, perhaps, it will be useful for mathematical natural sciences in the whole. Mathematical theory of *YY*-numbers gives new formal and conceptual apparatus to model phenomena of reproduction and self-organization in living nature.

It can be demonstrated easily that Yin-Yang algebras are the special generalization of the algebras of hypercomplex numbers in the form of "double-hypercomplex" numbers. Yin-Yang-numbers (*YY*-numbers) or bipolar numbers become the appropriate hypercomplex numbers in those cases when all their female (or male) coordinates are equal to zero. Traditional hypercomplex numbers can be represented as the "mono-sex" half (a Yin half or a Yang half) or "mono-polar" half of appropriate *YY*-numbers. The

Figure 7. The multiplication tables of the Yin-genoquaternion G_f (on the left side) and of Yang-genoquaternions G_m (on the left side)

	f_0	f_2	f_4	f_6			m_1	m_3	m_5	m_7
f_0	f_0	f_2	f_4	f_6		m_1	m_1	m_3	m_5	m_7
f_2	f_2	$-f_0$	$-f_6$	f_4		m_3	m_3	$-m_0$	$-m_6$	m_4
f_4	f_4	f_6	f_0	f_2		m_5	m_5	m_6	m_0	m_2
f_6	f_6	$-f_4$	$-f_2$	f_0		m_7	m_7	$-m_4$	$-m_2$	m_0

algorithm of such generalization will be described later. We denote Yin-Yang numbers by double letters (for example, *YY*) to distinguish them from traditional (complex and hypercomplex) numbers.

If all male coordinates are equal to 0 ($x_1 = x_3 = x_5 = x_7 = 0$), the numbers YY_8 become the Yin-genoquaternions $G_f = x_0*\mathbf{f}_0 + x_2*\mathbf{f}_2 + x_4*\mathbf{f}_4 + x_6*\mathbf{f}_6$, the multiplication table of which is shown on Figure 7 These Yin-quaternions can be called also as "pyrimidine quaternions" conditionally because their coordinates x_0, x_2, x_4, x_6 correspond to triplets with the pyrimidine suffixes C or U (Figure 4).

If all female coordinates are equal to 0 ($x_0 = x_2 = x_4 = x_6 = 0$), the numbers YY_8 become the Yang-genoquaternions $G_m = x_1*\mathbf{m}_1 + x_3*\mathbf{m}_3 + x_5*\mathbf{m}_5 + + x_7*\mathbf{m}_7$, the multiplication table of which is shown on Figure 7. These Yang-quaternions can be called also as "purine quaternions" conditionally because their coordinates x_1, x_3, x_5, x_7 correspond to triplets with the purine suffixes A or G (Figure 4).

These genetic quaternions G_f and G_m have the identical multiplication tables, which differ from the multiplication table of Hamilton quaternions (see Figure 2). Taking these facts into account, the octet genonumbers YY_8 can be named "the double genetic quaternions". It causes heuristic associations with a double helix of DNA, which is the bearer of genetic information. Just as the structure of three-dimensional physical space corresponds to the algebra of quaternions by Hamilton, so the structure of the genetic code corresponds to the algebra of the double genoquaternions.

The set of the basic elements of the YY_8-algebra forms a semi-group. Two squares are marked out by bold lines in the left upper corner of the multiplication table on Figure 6. The first two basic elements $\mathbf{f}_0$ and $\mathbf{m}_1$ are disposed in the smaller (2x2)-square of this table only. The greater (4x4)-square collects the four first basic elements $\mathbf{f}_0$, $\mathbf{m}_1$, $\mathbf{f}_2$, $\mathbf{m}_3$. These aspects say that sub-algebras YY_2 and YY_4 exist inside the algebra YY_8. We shall return to these sub-algebras later.

Each genetic triplet, which is disposed in the genomatrix $[C\ A; U\ G]^{(3)}$ on Figure 4 together with one of the female *YY*-coordinates x_0, x_2, x_4, x_6 in a mutual matrix cell, is named the female triplet or the Yin-triplet. The third position of all female triplets is occupied by the letter **γ**, which corresponds to the pyrimidine C or U/T. Thereby the female triplets can be named "pyrimidine triplets" as well. Each triplet, which is disposed in the genomatrix $[C\ A; U\ G]^{(3)}$ on Figure 4 together with one of the male *YY*-coordinates x_1, x_3, x_5, x_7 in a mutual matrix cell, is named the male triplet or the Yang-triplet. The third position of all male triplets is occupied by the letter **δ**, which corresponds to the purine A or G. Thereby the male triplets can be named "purine triplets". In such algebraic way the whole set of 64 triplets is divided into two sub-sets of Yin-triplets (or female triplets) and Yang-triplets (or male triplets). We shall demonstrate later that the set of 20 amino acids is divided into the sub-sets of "female amino acids", "male amino acids" and "androgenous amino acids" from this matrix viewpoint.

Later we will continue to describe significant mathematical properties of the octet Yin-Yang-matrices. But now let us consider the close connection of structures of the genetic code with the octet Yin-Yang-matrices in many aspects.

THE STRUCTURAL ANALOGIES BETWEEN THE GENOMATRIX [C A; U G]$^{(3)}$ AND THE MATRIX YY_8

The main interest of bioinformatics to the octet Yin-Yang-algebra is connected with a possibility of its use as an adequate model of the structure of the genetic code. This possibility depends on structural coincidences between the Yin-Yang matrix YY_8 and the genetic matrix [C A; U G]$^{(3)}$. A list of such non-trivial coincidences includes the following ones:

1. *The first coincidence.*

The black-and-white mosaics of thc Yin-Yang matrix YY_8 and the genetic matrix [C A; U G]$^{(3)}$ are identical. (By an unknown reason, nature has divided the set of the 64 genetic triplets into two subset of 32 black triplets and 32 white triplets, which are disposed in the cells of 32 positive coordinates and 32 negative coordinates of the Yin-Yang matrix YY_8).

2. *The second coincidence.*

In the Yin-Yang matrix YY_8, the pairs of the adjacent rows 0-1, 2-3, 4-5, 6-7 are identical to each other by the assortment and the disposition of numeric coordinates $x_0, x_1, x_2, x_3, x_4, x_5, x_6, x_7$.

In the genetic matrix [C A; U G]$^{(3)}$, the same pairs of adjacent rows 0-1, 2-3, 4-5, 6-7 are identical each to another by the assortment and the disposition of amino acids and stop-codons.

3. *The third coincidence.*

In the Yin-Yang matrix YY_8, the female coordinates x_0, x_2, x_4, x_6 occupy the columns with the even numbers 0, 2, 4, 6, and the male coordinates x_1, x_3, x_5, x_7 occupy the columns with the odd numbers 1, 3, 5, 7.

In the genetic matrix [C A; U G]$^{(3)}$, the triplets with pyrimidine C or U on their third positions occupy the columns with the even numbers 0, 2, 4, 6; and the triplets with purine A or G on their third positions occupy the columns with the odd numbers 1, 3, 5, 7.

4. *The fourth coincidence.*

In the Yin-Yang matrix YY_8, one half of the quantity of the numeric coordinates (x_0, x_1, x_2, x_3) exists in the two quadrants along the main diagonal only; the second half of the numeric coordinates (x_4, x_5, x_6, x_7) exists in the two quadrants along the second diagonal only.

In the genetic matrix [C A; U G]$^{(3)}$, one half of kinds of amino acids exists in the two quadrants along the main diagonal only (Ala, Arg, Asp, Gln, Glu, Gly, His, Leu, Pro, Val); the second half of kinds of amino acids exists in the two quadrants along the second diagonal only (Asn, Cys, Ile, Lys, Met, Phe, Ser,Thr, Trp, Tyr).

5. *The fifth coincidence.*

In the Yin-Yang matrix YY_8, those six kinds of different numeric matrices are generated by means of some kinds of permutations of columns and rows of this matrix, each of which possesses its own kind of the 8-dimensional Yin-Yang-algebra.

In the genetic matrix $[C\ A; U\ G]^{(3)}$, the same six kinds of permutations of columns and rows fit the six possible kinds of permutations of positions inside the 64 triplets (1-2-3, 2-3-1, 3-1-2, 3-2-1, 2-1-3, 1-3-2), which lead to the new genomatrices with symmetric and interrelated mosaics (see Chapter 2).

The fifth coincidence will be explained additionally below.

One should note that the black cells of the genomatrix $[C\ A; U\ G]_{123}^{(3)}$ contain the black *NN*-triplets, which encode the 8 high-degeneracy amino acids, and the coding meaning of which does not depend on the letter on their third position (see Chapter 2). The set of the 8 high-degeneracy amino acids contains those amino acids, each of which is encoded by 4 triplets or more: Ala, Arg, Gly, Leu, Pro, Ser, Thr, Val. The white cells of the genomatrix $[C\ A; U\ G]_{123}^{(3)}$ contain the white *NN*-triplets, the coding meaning of which depends on the letter on their third position; these triplets encode the 12 low-degeneracy amino acids together with stop-signals: Asn, Asp, Cys, Gln, Glu, His, Ile, Lys, Met, Phe, Trp, Tyr.

The described structural coincidences of two matrices YY_8 and $[C\ A; U\ G]_{123}^{(3)}$ allow one to consider the octet algebra YY_8 as the adequate model of the structure of the genetic code. One can postulate such an algebraic model and then deduce some peculiarities of the genetic code from this model. These results of the comparison analysis give the following answer to the question of mysterious principles of the degeneracy of the vertebrate mitochondrial genetic code from the viewpoint of the proposed algebraic model. The matrix disposition of the 20 amino acids and the stop-signals is determined by algebraic principles of the matrix disposition of the *YY*-coordinates. Moreover the disposition of the 32 black triplets and the high-degeneracy amino acids in this basic dialect of the genetic code is determined by the disposition of the *YY*-coordinates with the sign "+". And the disposition of the 32 white triplets, the low-degeneracy amino acids and stop-signals is determined by the disposition of the *YY*-coordinates with the sign "-". One can recall here that the division of the set of 20 amino acids into the two sub-sets of the 8 high-degeneracy amino acids and the 12 low-degeneracy amino acids is the invariant rule of all the dialects of the genetic code practically (see Chapter 3). The described structural coincidences between both matrices do not exhaust the interconnections between the genetic code systems and the Yin-Yang matrices.

THE SIX KINDS OF THE GENETIC OCTET YIN-YANG-ALGEBRAS CONNECTED WITH PERMUTATIONS OF POSITIONS IN TRIPLETS

Now we continue to study beautiful and unexpected mathematical properties of the octet Yin-Yang-algebras.

Chapter 2 has described the 6 variants of the mosaic genetic matrices, which have corresponded to the 6 possible kinds of permutation of positions in triplets: $[C\ A; U\ G]^{(3)}_{123}$, $[C\ A; U\ G]^{(3)}_{231}$, $[C\ A; U\ G]^{(3)}_{312}$, $[C\ A; U\ G]^{(3)}_{321}$, $[C\ A; U\ G]^{(3)}_{213}$, $[C\ A; U\ G]^{(3)}_{132}$. Each of these genetic matrices can be obtained from the initial matrix $[C\ A; U\ G]^{(3)}_{123}$ by an appropriate permutation of its columns and rows. One can make the same permutations of columns and rows in the Yin-Yang-matrix YY_8, which is marked in this paragraph as $(YY_8)_{123}$. By such way the appropriate matrices $(YY_8)_{123}$, $(YY_8)_{231}$, $(YY_8)_{312}$, $(YY_8)_{321}$, $(YY_8)_{213}$,

$(YY_8)_{132}$ arise. It is quite unexpected that not only the initial matrix $(YY_8)_{123}$ (Figure 3) but each of the other five matrices $(YY_8)_{231}$, $(YY_8)_{312}$, $(YY_8)_{321}$, $(YY_8)_{213}$, $(YY_8)_{132}$ is the matrix form of presentation of its own 8-dimensional Yin-Yang-algebra (another name is bipolar algebra). For example, Figure 8 shows the Yin-Yang-matrix $(YY_8)_{231}$, which corresponds to the genomatrix $[C\ A;\ U\ G]^{(3)}_{231}$, together with its multiplication table of the basic elements.

Figure 9 demonstrates the multiplication tables for other four Yin-Yang-matrices $(YY_8)_{312}$, $(YY_8)_{132}$, $(YY_8)_{213}$, $(YY_8)_{321}$. Thereby the degeneracy of the genetic code is connected with the bunch of six genetic Yin-Yang-algebras (Petoukhov, 2008a, 2008d).

Taking into account the multiplication tables on Figure 6, 8, and 9, the proper YY_8-numbers in the linear form of their presentation have the following expressions:

$$(YY_8)^{CAUG}_{123} = x_0*\mathbf{f}_0+x_1*\mathbf{m}_1+x_2*\mathbf{f}_2+x_3*\mathbf{m}_3+x_4*\mathbf{f}_4+x_5*\mathbf{m}_5+x_6*\mathbf{f}_6+x_7*\mathbf{m}_7$$

$$(YY_8)^{CAUG}_{231} = x_0*\mathbf{f}_0+x_1*\mathbf{f}_1+x_2*\mathbf{f}_2+x_3*\mathbf{f}_3+x_4*\mathbf{m}_4+x_5*\mathbf{m}_5+x_6*\mathbf{m}_6+x_7*\mathbf{m}_7$$

Figure 8. Above: the Yin-Yang-matrix $(YY_8)_{231}$, which corresponds to the genomatrix $[C\ A;\ U\ G]^{(3)}_{231}$. Below: its multiplication table of the 8 basic elements

CCC x_0	CAC $-x_2$	ACC x_4	AAC $-x_6$	CCA x_1	CAA $-x_3$	ACA x_5	AAA $-x_7$
CUC x_2	CGC x_0	AUC $-x_6$	AGC $-x_4$	CUA x_3	CGA x_1	AUA $-x_7$	AGA $-x_5$
UCC x_4	UAC $-x_6$	GCC x_0	GAC $-x_2$	UCA x_5	UAA $-x_7$	GCA x_1	GAA $-x_3$
UUC $-x_6$	UGC $-x_4$	GUC x_2	GGC x_0	UUA $-x_7$	UGA $-x_5$	GUA x_3	GGA x_1
CCU x_0	CAU $-x_2$	ACU x_4	AAU $-x_6$	CCG x_1	CAG $-x_3$	ACG x_5	AAG $-x_7$
CUU x_2	CGU x_0	AUU $-x_6$	AGU $-x_4$	CUG x_3	CGG x_1	AUG $-x_7$	AGG $-x_5$
UCU x_4	UAU $-x_6$	GCU x_0	GAU $-x_2$	UCG x_5	UAG $-x_7$	GCG x_1	GAG $-x_3$
UUU $-x_6$	UGU $-x_4$	GUU x_2	GGU x_0	UUG $-x_7$	UGG $-x_5$	GUG x_3	GGG x_1

	f_0	f_1	f_2	f_3	m_4	m_5	m_6	m_7
f_0	f_0	f_1	f_2	f_3	m_4	m_5	m_6	m_7
f_1	f_1	$-f_0$	$-f_3$	f_2	m_5	$-m_4$	$-m_7$	m_6
f_2	f_2	f_3	f_0	f_1	m_6	m_7	m_4	m_5
f_3	f_3	$-f_2$	$-f_1$	f_0	m_7	$-m_6$	$-m_5$	m_4
m_4	f_0	f_1	f_2	f_3	m_4	m_5	m_6	m_7
m_5	f_1	$-f_0$	$-f_3$	f_2	m_5	$-m_4$	$-m_7$	m_6
m_6	f_2	f_3	f_0	f_1	m_6	m_7	m_4	m_5
m_7	f_3	$-f_2$	$-f_1$	f_0	m_7	$-m_6$	$-m_5$	m_4

$$(YY_8)^{CAUG}{}_{312} = x_0*\mathbf{f}_0+x_1*\mathbf{f}_1+x_2*\mathbf{m}_2+x_3*\mathbf{m}_3+x_4*\mathbf{f}_4+x_5*\mathbf{f}_5+x_6*\mathbf{m}_6+x_7*\mathbf{m}_7$$

$$(YY_8)^{CAUG}{}_{132} = x_0*\mathbf{f}_0+x_1*\mathbf{f}_1+x_2*\mathbf{m}_2+x_3*\mathbf{m}_3+x_4*\mathbf{f}_4+x_5*\mathbf{f}_5+x_6*\mathbf{m}_6+x_7*\mathbf{m}_7$$

Figure 9. The multiplication tables of the basic elements of the octet Yin-Yang-algebras $(YY_8)_{312}$, $(YY_8)_{132}$, $(YY_8)_{213}$, $(YY_8)_{321}$

	f_0	f_1	m_2	m_3	f_4	f_5	m_6	m_7
f_0	f_0	f_1	m_2	m_3	f_4	f_5	m_6	m_7
f_1	f_1	f_0	m_3	m_2	f_5	f_4	m_7	m_6
m_2	f_0	f_1	m_2	m_3	f_4	f_5	m_6	m_7
m_3	f_1	f_0	m_3	m_2	f_5	f_4	m_7	m_6
f_4	f_4	$-f_5$	m_6	$-m_7$	$-f_0$	f_1	$-m_2$	m_3
f_5	f_5	$-f_4$	m_7	$-m_6$	$-f_1$	f_0	$-m_3$	m_2
m_6	f_4	$-f_5$	m_6	$-m_7$	$-f_0$	f_1	$-m_2$	m_3
m_7	f_5	$-f_4$	m_7	$-m_6$	$-f_1$	f_0	$-m_3$	m_2

	f_0	f_1	m_2	m_3	f_4	f_5	m_6	m_7
f_0	f_0	f_1	m_2	m_3	f_4	f_5	m_6	m_7
f_1	f_1	$-f_0$	m_3	$-m_2$	$-f_5$	f_4	$-m_7$	m_6
m_2	f_0	f_1	m_2	m_3	f_4	f_5	m6	m_7
m_3	f_1	$-f_0$	m_3	$-m_2$	$-f_5$	f_4	$-m_7$	m_6
f_4	f_4	f_5	m_6	m_7	f_0	f_1	m_2	m_3
f_5	f_5	$-f_4$	m_7	$-m_6$	$-f_1$	f_0	$-m_3$	m_2
m_6	f_4	f_5	m_6	m_7	f_0	f_1	m_2	m_3
m_7	f_5	$-f_4$	m_7	$-m_6$	$-f_1$	f_0	$-m_3$	m_2

	f_0	m_1	f_2	m_3	f_4	m_5	f_6	m_7
f_0	f_0	m_1	f_2	m_3	f_4	m_5	f_6	m_7
m_1	f_0	m_1	f_2	m_3	f_4	m_5	f_6	m_7
f_2	f_2	m_3	f_0	m_1	f_6	m_7	f_4	m_5
m_3	f_2	m_3	f_0	m_1	f_6	m_7	f_4	m_5
m_5	f_4	m_5	$-f_6$	$-m_7$	$-f_0$	$-m_1$	f_2	m_3
m_5	f_4	m_5	$-f_6$	$-m_7$	$-f_0$	$-m_1$	f_2	m_3
f_6	f_6	m_7	$-f_4$	$-m_5$	$-f_2$	$-m_3$	f_0	m_1
m_7	f_6	m_7	$-f_4$	$-m_5$	$-f_2$	$-m_3$	f_0	m_1

	f_0	f_1	f_2	f_3	m_4	m_5	m_6	m_7
f_0	f_0	f_1	f_2	f_3	m_4	m_5	m_6	m_7
f_1	f_1	f_0	f_3	f_2	m_5	m_4	m_7	m_6
f_2	f_2	$-f_3$	$-f_0$	f_1	m_6	$-m_7$	$-m_4$	m_5
f_3	f_3	$-f_2$	$-f_1$	f_0	m_7	$-m_6$	$-m_5$	m_4
m_4	f_0	f_1	f_2	f_3	m_4	m_5	m_6	m_7
m_5	f_1	f_0	f_3	f_2	m_5	m_4	m_7	m_6
m_6	f_2	$-f_3$	$-f_0$	f_1	m_6	$-m_7$	$-m_4$	m_5
m_7	f_3	$-f_2$	$-f_1$	f_0	m_7	$-m_6$	$-m_5$	m_4

$(YY_8)^{CAUG}{}_{213} = x_0*\mathbf{f_0}+x_1*\mathbf{m_1}+x_2*\mathbf{f_2}+x_3*\mathbf{m_3}+x_4*\mathbf{f_4}+x_5*\mathbf{m_5}+x_6*\mathbf{f_6}+x_7*\mathbf{m_7}$

$(YY_8)^{CAUG}{}_{321} = x_0*\mathbf{f_0}+x_1*\mathbf{f_1}+x_2*\mathbf{f_2}+x_3*\mathbf{f_3}+x_4*\mathbf{m_4}+x_5*\mathbf{m_5}+x_6*\mathbf{m_6}+x_7*\mathbf{m_7}$ (2)

All these Yin-Yang matrices have secret connections with Hadamard matrices: when all their coordinates are equal to the real unit 1 ($x_0 = x_1 = \ldots = x_7 = 1$) and when the signs of components of the matrices are changed by means of the U-algorithm described in Chapter 6, then all these Yin-Yang octet matrices become the Hadamard matrices. In necessary cases biological computers of organisms can transform these Yin-Yang matrices into the Hadamard matrices to operate with systems of orthogonal vectors. One can add that for the case, when all their coordinates are equal to 1 ($x_0 = x_1 = \ldots = x_7 = 1$), all these six Yin-Yang matrices $(YY_8)^{CAUG}{}_{123}$, $(YY_8)^{CAUG}{}_{231}$, …, $(YY_8)^{CAUG}{}_{321}$ possess the property of their tetra-reproduction which is described below and which evokes the tetra-reproduction of gametal cells in the biological process of meiosis.

Two facts can be mentioned as well. The complementary triplets (codon and anti-codon) play an essential role in the genetic code systems. One can replace each codon by its anti-codon in the genomatrices $[C\ A;\ U\ G]_{123}{}^{(3)}$, $[C\ A;\ U\ G]_{231}{}^{(3)}$, $[C\ A;\ U\ G]_{312}{}^{(3)}$, $[C\ A;\ U\ G]_{132}{}^{(3)}$, $[C\ A;\ U\ G]_{213}{}^{(3)}$, $[C\ A;\ U\ G]_{321}{}^{(3)}$. The new six genomatrices appear in this case. Have they any connection with Yin-Yang algebras? This question has the positive answer. The multiplication tables for the basic elements of Yin-Yang matrices, connected with these new genomatrices, are identical to the multiplication tables for the initial genomatrices. In other words, the "complementary" transformations of the genomatrices $[C\ A;\ U\ G]_{123}{}^{(3)}$, $[C\ A;\ U\ G]_{231}{}^{(3)}$, $[C\ A;\ U\ G]_{312}{}^{(3)}$, $[C\ A;\ U\ G]_{132}{}^{(3)}$, $[C\ A;\ U\ G]_{213}{}^{(3)}$, $[C\ A;\ U\ G]_{321}{}^{(3)}$ change the matrix forms of presentation of the initial YY_8-numbers only but do not change the Yin-Yang algebras of the genomatrices. But if we consider the transposed matrices, which are generated from the matrices $(YY_8)^{CAUG}{}_{123}$, $(YY_8)^{CAUG}{}_{231}$, etc., they correspond to new octet Yin-Yang-algebras.

THE GENETIC YIN-YANG OCTETS AS "DOUBLE GENOQUATERNIONS"

Taking into account the described fact of existence of many octet Yin-Yang-algebras and correspondingly many kinds of octet genonumbers, we shall name any numbers with 8 items $x_0*\mathbf{i_0}+x_1*\mathbf{i_1}+\ldots x_7*\mathbf{i_7}$ by the name "octets" independently of multiplication tables of their basic elements. We shall name numbers with 4 items $x_0*\mathbf{i_0}+x_1*\mathbf{i_1}+x_2*\mathbf{i_2}+x_3*\mathbf{i_3}$ by the name "quaternions" independently of multiplication tables of their basic elements (quaternions by Hamilton are the special case of quaternions). Let us analyze the expression (1) of the genetic octet YY_8 together with its multiplication table (Figure 6). If all male coordinates are equal to zero ($x_1 = x_3 = x_5 = x_7 = 0$), this genetic octet YY_8 becomes the genetic Yin-quaternion G_f (or the Yin-genoquaternion):

$$G_f = x_0*\mathbf{f_0} + x_2*\mathbf{f_2} + x_4*\mathbf{f_4} + x_6*\mathbf{f_6} \qquad (3)$$

The proper multiplication table for this quaternion is shown on Figure 10 (on the left side). This table is generated from the multiplication table for the algebra YY_8 (Figure 5) by nullification (or by excision) of the columns and rows, which have the male basic elements. Taking into account that the basic element $\mathbf{f_0}$ possesses the multiplication properties of the real unit relative to all female basic elements, one can rewrite the expression (3) in the following form:

Figure 10. The multiplication tables for the genetic Yin-quaternions G_f (on the left side)

	$\mathbf{f_0}$	$\mathbf{f_2}$	$\mathbf{f_4}$	$\mathbf{f_6}$
$\mathbf{f_0}$	$\mathbf{f_0}$	$\mathbf{f_2}$	$\mathbf{f_4}$	$\mathbf{f_6}$
$\mathbf{f_2}$	$\mathbf{f_2}$	$\mathbf{-f_0}$	$\mathbf{-f_6}$	$\mathbf{f_4}$
$\mathbf{f_4}$	$\mathbf{f_4}$	$\mathbf{f_6}$	$\mathbf{f_0}$	$\mathbf{f_2}$
$\mathbf{f_6}$	$\mathbf{f_6}$	$\mathbf{- f_4}$	$\mathbf{- f_2}$	$\mathbf{f_0}$

	$\mathbf{m_1}$	$\mathbf{m_3}$	$\mathbf{m_5}$	$\mathbf{m_7}$
$\mathbf{m_1}$	$\mathbf{m_1}$	$\mathbf{m_3}$	$\mathbf{m_5}$	$\mathbf{m_7}$
$\mathbf{m_3}$	$\mathbf{m_3}$	$\mathbf{-m_1}$	$\mathbf{-m_7}$	$\mathbf{m_5}$
$\mathbf{m_5}$	$\mathbf{m_5}$	$\mathbf{m_7}$	$\mathbf{m_1}$	$\mathbf{m_3}$
$\mathbf{m_7}$	$\mathbf{m_7}$	$\mathbf{- m_5}$	$\mathbf{- m_3}$	$\mathbf{m_1}$

$$G_f = x_0*\mathbf{1} + x_2*\mathbf{f_2} + x_4*\mathbf{f_4} + x_6*\mathbf{f_6} \tag{4}$$

If all female coordinates are equal to zero ($x_0 = x_2 = x_4 = x_6 = 0$), this genetic octet YY_8 becomes the genetic Yang-quaternion G_m (or the Yang-genoquaternion):

$$G_m = x_1*\mathbf{m_1} + x_3*\mathbf{m_3} + x_5*\mathbf{m_5} + x_7*\mathbf{m_7} \tag{5}$$

The appropriate multiplication table for this quaternion is shown on Figure 4 (on the right side). Taking into account that the basic element $\mathbf{m_1}$ possesses the multiplication properties of the real unit relative to all male basic elements, one can rewrite the expression (5) in the following form:

$$G_m = x_1*\mathbf{1} + x_3*\mathbf{m_3} + x_5*\mathbf{m_5} + x_7*\mathbf{m_7} \tag{6}$$

and for the genetic Yang-quaternions G_m (on the right side).

The quaternions G_f and G_m are similar to each other by the structure of their multiplication tables, which differ from the multiplication table of quaternions by Hamilton (Figure 2). The quaternions G_f and G_m can be expressed in the following general form:

$$G = y_0*\mathbf{1} + y_1*\mathbf{i_1} + y_2*\mathbf{i_2} + y_3*\mathbf{i_3} \tag{7}$$

The system of quaternions by Hamilton has many useful properties and applications in mathematics and physics. The system of genoquaternions possesses many analogical properties, which permits one to think about its useful applications in bioinformatics, mathematical biology, etc. For example, the numeric system of genoquaternions is the system with the operation of division and it possesses the associative property, the notions of the "norm of genoquaternion" and of the "inverse genoquaternion", etc. Figure 11 demonstrates some analogies between both types of quaternions.

In view of these materials, one can name the genetic octet $x_0*\mathbf{i_0}+x_1*\mathbf{i_1}+\ldots x_7*\mathbf{i_7}$ (with its individual multiplication table on Figure 6) as "the double genoquaternion" conditionally. This name generates heuristic associations with the famous name "the double spiral" of DNA.

Figure 11. The comparison of some properties between the systems of quaternions by Hamilton (on the left side) and of genoquaternions (on the right side)

Quaternions by Hamilton	Genoquaternions
$q = x_0*\mathbf{1} + x_1*\mathbf{i}_1 + x_2*\mathbf{i}_2 + x_3*\mathbf{i}_3$	$G = x_0*\mathbf{1} + x_1*\mathbf{i}_1 + x_2*\mathbf{i}_2 + x_3*\mathbf{i}_3$
$(q_1*q_2)*q3 = q_1*(q_2*q3)$	$(G_1*G_2)*G_3 = G_1*(G_2*G_3)$
Conjugate quaternion $q_s = x_0*\mathbf{1} - x_1*\mathbf{i}_1 - x_2*\mathbf{i}_2 - x_3*\mathbf{i}_3$	Conjugate genoquaternion $G_s = x_0*\mathbf{1} - x_1*\mathbf{i}_1 - x_2*\mathbf{i}_2 - x_3*\mathbf{i}_3$
To the norm of quaternions: $\vert q\vert^2 = q*q_s = q_s*q = x_0^2 + x_1^2 + x_2^2 + x_3^2$	To the norm of genoquaternions: $\vert G\vert^2 = G*G_s = G_s*G = x_0^2 + x_1^2 - x_2^2 - x_3^2$
The inverse quaternion exists: $q^{-1} = q_s/\vert q\vert^2$	The inverse genoquaternion exists: $G^{-1} = G_s/\vert G\vert^2$
$(q_1 + q_2)_s = (q_1)_s + (q_2)_s$	$(G_1 + G_2)_s = (G_1)_s + (G_2)_s$
$(q_1*q_2)_s = (q_2)_s * (q_1)_s$	$(G_1*G_2)_s = (G_2)_s * (G_1)_s$

THE COMPARISON BETWEEN THE CLASSICAL VECTOR CALCULATION AND THE GENOVECTOR CALCULATION

Let us recall about one of the famous applications of quaternions by Hamilton, which concerns the beautiful connection between these quaternions $q = x_0*\mathbf{1}+x_1*\mathbf{i}_1+x_2*\mathbf{i}_2+ x_3*\mathbf{i}_3$ and the classical vector calculation developed by J. Gibbs. One can take two vectors $\underline{a}$ and $\underline{b}$, which belong to the plane $(\mathbf{i}_v, \mathbf{i}_w)$, where $v < w$, $v = 1, 2$; $w = 2, 3$; $\underline{a} = a_1*\mathrm{i}_v + a_2*\mathbf{i}_w$, $\underline{b} = b_1*\mathbf{i}_1 + b_2*\mathbf{i}_2$. These vectors can be presented in the following usual form:

$$\underline{a} = |\underline{a}|*(\mathbf{i}_v*cos\ \alpha + \mathbf{i}_w*sin\ \alpha),\ \underline{b} = |\underline{b}|*(\mathbf{i}_v*cos\ \beta + \mathbf{i}_w*sin\ \beta), \quad (8)$$

where α and β are appropriate angles between these vectors and the axises $\mathbf{i}_v$ and $\mathbf{i}_w$ in the orthogonal system of the basic vectors $(\mathbf{i}_1, \mathbf{i}_2, \mathbf{i}_3)$. If we multiply together these vectors as Hamilton's quaternions in accordance with the multiplication table on Figure 2, the following equation arises:

$$\underline{a}*\underline{b} = - |\underline{a}|*|\underline{b}|*cos(\alpha - \beta) + |\underline{a}|*|\underline{b}|*sin(\alpha - \beta)*\mathbf{i}_{vw}, \quad (9)$$

where $\mathbf{i}_{vw}$ is the third basic vector, which is orthogonal to the basic vectors $\mathbf{i}_v$ and $\mathbf{i}_w$.

The equation (9) shows that the quaternion multiplication of two vectors contains two parts: the scalar part and the vector part. The scalar part $|\underline{a}|*|\underline{b}|*cos(\alpha - \beta)$ is famous under the name "the scalar product" and the vector part $\underline{a}|*|\underline{b}|*sin(\alpha - \beta)*\mathbf{i}_3$ is famous under the name "the vector product" in the classical vector calculation. This vector calculation is utilized widely in mechanics to describe movements of hard bodies in our physical space, etc. Mechanics of bodies in the usual physical space fits this vector calculation. From the viewpoint of this vector calculation, space is isotropic because the expression (5) with its scalar and vector parts is the same for each pair of vectors, which belong to the planes $(\mathbf{i}_1, \mathbf{i}_2)$, $(\mathbf{i}_1, \mathbf{i}_3)$, $(\mathbf{i}_2, \mathbf{i}_3)$, and the scalar products and the vectors product possess the analogical forms for all three cases of the planes.

But what results arise, if we multiply together the vectors $\underline{a}$ and $\underline{b}$ (8) as genoquaternions in accordance with their multiplication table (Figure 2)? Let us consider the following three cases, each of which contains a scalar part and a vector part in the final expressions (10), (11), (12), but in different forms.

The case 1. The vectors $\underline{a}$ and $\underline{b}$ belong to the plane (i_1, i_2). They can be expressed in the following form: $\underline{a} = |\underline{a}|*(i_1*cos\ \alpha + i_2*sin\ \alpha)$, $\underline{b} = |\underline{b}|*(i_1*cos\ \beta + i_2*sin\ \beta)$. If we multiply together these vectors as genoquaternions (Figure 14, in the right side), the result arises:

$$a*b = |a|*|b|*(i_1*cos\ \alpha+i_2*sin\ \alpha)*(i_1*cos\ \beta+i_2*sin\ \beta) =$$

$$= -|a|*|b|*cos(\alpha+\beta)+|a|*|b|*sin(\alpha-\beta)*i_3 \qquad (10)$$

The equation (10) of the genovector calculation differs from the equation (9) of the classical vector calculation in the scalar part only (by the value cos(α+β)).

The case 2. The vectors $\underline{a}$ and $\underline{b}$ belong to the plane (i_1, i_3): $\underline{a} = |\underline{a}|*(i_1*cos\ \alpha + i_3*sin\ \alpha)$, $\underline{b} = |\underline{b}|*(i_1*cos\ \beta + i_3*sin\ \beta)$. The product of these two vectors as genoquaternions gives the following result:

$$a*b = |a|*|b|*(i_1*cos\ \alpha+i_3*sin\ \alpha)*(i_1*cos\ \beta+i_3*sin\ \beta)$$

$$= -|a|*|b|*cos(\alpha+\beta)-|a|*|b|*sin(\alpha-\beta)*i_2 \qquad (11)$$

The equation (11) of the genovector calculation differs from the classical equation (9) in the scalar part (by the value *cos*(α+β)) and in the vector part (by the opposite sign).

The case 3. The vectors $\underline{a}$ and $\underline{b}$ belong to the plane (i_2, i_3): $\underline{a} = |\underline{a}|*(i_2*cos\ \alpha + i_3*sin\ \alpha)$, $\underline{b} = |\underline{b}|*(i_2*cos\ \beta + i_3*sin\ \beta)$. The product of these two vectors as genoquaternions gives

$$a*b=|a|*|b|*(i_2*cos\ \alpha+i_3*sin\ \alpha)*(i_2*cos\ \beta+i_3*sin\ \beta) =$$

$$= +|a|*|b|*cos(\alpha-\beta)-|a|*|b|*sin(\alpha-\beta)*i_1 \qquad (12)$$

The equation (12) of the genovector calculation differs from the classical equation (9) by the opposite sign in the scalar part and in the vector part.

We name vectors, which are considered as qenoquaternions (with applications of the rules of genoquaternion operations to them), as "genovectors". It is obvious that the genovector calculation fits the case of an anisotropic space because the results of multiplication of arbitrary vectors $\underline{a}$ and $\underline{b}$ depend on the plane, to which these vectors belong. The spaces of biological phenomena of morphogenesis, growth, etc. have anisotropic characters as well. Since the genovector calculation was developed from the genetic code features, it seems that this calculation (and its generalization for the system of Yin-Yang genooctets) can be adequate to model anisotropic processes in biological spaces including processes of bioinformatics and of biological morphogenesis on different levels of each united organism.

Many mathematical formalisms and notions, which were convinced in the theory of quaternions by Hamilton and which were utilized in many scientific branches, have their analogies in the theory of genoquaternions (Petoukhov, 2008a, 2008d) and in the theory of genetic tetrions described below.

THE PARAMETRIC REDUCTION OF THE GENETIC OCTET YIN-YANG ALGEBRA TO THE 4-DIMENSIONAL ALGEBRA OF TETRIONS

This paragraph shows the special case of the parametric reduction of the genetic octet Yin-Yang-algebra to one of 4-dimensional algebras. This case relates to alphabetic peculiarities of the genetic code.

The previous paragraphs have considered the numeric system $YY_8 = x_0*\mathbf{f}_0+x_1*\mathbf{m}_1+x_2*\mathbf{f}_2+x_3*\mathbf{m}_3+x_4*\mathbf{f}_4+x_5*\mathbf{m}_5+x_6*\mathbf{f}_6+x_7*\mathbf{m}_7$ (equation (1)) with the 8 arbitrary coordinates $x_0, x_1, \ldots, x_7$. But in accordance with the matrix on Figure 4 all these 8 coordinates are expressed by means of 4 parameters α, β, γ, δ:

$$x_0 = \alpha\alpha\gamma;\ x_1 = \alpha\alpha\delta;\ x_2 = \alpha\beta\gamma;\ x_3 = \alpha\beta\delta;\ x_4 = \beta\alpha\gamma;\ x_5 = \beta\alpha\delta;\ x_6 = \beta\beta\gamma;\ x_7 = \beta\beta\delta \text{ equation} \quad (13)$$

Hence these 8 coordinates are not independent of each other and they are interconnected by the following expressions:

$$x_1 = x_0*\delta/\gamma;\ x_3 = x_2*\delta/\gamma;\ x_5 = x_4*\delta/\gamma;\ x_7 = x_6*\delta/\gamma \text{ equations} \quad (14)$$

One can see from the expression (13) that the coordinates belong to the female (male) type if they have the symbol γ (δ correspondingly) on their third position. The expressions (14) show the existence of the pairs of "complementary" male and female coordinates, which differ by the coefficient δ/γ only: x_1 and x_0; x_3 and x_2; x_5 and x_4; x_7 and x_6. These interconnections of coordinates lead to the particular form of the octet number YY_8, where the female coordinates x_0, x_2, x_4, x_6 exist only (another possible form has the male coordinates x_1, x_3, x_5, x_7 only):

$$T = x_0*(\mathbf{f}_0+\delta/\gamma*\mathbf{m}_1) + x_2*(\mathbf{f}_2+\delta/\gamma*\mathbf{m}_3) + x_4*(\mathbf{f}_4+\delta/\gamma*\mathbf{m}_5) + x_6*(\mathbf{f}_6+\delta/\gamma*\mathbf{m}_7) =$$

$$= \alpha\alpha\gamma*(\mathbf{f}_0+\delta/\gamma*\mathbf{m}_1) + \alpha\beta\gamma*(\mathbf{f}_2+\delta/\gamma*\mathbf{m}_3) + \beta\alpha\gamma*(\mathbf{f}_4+\delta/\gamma*\mathbf{m}_5) + \beta\beta\gamma*(\mathbf{f}_6+\delta/\gamma*\mathbf{m}_7) =$$

Each of these four matrices $(\mathbf{f}_0+\delta/\gamma*\mathbf{m}_1)$, $(\mathbf{f}_2+\delta/\gamma*\mathbf{m}_3)$, $(\mathbf{f}_4+\delta/\gamma*\mathbf{m}_5)$, $(\mathbf{f}_6+\delta/\gamma*\mathbf{m}_7)$ on the Figure 12 is constructed by means of the fusion of appropriate male and female matrices of the complementary pairs into united object. It is interesting that these four matrices form their own closed set relative to multiplication. Figure 13 shows the table of multiplication of these matrices.

In view of these facts the expression T on Figure 12 with all possible values of real numbers α, β, γ, δ represents the new system of 4-dimensional numbers, which are named "genetic tetrions" (or genotetrions) to distinguish them from 4-dimensional hypercomplex numbers called "quaternions" traditionally (including genoquaternions described above). If quaternions and other hypercomplex numbers have the real unit among their basic elements, tetrions have not the real unit among their basic elements at all. The first basic element $(\mathbf{f}_0+\delta/\gamma*\mathbf{m}_1)$ of the tetrions (Figure 12) is the matrix presentation of the real number (1+δ/γ). This basic element possesses the commutative property relative to all these basic elements. The first item $x_0*(\mathbf{f}_0+\delta/\gamma*\mathbf{m}_1)$ is considered as the scalar part of tetrions, and other three items $x_2*(\mathbf{f}_2+\delta/\gamma*\mathbf{m}_3) + x_4*(\mathbf{f}_4+\delta/\gamma*\mathbf{m}_5) + x_6*(\mathbf{f}_6+\delta/\gamma*\mathbf{m}_7)$ are considered as the vector part of tetrions.

The square of any basic element of the tetrions T is equal to $(1+\delta/\gamma)*(\mathbf{f}_0+\delta/\gamma*\mathbf{m}_1)$ with the sign "+" or "-". This peculiarity is demonstrated on Figure 13 in the cells (marked by bold borders) along the main diagonal. So instead of the real unit, tetrions have the real number (1+*v*), where "*v*" is the real number, which is equal to δ/γ in the case of the genetic tetrions *T*. One can consider such tetrions as the special

Figure 12. The presentation of the matrix T as the sum of the superposition of the matrices $(f_0+\delta/\gamma*m_1)$, $(f_2+\delta/\gamma*m_3)$, $(f_4+\delta/\gamma*m_5)$, $(f_6+\delta/\gamma*m_7)$

$$T = x_0*(\mathbf{f_0}+\delta/\gamma*\mathbf{m_1}) + x_2*(\mathbf{f_2}+\delta/\gamma*\mathbf{m_3}) + x_4*(\mathbf{f_4}+\delta/\gamma*\mathbf{m_5}) + x_6*(\mathbf{f_6}+\delta/\gamma*\mathbf{m_7}) = \alpha\alpha\gamma*(\mathbf{f_0}+\delta/\gamma*\mathbf{m_1}) + \alpha\beta\gamma*(\mathbf{f_2}+\delta/\gamma*\mathbf{m_3}) + \beta\alpha\gamma*(\mathbf{f_4}+\delta/\gamma*\mathbf{m_5}) + \beta\beta\gamma*(\mathbf{f_6}+\delta/\gamma*\mathbf{m_7}) =$$

$$= \alpha\alpha\gamma * \begin{vmatrix} 1 & \delta/\gamma & 0 & 0 & 0 & 0 & 0 & 0 \\ 1 & \delta/\gamma & 0 & 0 & 0 & 0 & 0 & 0 \\ 0 & 0 & 1 & \delta/\gamma & 0 & 0 & 0 & 0 \\ 0 & 0 & 1 & \delta/\gamma & 0 & 0 & 0 & 0 \\ 0 & 0 & 0 & 0 & 1 & \delta/\gamma & 0 & 0 \\ 0 & 0 & 0 & 0 & 1 & \delta/\gamma & 0 & 0 \\ 0 & 0 & 0 & 0 & 0 & 0 & 1 & \delta/\gamma \\ 0 & 0 & 0 & 0 & 0 & 0 & 1 & \delta/\gamma \end{vmatrix}$$

$$+ \alpha\beta\gamma * \begin{vmatrix} 0 & 0 & 1 & \delta/\gamma & 0 & 0 & 0 & 0 \\ 0 & 0 & 1 & \delta/\gamma & 0 & 0 & 0 & 0 \\ -1 & -\delta/\gamma & 0 & 0 & 0 & 0 & 0 & 0 \\ -1 & -\delta/\gamma & 0 & 0 & 0 & 0 & 0 & 0 \\ 0 & 0 & 0 & 0 & 0 & 0 & 1 & \delta/\gamma \\ 0 & 0 & 0 & 0 & 0 & 0 & 1 & \delta/\gamma \\ 0 & 0 & 0 & 0 & -1 & -\delta/\gamma & 0 & 0 \\ 0 & 0 & 0 & 0 & -1 & -\delta/\gamma & 0 & 0 \end{vmatrix}$$

$$+ \beta\alpha\gamma * \begin{vmatrix} 0 & 0 & 0 & 0 & -1 & -\delta/\gamma & 0 & 0 \\ 0 & 0 & 0 & 0 & -1 & -\delta/\gamma & 0 & 0 \\ 0 & 0 & 0 & 0 & 0 & 0 & 1 & \delta/\gamma \\ 0 & 0 & 0 & 0 & 0 & 0 & 1 & \delta/\gamma \\ -1 & -\delta/\gamma & 0 & 0 & 0 & 0 & 0 & 0 \\ -1 & -\delta/\gamma & 0 & 0 & 0 & 0 & 0 & 0 \\ 0 & 0 & 1 & \delta/\gamma & 0 & 0 & 0 & 0 \\ 0 & 0 & 1 & \delta/\gamma & 0 & 0 & 0 & 0 \end{vmatrix} +$$

$$+ \beta\beta\gamma * \begin{vmatrix} 0 & 0 & 0 & 0 & 0 & 0 & -1 & -\delta/\gamma \\ 0 & 0 & 0 & 0 & 0 & 0 & -1 & -\delta/\gamma \\ 0 & 0 & 0 & 0 & -1 & -\delta/\gamma & 0 & 0 \\ 0 & 0 & 0 & 0 & -1 & -\delta/\gamma & 0 & 0 \\ 0 & 0 & -1 & -\delta/\gamma & 0 & 0 & 0 & 0 \\ 0 & 0 & -1 & -\delta/\gamma & 0 & 0 & 0 & 0 \\ -1 & -\delta/\gamma & 0 & 0 & 0 & 0 & 0 & 0 \\ -1 & -\delta/\gamma & 0 & 0 & 0 & 0 & 0 & 0 \end{vmatrix}$$

generalization of appropriate hypercomplex numbers by means of utilizing any kind of real numbers in the role of their first basic element instead of utilizing the real unit in this role in the case of traditional hypercomplex numbers.

The system of tetrions T (Figure 12) possesses the commutative and associative properties. It is the system with operation of division from the left side and from the right side (by analogy with the division in the system of quaternions). By definition the conjugate tetrion T_S is presented by the expression:

Figure 13. The table of multiplication of the matrices $(\mathbf{f_0}+\delta/\gamma*\mathbf{m_1})$, $(\mathbf{f_2}+\delta/\gamma*\mathbf{m_3})$, $(\mathbf{f_4}+\delta/\gamma*\mathbf{m_5})$, $(\mathbf{f_6}+\delta/\gamma*\mathbf{m_7})$, *which are basic elements of the genetic tetrions*

	$f_0+\delta/\gamma*m_1$	$f_2+\delta/\gamma*m_3$	$f_4+\delta/\gamma*m_5$	$f_6+\delta/\gamma*m_7$
$f_0+\delta/\gamma*m_1$	$(1+\delta/\gamma)*(f_0+\delta/\gamma*m_1)$	$(1+\delta/\gamma)*(f_2+\delta/\gamma*m_3)$	$(1+\delta/\gamma)*(f_4+\delta/\gamma*m_5)$	$(1+\delta/\gamma)*(f_6+\delta/\gamma*m_7)$
$f_2+\delta/\gamma*m_3$	$(1+\delta/\gamma)*(f_2+\delta/\gamma*m_3)$	$-(1+\delta/\gamma)*(f_0+\delta/\gamma*m_1)$	$-(1+\delta/\gamma)*(f_6+\delta/\gamma*m_7)$	$(1+\delta/\gamma)*(f_4+\delta/\gamma*m_5)$
$f_4+\delta/\gamma*m_5$	$(1+\delta/\gamma)*(f_4+\delta/\gamma*m_5)$	$(1+\delta/\gamma)*(f_6+\delta/\gamma*m_7)$	$(1+\delta/\gamma)*(f_0+\delta/\gamma*m_1)$	$(1+\delta/\gamma)*(f_2+\delta/\gamma*m_3)$
$f_6+\delta/\gamma*m_7$	$(1+\delta/\gamma)*(f_6+\delta/\gamma*m_7)$	$-(1+\delta/\gamma)*(f_4+\delta/\gamma*m_5)$	$-(1+\delta/\gamma)*(f_2+\delta/\gamma*m_3)$	$(1+\delta/\gamma)*(f_0+\delta/\gamma*m_1)$

$$T_S = x_0*(\mathbf{f_0}+\delta/\gamma*\mathbf{m_1}) - x_2*(\mathbf{f_2}+\delta/\gamma*\mathbf{m_3}) - x_4*(\mathbf{f_4}+\delta/\gamma*\mathbf{m_5}) - x_6*(\mathbf{f_6}+\delta/\gamma*\mathbf{m_7}) \quad (15)$$

The following expressions for two tetrions T_1 and T_2 hold true:

$$(T_1 + T_2)_S = (T_1)_S + (T_2)_S\,;\; (T_1*T_2)_S = (T_2)_S * (T_1)_S \quad (16)$$

The square of the module of tetrions is listed below.

$$|T|^2 = T*T_S = T_S*T = (1+\delta/\gamma)*(x_0^2+x_2^2-x_4^2-x_6^2)$$

$$= (1+\delta/\gamma)*[(\alpha\alpha\gamma)^2+(\alpha\beta\gamma)^2-(\beta\alpha\gamma)^2-(\beta\beta\gamma)^2] \quad (17)$$

The inverse genotetrion exists: $T^{-1} = T_S/|T|^2$. It allows defining the operation of division traditionally by means of multiplication by the inverse genotetrion. One can see that these properties of the genetic tetrions are similar to the properties of genoquaternions considered above (Figure 11) and that the genotetrion's multiplication table and genoquaternion's multiplication table are similar to each other by the disposition of the signs "+" and "-" (Figure 10 and Figure 13).

The system of genetic tetrions leads to a special kind of vector calculation. By analogy with the expressions (10-12) for genoquaternions, one can arrive at the similar expressions (18-20) of vector calculation for genotetrions. Let us analyze the multiplication of two vectors $\underline{\boldsymbol{a}}$ and $\underline{\boldsymbol{b}}$ (equation 8) as tetrions in accordance with the multiplication table (Figure 13) in the same three cases which were described for the expressions (10-12). In the result we arrive at the following equations (18-20).

The case 1. The vectors $\underline{\boldsymbol{a}}$ and $\underline{\boldsymbol{b}}$ belong to the plane of the basic vectors $(\mathbf{f_2}+\delta/\gamma*\mathbf{m_3}, \mathbf{f_4}+\delta/\gamma*\mathbf{m_5})$. Then

$$\boldsymbol{a}*\boldsymbol{b} = -|\boldsymbol{a}|*|\boldsymbol{b}|*(1+\delta/\gamma)^2*cos(\alpha+\beta) + |\boldsymbol{a}|*|\boldsymbol{b}|*sin(\alpha-\beta)*(1+\delta/\gamma)*(\mathbf{f_6}+\delta/\gamma*\mathbf{m_7}). \quad (18)$$

The case 2. The vectors $\underline{\boldsymbol{a}}$ and $\underline{\boldsymbol{b}}$ belong to the plane $(\mathbf{f_2}+\delta/\gamma*\mathbf{m_3}, \mathbf{f_6}+\delta/\gamma*\mathbf{m_7})$. Then

$$\boldsymbol{a}*\boldsymbol{b} = -|\boldsymbol{a}|*|\boldsymbol{b}|*(1+\delta/\gamma)^2*cos(\alpha+\beta) - |\boldsymbol{a}|*|\boldsymbol{b}|*sin(\alpha-\beta)*(1+\delta/\gamma)*(\mathbf{f_4}+\delta/\gamma*\mathbf{m_5}). \quad (19)$$

The case 3. The vectors $\underline{a}$ and $\underline{b}$ belong to the plane ($\mathbf{f}_2+\delta/\gamma*\mathbf{m}_3$, $\mathbf{f}_6+\delta/\gamma*\mathbf{m}_7$). Then

$$a*b = +|a|*|b|*(1+\delta/\gamma)^2*cos(\alpha-\beta) - |a|*|b|*sin(\alpha-\beta)*(1+\delta/\gamma)*(\mathbf{f}_2+\delta/\gamma*\mathbf{m}_3). \quad (20)$$

It is obvious that the vector calculation of genetic tetrions fits the case of an anisotropic space because the results of multiplication of arbitrary vectors $\underline{a}$ and $\underline{b}$ depend on the plane, to which these vectors belong. Can the scalar and vector parts of genetic tetrions be considered correspondingly as the time coordinate and the space coordinates in the theory of the genetic space-time? This and other interesting questions are under investigation now.

In the described approach, the genetic code is presented as the replica of the tetrions in their matrix form. It permits one to consider the algebra of genetic tetrions as a candidate for the role of the mathematical system of genetic preceding code (the "pre-code" or the more fundamental code) relative to the genetic code. Really, from a traditional viewpoint, a code is an aggregate of symbols which corresponds to elements of information. In our algebraic case, the discussion is about the matrix system, the symbols of which can be confronted with triplets and with other elements of the genetic code. In other words, the genetic code can be encoded itself by symbols of elements of the tetrion numerical system. Such tetrion pre-code has its own pre-code alphabet, which consists of the four letters α, β, γ, δ in contrast to the usual genetic alphabet A, C, G, U/T. This set of the letters α, β, γ, δ, which present the molecular parameters of the letters of the genetic alphabet, can be named as the alphabet of genetic algebras or as the algebraic alphabet of the genetic code as well. Revealing such a tetrion pre-code as a new numeric system can help with sorting, ordering and a deeper understanding of genetic informatics. It can also help to develop new effective methods of processing and transfer of information in many applied problems. Mathematical features of such pre-code can explain evolutionary features of the genetic code. One should emphasize that not only the (8x8)-matrix YY_8 (Figure 3 and Figure 4), but each of its (4x4)-quadrants and each of its (2x2)-subquadrants defines its own special algebras, if we take into account the coordinates $x_0, x_1, \ldots, x_7$ and the algebraic alphabet α, β, γ, δ. It means that the genetic code is an ensemble of special multidimensional algebras from such a matrix viewpoint.

ABOUT GENETIC MECHANICS AND THE IDEA BY PYTHAGORAS

In the beginning of the XIX century the following opinion existed: the world possesses the single real geometry (Euclidean geometry) and the single arithmetic. But this opinion was neglected after the discovery of non-Euclidean geometries and of quaternions by Hamilton. Science understood that different natural systems can possess their own individual geometries and their own individual algebras (see this theme in the book (Kline, 1980)). The example of Hamilton, who wasted 10 years in his attempts to solve the task of description of transformations of 3D space by means of 3-dimensional algebras without success, is a very demonstrative one. This example says that if a scientist does not guess correctly what types of algebras are adequate for the natural system, which is investigated by him, he can waste many years without any result by analogy with Hamilton. One can add that geometrical and physical-geometrical properties of separate natural systems (including laws of conservation, theories of oscillations and waves, theories of potentials and fields, etc.) can depend on the type of algebras which are adequate for them.

The fact that the genetic code has led us to the algebra of genetic tetrions (which can be interpreted as a special case of the genetic octet Yin-Yang-algebra) shows the importance of this algebra for each united organism. It seems that many difficulties of modern science to understand genetic and biological systems are determined by approaches to these systems from the viewpoint of non-adequate algebras, which were developed formerly for other systems only. In particular, the classical vector calculation, which plays the role of the important tool in classical mechanics and which fits geometrical properties of our physical space, can be inappropriate for important biological phenomena.

In view of described materials, the hypothesis can be put forward that a very special mechanics of biogenetic systems exists, which is connected with the vector calculation of genetic tetrions and with their generalization in the form of Yin-Yang octets (Petoukhov, 2008a, 2008d, c). It can be named "genetic mechanics" because of its relation with the genetic code. Modern biomechanics is the set of applications of classical mechanics for modeling some properties of living matter. In our opinion, such traditional biomechanics are not adequate to many biological phenomena and it will be replaced in many aspects by genetic mechanics in future. We think that living matter lives in its own biological (bioinformation) space which has specific algebraic and geometric properties.

The hypothesis of a non-Euclidean geometry of living nature exists long ago (Vernadsky, 1965) but without any concrete definition of the type of such geometry. And how one can construct such geometry if biological organisms – bacteria, birds, fishes, plants, etc. - differ from each other so significantly in their morphogenetic and many other features? The discovery of the genetic code, the basic elements of which are general for all biological organisms, has allowed hoping that such geometric and algebraic tasks can be solved by means of investigation of genetic code structures. Some results of such investigation are presented in our book.

It happens frequently, that mathematicians construct a new beautiful abstract mathematics and then they search for opportunities of its application in different areas of natural sciences. On the contrary, in our case the phenomenology of the genetic code has led unexpectedly to the new mathematics of tetrions and Yin-Yang-octets. And we investigate formal features of this mathematics on the second stage only. The genetic code is the result of a gigantic experiment of nature. This molecular code bears the imprint of a great set of known and unknown laws of nature. In this connection, algebraic features of genetic structures are very essential to guess right a perspective direction of development of algebraic bases of mathematical natural sciences in the future. In our opinion, the tetrion algebra, the Yin-Yang-algebra and their geometries can be useful not only for biology, but also for other fields of mathematical natural sciences and for applied sciences (signals processing, mathematical economy, etc.). For example, they allow developing new algorithms and methods of digital signal processing.

It is important to discuss about the following as well. We have noted already that the notion of "number" is the main notion of mathematics and mathematical natural sciences. Pythagoras has formulated the famous idea: "All things are numbers".

Such known slogans of Pythagoreans as "numbers operate the world", "the world is number" reflect the representations of Pythagoreans. For Pythagoreans the systems of numbers expressed "essence" of everything. In view of this idea, natural phenomena should be explained by means of systems of numbers; the systems of numbers play a role of the beginning for uniting all things and for expressing the harmony of nature (Kline, 1980, p. 21, 24). Many prominent scientists and thinkers were supporters of this viewpoint or of one similar to it. Not without reason B. Russell (1945) noted that he did not know any other person who could exert such influence on the thinking of people as Pythagoras. From this viewpoint, there is no more fundamental scientific idea in the world, than this idea. C.Gauss, J.Argand and C. Wessel have demonstrated that a plane with its properties fits 2-dimensional complex numbers. W.Hamilton has proved that the properties of our 3-dimensional physical space fit mathematical properties of the special quaternions. The materials, which are described in this chapter, show that the genetic code is connected with "double genoquaternions" by analogy with the fact that the physical 3D-space fits Hamilton's quaternions. The described results give new materials to the great idea by Pythagoras in its possible modernized formulation: "All things are multi-dimensional number".

As a result of the matrix investigation of the genetic code, which is the basis of biological organisms, we find ourselves unexpectedly in area of the bases of mathematics and mathematical natural science, since number is the main notion there. One can note that mathematical natural sciences were created for putting in good order of information about the world and so they are information sciences. They utilize the notion of multi-dimensional number as the main notion. But genetic information is based on the multi-dimensional numbers also as the described results and models reveal. So the mathematical natural sciences can be considered as a continuation of bio-informatics principles, in accordance with which we are constructed genetically.

Such construction of science in its information essence reminds one of the constructions of instincts of biological organisms, according to which they build the dwellings by utilizing those genetic-information mechanisms and principles, on which their biological bodies are constructed. Figuratively speaking, the viable mathematical natural sciences are a continuation of our body, which is coordinated structurally with genetic bases of the body (the problem of anthropomorphism of development of mathematical natural sciences arises here).

WHAT IS LIFE FROM THE VIEWPOINT OF ALGEBRA? THE PROBLEM OF ALGEBRAIZATION OF BIOINFORMATICS AND BIOLOGY

Taking into account the great meaning of the genetic code for biological organisms, the described discovery of algebraic properties of the genetic code gives the basis for investigation of biological organizations from the algebraic viewpoint. Modern algebra is the wide branch of mathematics. It possesses many theorems, applications of which to genetic systems can give new vision in the field of theoretical biology. It is essential that algebra plays a great role in the modern theory of information encoding and of signal processing. It seems important, that the matrix forms of presentation of elements of the genetic octet Yin-Yang-algebra are connected with Hadamard matrices by means of the simple U-algorithm (see Chapter 6). Hadamard matrices play a significant role in the theory of quantum computers and of quantum mechanics, in particular. For this reason such connection can lead to possible understanding of the systems of the genetic code as quantum mechanical or quantum computer systems. Revealed algebraic properties of the genetic code present the opportunity to put forward the interesting problem of algebraization of bioinformatics on the basis of the algebras of the genetic code.

All these facts provoke the high interest to the question: what is life from the viewpoint of algebra? This question exists now in parallel with the old question from the famous book by E.Schrodinger: what is life from the viewpoint of physics? One can add that attempts are known in modern theoretical physics to reveal information bases of physics; in these attempts information principles are considered as the most fundamental.

Here one can mention as well the known problem of geometrization of physics that is the problem of creation and interpretation of physical theories in a form of theories of invariants of groups of transformations (see for example (Lochak, 1994)). Such general approach to different physical theories was very fruitful. One can hope that the problem of algebraization of bioinformatics (and of biology, which is connected closely with bioinformatics), that is understanding phenomena of bioinformatics from the viewpoint of algebras of the genetic code, will be useful as well.

One of the main questions in this field is the question about geometrical properties of vector spaces of bioinformatics, including various physiological spaces of sensory perception. Human organism encodes not only genetic information but also information about external world systematically. For instance, when a bright point of external picture is projected on retina of our eyes, an ophthalmic nerve delivers into nervous system not original information about a brightness of this point but encoded information already about a logarithm of this brightness.

In view of this, our organism is a machine for processing of flows of encoded information, principles of coding of which are inherited and are related with mathematics of genetic coding. What are possible geometries of such bioinformation spaces from the viewpoint of described Yin-Yang-algebra of the genetic code? In this question, one can utilize an analogy with quaternions by Hamilton $Q = x_0*\mathbf{1} + x_1*\mathbf{i}_1 + x_2*\mathbf{i}_2 + x_3*\mathbf{i}_3$ (Figure 2), where the first coordinate $x_0*\mathbf{1}$ is a scalar coordinate and three others $x_1*\mathbf{i}_1$, $x_2*\mathbf{i}_2$, $x_3*\mathbf{i}_3$ are vector coordinates. Quaternions by Hamilton correspond to properties of 3-dimensional vector space of physical world. By analogy one can suppose that each of two types of genetic quaternions $G_f = x_0*\mathbf{f}_0 + x_2*\mathbf{f}_2 + x_4*\mathbf{f}_4 + x_6*\mathbf{f}_6$ and $G_m = x_1*\mathbf{m}_1 + x_3*\mathbf{m}_3 + x_5*\mathbf{m}_5 + x_7*\mathbf{m}_7$ (see equations 3-7) correspond to properties of their own 3-dimensional vector space of bioinformatics. Each of them has one scalar coordinate ($x_0*\mathbf{f}_0$ or $x_1*\mathbf{m}_1$) and three vector coordinates ($x_2*\mathbf{f}_2$, $x_4*\mathbf{f}_4$, $x_6*\mathbf{f}_6$ or $x_3*\mathbf{m}_3$, $x_5*\mathbf{m}_5$, $x_7*\mathbf{m}_7$). Then octet genetic Yin-Yang-numbers $YY_8 = x_0*\mathbf{f}_0 + x_1*\mathbf{m}_1 + x_2*\mathbf{f}_2 + x_3*\mathbf{m}_3 + x_4*\mathbf{f}_4 + x_5*\mathbf{m}_5 + x_6*\mathbf{f}_6 + x_7*\mathbf{m}_7$ (equation 1) have two scalar coordinates ($x_0*\mathbf{f}_0$ or $x_1*\mathbf{m}_1$) and six vector coordinates ($x_2*\mathbf{f}_2$, $x_4*\mathbf{f}_4$, $x_6*\mathbf{f}_6$ and $x_3*\mathbf{m}_3$, $x_5*\mathbf{m}_5$, $x_7*\mathbf{m}_7$).

Correspondingly these Yin-Yang-numbers fit 6-dimensional bioinformation vector space, which unites two 3-dimensional bioinformation vector spaces of oppositional types (Yin and Yang) in a special cross-manner. This viewpoint is in a good agreement with a biological phenomenology: with existence of two oppositional cerebral hemispheres, which differ each from another by their functions and be the left-right morphology; etc. R. Penrose (1989) has emphasized at his analysis of phenomenon of thinking, that sharp functional distinguish exists between both cerebral hemispheres and that these hemispheres are related with halves of human body by means of cross-connections as well (see Figure 6 of Chapter 1). Each person has two eyes, two ears, etc.

Such data shows the existence of two bioinformation spaces (the right space and the left space) as sub-spaces of the whole bioinformation space of our organism. This theme of double (or twin) bioinformation sub-spaces continues a theme of double objects on a level of molecular-genetic structures: a double helix of DNA, a double configuration of chromosomes, etc. One of interesting examples is received in experiments with human vestibular disorders. The work (Petoukhov, 1975) has revealed an interesting class of human vestibular-visual illusions at observation of a single shining filament of a small light bulb in the dark: in experiments with oscillation of their head in the dark, after a certain latent period behind the end of oscillation many people experience a process of a development of a physiological phenomenon of double vision of this filament in a form of a smooth symmetrical divergence of positions of the "two" filaments on significant distance each from another (Figure 14).

This phenomenon shows additionally the existence of two 3-dimensional physiological spaces of perception, a joint coordination of which can be broken in some circumstances. Similar phenomena of double vision are known in a case of drunken persons and in some other cases. Such non-coordination of two 3-dimensional spaces of perception can lead not only to spatial illusions, but to a non-coordination

Figure 14. The phenomenon of double vision

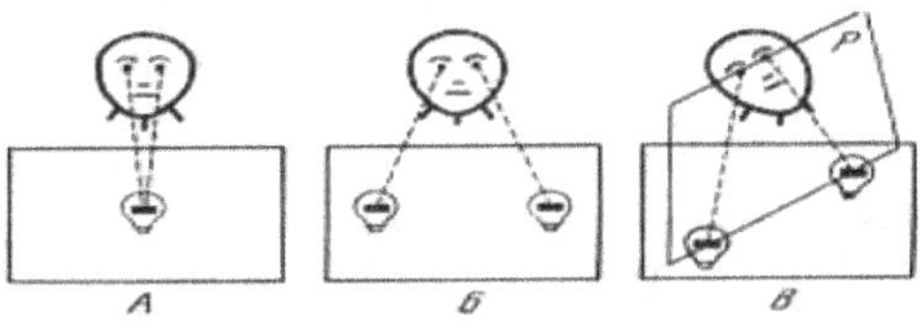

of movements, to nausea, to giddiness and to motion sickness. In view of this coincidence of biological phenomenology and genetic mathematics, knowledge of physiologic meaning of genetic Yin-Yang-algebras allows studying and modeling not only properties of molecular-genetic ensembles but genetic inherited macro-physiological systems and phenomena as well including illusions of perception, etc.

But how can bioinformation spaces with their genetic fundamentals lead to a realization of their genetic vector constructions in a form of material constructions from biochemical molecules? Why do biological atomic-molecular elements, which belong to the world of quantum mechanics, require mathematical constructions of bioinformatics? How are the quantum mechanics with its complex numbers, unitary operators and other mathematical formalisms interfaced with mosaic matrices of matrix genetics? Is there any connection of matrix genetics with matrix mechanics of Heisenberg?

In view of such important questions, one should emphasize a deep connection of matrix presentations of genetic systems with Hadamard matrices which play so significant role in a set of unitary operators of quantum mechanics, in logical gates of quantum computers, etc. A set of genetic Yin-Yang-matrices, which are presented in this Chapter and in Chapter 11, is transformed into a relevant set of Hadamard matrices by means of the same U-algorithm (see Chapter 6). Thereby all sets of considered genetic Yin-Yang-algebras become relevant sets of Hadamard matrices at action of such U-algorithm, and vice versa. Many genetic matrices, which were revealed and analyzed in matrix genetics, can be received algorithmically from relevant Hadamard matrices as initial matrices; this fact can be useful for future theory of connection of quantum mechanics with matrix genetics.

In addition, all genetic Hadamard matrices are block matrices, components of which are related to the complex number **Z** = **1+i** in its matrix form of presentation (Figure 15). For example, the genetic Hadamard matrix from Figure 4 of Chapter 6, which corresponds to the genomatrix $[C\ A;\ U\ G]^{(3)}$, can be expressed through **Z** in a following form:

SOLUTIONS AND RECOMMENDATIONS

Let us discuss the specificity of our approach to the question of the essence of the genetic code. From the scientific viewpoint, an explanation of something or understanding of something in a natural phenomenon means a substitution of categories, which characterize this phenomenon, by the more fundamental scientific categories. For example, physics explains the phenomenon of spontaneous movement of a ball from an edge of a pit into its bottom by means of the statement that the ball will have a minimum of potential energy on the pit bottom. This explanation substitutes for the initial question about the natural phenomenon by the new question about such fundamental physical category as a minimum of potential energy. This explanation is physical because it is based on physical law and it uses the physical notion of potential energy. But natural sciences utilize not only physical explanations but mathematical kinds

Figure 15. The example of a presentation of one of the genetic Hadamard matrices (in the middle) as a block matrix with components in a form of matrix Z (on the right side), which is the matrix form of presentation of the complex number **Z = 1+i** *(on the left side). Black (white) cells in two left matrices mean elements "+1" ("-1")*

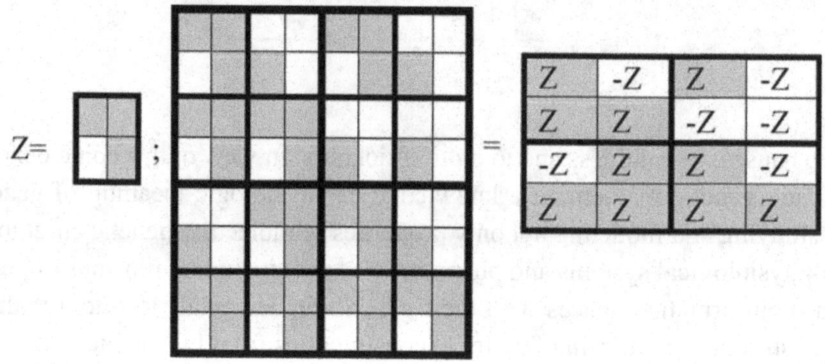

of explanations as well. For example, an explanation and a modeling of properties of elementary particles are based on mathematical theory of group presentation; properties of chemical compounds are explained on the basis of the periodic table by Mendeleev, etc. The algebraic model, which is described in our book, interprets the peculiarities of alphabetical systems of the genetic code on the mathematical base and moreover on the base of the main mathematical notion of "number" (or of "numeric system"). It means that this model and explanation belong to the mathematical and meta-mathematical kinds of explanations.

Matrix genetics reveals that other numeric systems and other good ordering systems govern in living matter in comparison with those, which mathematical natural sciences utilize traditionally. Our book proposes the new kind of generalization of real and hypercomplex numbers in the form of Yin-Yang (or bipolar) numbers. Starting from the extraordinary importance of genetic coding for biological organisms and from the bipolar character of structures of the genetic code, one can think that mathematization of all biology will be connected with using this Yin-Yang (bipolar or bisex) mathematics and its language.

In our opinion, the knowledge about the Yin-Yang-algebraic character of the genetic code is necessary for deep understanding of genetic coding and phenomena of reproduction, self-organization and self-developing of living matter on the whole. Yin-Yang-algebras are a comfortable instrument to analyze and to model many properties of hierarchical systems of biological organisms. Yin-Yang numeric systems are the candidate to play a role of numeric system in putting in order living matter.

It seems that many difficulties of modern bioinformatics are connected with utilizing inadequate algebras, which were developed for completely different natural systems. Hamilton had similar difficulties in his ten-year attempts to describe 3D-space transformations by means of algebras of 3-dimensional numbers while this description needs the algebra of 4-dimensional quaternions. (Hamilton considered the discovery of algebra of his quaternions as the major achievement of his life). All the history of development of the notion "number" can be considered as a process of gradual overcoming of inadequacy of numerical systems for those or other problems. The genetic code, as the information basis of all living matter, leads to the next overcoming of an inadequacy of existing numerical systems and to a transition into a new numerical era with a new category of the generalized numbers. In general the discussed situation is reflected in the following phrase: living matter is structured on the basis of its own numerical systems of order, which were unknown in mathematical natural sciences till now.

E. Schrodinger considered gaining knowledge about a "stream of order" in living matter as the very important task and he wrote in his book (Schrodinger, 1944, Chapter VII):

What I wish to make clear in this last chapter is, in short, that from all we have learnt about the structure of living matter, we must be prepared to find it working in a manner that cannot be reduced to the ordinary laws of physics. And that not on the ground that there is any "new force" or what not, directing the behavior of the single atoms within a living organism, but because the construction is different from anything we have yet tested in the physical laboratory... The unfolding of events in the life cycle of an organism exhibits an admirable regularity and orderliness, unrivalled by anything we meet with in inanimate matter... To put it briefly, we witness the event that existing order displays the power of maintaining itself and of producing orderly events... We must be prepared to find a new type of physical law prevailing in it /living matter/.

Molecular genetics puts forward the question about the origin of the genetic code. Usually the following three versions are considered in discussions about stochastic process of biological evolution (Ratner, 2002, p. 199-202): 1) the structural properties of the genetic code were set preliminarily (were preset) by physical-chemical conditions of components and conditions; 2) they were picked out as adaptive among other alternative variants; 3) they were fixed accidentally. For example, the famous hypothesis by F.Crick (1968) about "the frozen accident" supposed that the first accidental system of coding, which possessed satisfactory features, was reproduced with its further evolutionary improvements for accelerated reproductions.

Matrix genetics yields new materials to this question of the origin of the genetic code by the revealing that the bases of the genetic code are connected with the multi-dimensional algebra, which generalized the notion of hypercomplex numbers. Any algebra, which is an abstract essence, does not depend on time and space or it exists outside time and space as the member of the mathematical world of Plato (Penrose, 1989). According to Plato, mathematical ideas have their own existence and they live in an ideal world, the entrance into which is possible by means of our intellect. So, algebras do not depend on evolutionary processes on the Earth.

But the correspondence of the genetic code to the specificity of genetic algebras can provide evolutionary advantages for living matter. Evolutionary biology explains any separate property of biological organisms by its evolutionary usefulness. From this viewpoint of evolutionary biology, the structuredness of the genetic code in accordance with the octet Yin-Yang-algebra can be explained, for example, by the opportunity of processing two different streams of information in parallel manner for evolutionary advantages. Here we have a certain coincidence with the works (Geodakian, 1999), which connect the existence of two sexes with two different (operative and conservative) streams of information.

One should take into account the following additional circumstance. The matrix forms of presentation of elements of the genetic octet Yin-Yang-algebra are connected with Hadamard matrices by means of the simple U-algorithm (Petoukhov, 2008a-d). Hadamard matrices play a significant role in the theory of quantum computers and of quantum mechanics, in particular. For this reason such a connection seems to be important for a possible understanding of the systems of the genetic code as quantum mechanical system or quantum computer systems.

One can note additionally that the binary opposition "male-female" is connected with the binary opposition "the left side and the right side" in the history of various nations. More precisely, in accordance with many biological, ethnographical and mythological materials, the left side of human body correlates with the female beginning, and the right side of human body correlates with the male beginning (Ivanov, 1990, p.506-508). It is known that mirror symmetry of "left-and-right" is broken in bio-organic world. L. Paster, who has discovered this fact, put forward the hypothesis that this property of dissymmetry is the line of demarcation between living matter and inanimate matter. The origin of this dissymmetry phenomenon is not understood till now. In our opinion, the Yin-Yang-algebra of the genetic code can lead to new approaches in understanding this dissymmetry phenomenon.

The described Yin-Yang-algebraic model and its language are the parts of the general process of mathematization of science. It is known that an appearance of mathematical models in any field of science shows that a system of notions in this field becomes precise in high degree to allow rigorous and abstract analyses by means of mathematical instruments. Mathematical models are defined frequently in a form of a special "language" for a description of appropriate phenomena. For example, differential calculus and integral calculus have arisen in the XVII century in such forms. Application of Yin-Yang-algebras for modeling the genetic code brings a new language as well.

One can think that various genetic algebras, which are connected with various parameters of ensembles of genetic molecules, correspond to the various information channels in multi-channel informatics of organism. Many thinkers spoke about a harmony of living nature. The genetic algebras, which are described in our book, give valuable opportunities to analyze this harmony.

Understanding the fact of existence of the genetic code was the most difficult thing in a problem of a genetic code. The whole century was required for it. When it has been understood, ten years were needed only to know details (Ycas, 1969). By analogy with it, the understanding the existence of the special and new algebra for modeling the degeneracy of the genetic code was the most difficult thing in described genetic researches, where many other – biophysical, biochemical, mathematical – variants of modeling were tried (Petoukhov, 1999-2008).

The mathematical part of the materials described in this chapter proposes some interesting prolongations in various aspects, which can be recommended for further investigations and applications. One of them is a generalization of hypercomplex numbers into a form of appropriate Yin-Yang-algebras. This kind of generalization should be taken into account in developing the theory of multi-dimensional numbers. Let us stop on it for more details.

As it was mentioned above, Yin-Yang numbers (*YY*-numbers) can be considered as the generalization of hypercomplex numbers. Each of 2^{n-1}-dimensional hypercomplex numbers can be transformed into the 2^n-dimensional *YY*-number by a special algorithm. An inverse application of this algorithm to a 2^n-dimensional *YY*-number generates the appropriate 2^{n-1}-dimensional hypercomplex number. According to this algorithm, if we have a $(2^n x 2^n)$-matrix, which represents a 2^n-dimensional hypercomplex number, we should replace each component of this matrix by the (2x2)-matrix $[x_\kappa\, x_{\kappa+1};\, x_\kappa\, x_{\kappa+1}]$. By this algorithm we have the tetra-reproduction of matrix components, which reminds the tetra-reproduction of gametal cells in the process of meiosis. For such a reason this algorithm has the conditional name "the meiosis algorithm".

For example, if we have the (2x2)-matrix of the presentation of complex numbers, this meiosis algorithm transforms it into the (4x4)-matrix of the presentation of 4-dimensional "Yin-Yang-complex" numbers KK_4, which fit the special multiplication table of the appropriate 4-dimensional YY_4-algebra (Figure 16). Really, according to this algorithm, each component x_0 and x_1 of the initial matrix is replaced

Figure 16. The matrix forms of presentation of complex numbers (on the left side) and of YY-complex numbers (in the middle). On the right side: the multiplication table for the basic elements of the YY-complex number

x_0	$-x_1$
x_1	x_0

;

y_0	y_1	$-y_2$	$-y_3$
y_0	y_1	$-y_2$	$-y_3$
y_2	y_3	y_0	y_1
y_2	y_3	y_0	y_1

	$\mathbf{f}_0$	$\mathbf{m}_1$	$\mathbf{f}_2$	$\mathbf{m}_3$
$\mathbf{f}_0$	$\mathbf{f}_0$	$\mathbf{m}_1$	$\mathbf{f}_2$	$\mathbf{m}_3$
$\mathbf{m}_1$	$\mathbf{f}_0$	$\mathbf{m}_1$	$\mathbf{f}_2$	$\mathbf{m}_3$
$\mathbf{f}_2$	$\mathbf{f}_2$	$\mathbf{m}_3$	$-\mathbf{f}_0$	$-\mathbf{m}_1$
$\mathbf{m}_3$	$\mathbf{f}_2$	$\mathbf{m}_3$	$-\mathbf{f}_0$	$-\mathbf{m}_1$

by the (2x2)-matrix of the mentioned type: $x_0=[y_0y_1; y_0y_1]$, $x_1=[y_2y_3; y_2y_3]$. In the result we have *YY*-complex numbers $KK_4 = y_0*\mathbf{f}_0+y_1*\mathbf{m}_1+y_2*\mathbf{f}_2+y_3*\mathbf{m}_3$, where $\mathbf{f}_0$ and $\mathbf{m}_1$ are the female and male quasi-real units; $\mathbf{f}_2$ and $\mathbf{m}_3$ are the female and male imaginary units with the properties $(\mathbf{f}_2)^2 = -\mathbf{f}_0$, $(\mathbf{m}_3)^2 = -\mathbf{m}_1$.

By inverse application of this algorithm, one can arrive at the appropriate 4-dimensional hypercomplex number from the genetic *YY*-number YY_8. The *YY*-matrix YY_8 (Figure 3) contains the 4 kinds of (2x2)-sub-quadrants, each of which has one of the pairs of coordinates: x_0 and x_1; x_2 and x_3; x_4 and x_5; x_6 and x_7. One can replace each such sub-quadrant by a separate coordinate: $[x_0x_1; x_0x_1] = y_0$; $[x_2x_3; x_2x_3] = y_1$; $[x_4x_5; x_4x_5] = y_2$; $[x_6x_7; x_6x_7] = y_3$. As a result the (4x4)-matrix Q appears, which represents the genoquaternion $Q = y_0*\mathbf{1} + y_1*\mathbf{i}_1 + y_2*\mathbf{i}_2 + y_3*\mathbf{i}_3$, which was considered above and which has $\mathbf{i}_1^2 = -1$, $\mathbf{i}_2^2 = \mathbf{i}_3^2 = +1$. Figure 17 shows the matrix Q and the multiplication table for this genoquaternion. The genoquaternion Q suggests coquaternions (or split-quaternions, or para-quaternions, or hyperbolic quaternions), introduced by J.Cockle in 1849 year (http://en.Qikipedia.org/Qiki/Coquaternion), but their multiplication tables have differences. We name the number Q "genoquaternion of the first type". (A genoquaternion of the second type is produced by the special permutation of columns of the matrix Q, which is connected with the permutation of positions in genetic duplets (Petoukhov, 2008a, p.203)).

Let us pay some attention to the two squares, which are marked out by bold lines in the left top corner of the multiplication table on Figure 6 for the case of the Yin-Yang matrix $(YY_8)_{123}$. These two squares are connected with the 2-dimensional sub-algebra YY_2 and the 4-dimensional sub-algebra YY_4 of the 8-dimensional algebra YY_8.

The first of these squares with its size (2x2) is the multiplication table of the basic elements of the 2-dimensional Yin-Yang algebra YY_2. Figure 18 shows two matrix forms of presentation of appropriate Yin-Yang numbers YY_2. One of these forms $[z_0z_1; z_0z_1]$ coincides with the structure of each (2x2)-sub-quadrant of the genomatrices on Figure 2 in chapter 2, and Figures 3 and4 of this chapter, in relation of

Figure 17. The matrix form of presentation of the hypercomplex number Q (on the left side); its multiplication table is shown on the right side

$Q =$

y_0	$-y_1$	y_2	$-y_3$
y_1	y_0	$-y_3$	$-y_2$
y_2	$-y_3$	y_0	$-y_1$
$-y_3$	$-y_2$	y_1	y_0

	1	$\mathbf{i}_1$	$\mathbf{i}_2$	$\mathbf{i}_3$
1	**1**	$\mathbf{i}_1$	$\mathbf{i}_2$	$\mathbf{i}_3$
$\mathbf{i}_1$	$\mathbf{i}_1$	**-1**	$-\mathbf{i}_3$	$\mathbf{i}_2$
$\mathbf{i}_2$	$\mathbf{i}_2$	$\mathbf{i}_3$	**1**	$\mathbf{i}_1$
$\mathbf{i}_3$	$\mathbf{i}_3$	$-\mathbf{i}_2$	$-\mathbf{i}_1$	**1**

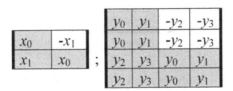

Figure 18. Two matrix forms of a presentation of the 2-dimensional numbers YY_2 (on the left side); the multiplication table of the basic elements of the Yin-Yang algebra YY_2

$$\begin{vmatrix} z_0 & z_1 \\ z_0 & z_1 \end{vmatrix} ; \begin{vmatrix} z_0 & -z_1 \\ -z_0 & z_1 \end{vmatrix} ;$$

	f_0	m_1
f_0	f_0	m_1
m_1	f_0	m_1

the disposition of the *YY*-coordinates $x_0, x_1, \ldots, x_7$ and of amino acids with stop-signals. It shows that the algebra YY_2 participates in the structural organization of the genetic code.

The second square with its size (4x4) on Figure 6 is the multiplication table of the 4-dimensional Yin-Yang algebra YY_4. The appropriate Yin-Yang numbers *YY*4 possess the following vector form: $YY_4 = z_0{*}f_0 + z_1{*}m_1 + z_2{*}f_2 + z_3{*}m_3$ and these numbers coincide with the Yin-Yang generalization of complex numbers (Figure 14).

The case of the 2-dimensional algebra YY_2 should be considered additionally. It is known that complex numbers have been widely recognized only after finding their geometrical interpretation on the geometric plane of complex variables. This plane was named "Gauss-Argand plane" according to the names of the mathematicians who have introduced such a plane. Is it possible to offer a substantial geometrical interpretation of the 2-dimensional Yin-Yang numbers YY_2? Yes, it is possible (Petoukhov, 2008a,d). For this purpose one can introduce the plane of Yin-Yang variables (or *YY*-plane). It is an ordinary plane with the Yin-Yang system of Cartesian coordinates. This Yin-Yang system (or *YY*-system) has the coordinate axes **f** and **m**, which play the role of female and male axes. By analogy with the case of complex numbers, each 2-dimensional *YY*-number is denoted on this *YY*-plane by the point or by the vector. A product *XX***ZZ* of two Yin-Yang vectors, where $\mathbf{XX} = x_0{*}\mathbf{f_0} + x_1{*}\mathbf{m_1}$ and $\mathbf{ZZ} = z_0{*}\mathbf{f_0} + z_1{*}\mathbf{m_1}$, possesses a geometric sense on such a plane. Really, the result of non-commutative multiplication of such two *YY*-vectors is equal to the second vector with the scale coefficient, which is equal to the sum of coordinates of the first vector (Figure 19, on the left side). The same first vector-factor at multiplication with all other vectors of the plane or of a geometric figure leads to their identical scaling (Figure 19, on the right side).

It associates with the known biological phenomenon of volumetric growth of living bodies, observed at the most different lines and branches of biological evolution. Biological bodies are capable of mysterious volumetric growth, occurring in the cooperative way in all the volume of the body or of its growing part. It is one of the sharp differences between living bodies and crystals, the surfaces of which grow by means of a local addition of new portions of substance to the surface of the crystal. By this connection, the Yin-Yang geometry is one of the candidates for the role of the geometry of biological volumetric

Figure 19. The non-commutative multiplication of two Yin-Yang vectors (on the left side). A scaling of a geometric figure on the Yin-Yang plane (on the right side)

$$\mathbf{XX}{*}\mathbf{ZZ} = (x_0{*}\mathbf{f_0} + x_1{*}\mathbf{m_1}){*}(z_0{*}\mathbf{f_0} + z_1{*}\mathbf{m_1}) = (x_0 + x_1){*}(z_0{*}\mathbf{f_0} + z_1{*}\mathbf{m_1})$$
$$\mathbf{ZZ}{*}\mathbf{XX} = (z_0{*}\mathbf{f_0} + z_1{*}\mathbf{m_1}){*}(x_0{*}\mathbf{f_0} + x_1{*}\mathbf{m_1}) = (z_0 + z_1){*}(x_0{*}\mathbf{f_0} + x_1{*}\mathbf{m_1})$$

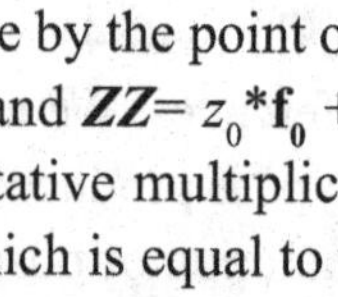

Figure 20. The matrix form of presentation of 3-polar numbers and the multiplication table of their basic elements

$$\begin{vmatrix} x_0 & x_1 & x_2 \\ x_0 & x_1 & x_2 \\ x_0 & x_1 & x_2 \end{vmatrix};$$

	i_0	i_1	i_2
i_0	i_0	i_1	i_2
i_1	i_0	i_1	i_2
i_2	i_0	i_1	i_2

growth. In our opinion, interesting branches of generalized crystallography can be developed by using Yin Yang algebras. We recommend paying attention to these new opportunities connected with application of methods of symmetry and with production of new patterns.

We recommend the further wide development of this Pythagorean approach to the genetic and other genetically heritable biological systems.

One additional aspect should be noted as well. It is known that mathematics deals not only with algebras of numbers but with algebras of operators also (see historical remarks in the book (Kline,1980, Chapter VIII)). G. Boole has published in 1854 his brilliant work about investigations of laws of thinking. He has proposed Boole's algebra of logics (or logical operators). Boole tried to construct an operator algebra which would reflect basic properties of human thinking. Boole's algebra plays a great role in modern science because of its connections with many scientific branches: mathematical logic, the problem of artificial intelligence, computer technologies, bases of the theory of probability, etc. In our opinion, the genetic algebras, which are described in this chapter, can be considered not only as the algebras of the numeric systems but also as the algebra of proper logical operators of genetic systems. This direction of thought can lead us to a deeper understanding of the logic of biological systems including an advanced variant of the idea of Boole (and by some other scientists) on the development of the algebraic theory of laws of thinking.

One of the possible applications of the genetic Yin-Yang algebra in the field of formal logic is a new possible approach to situations with the simultaneous presence of two kinds of logic, which correspond to the famous expression "the male logic and the female logic". Such applications are possible in analyses of a behavioral logic in groups of men and women or in systems, parts of which are under various, but interconnected variants of logic.

Biological organisms have famous possibilities to utilize the same structures in multi-purpose destinations. And the genetic algebras can be also utilized by biological organisms in different purposes.

FUTURE TRENDS AND CONCLUSION

It should be noted that the names "bipolar algebra" and "bipolar geometry", "bisex algebra" and "bisex geometry", "bipolar numbers", etc. can be utilized as the synonyms of the names "Yin-Yang algebra", "Yin-Yang geometry", "Yin-Yang numbers". In some cases the utilization of these names can be more comfortable but it depends on situations. For example it is comfortable in the question about algebras with many quasi-real units. Such algebras can be named "multi-polar algebras" (or "*n*-polar algebras").

Really bipolar algebras can be interpreted as a particular case of n-polar algebras, each of which possesses a set of their basic elements with "n" quasi-real units but without the real unit. Figure 20 shows

Figure 21. The symbol Yin-Yang and the symbol tomoe

the simplest example of 3-polar numbers $x_0*\mathbf{i}_0+x_1*\mathbf{i}_1+x_2*\mathbf{i}_2$ (in the matrix form of their presentation), which contain three quasi-real numbers only. The basic elements $\mathbf{i}_0$, $\mathbf{i}_1$, $\mathbf{i}_2$ of these 3-polar numbers have their matrix forms of presentation: $\mathbf{i}_0$=[1 0 0; 1 0 0; 1 0 0], $\mathbf{i}_1$=[0 1 0; 0 1 0; 0 1 0], $\mathbf{i}_2$=[0 0 1; 0 0 1; 0 0 1]. Their multiplication table is shown on Figure 20.

Bipolar numbers and tripolar numbers can be considered as numeric analogies of the famous symbols Yin-Yang and tomoe (Figure 21). Details about the Japanese tomoe symbol are given at the site http://altreligion.about.com/library/glossary/symbols/bldefstomoe.htm.

Multiplication of two 3-polar numbers gives the result, which is similar to the described case of multiplication of two bipolar numbers: the result is the 3-polar number, which is equal to the second factor increased by the sum of coordinates of the first factor (Figure 22). The 3-polar geometry is a candidate to play the role of the geometry of the volumetric biological growth in the case of 3D-space (by analogy with the bipolar geometry in the case of a plane).

Figure 23 shows another example of multi-polars: the matrix form of presentation of 8-dimensional 4-polar numbers $x_0*\mathbf{i}_0+x_1*\mathbf{i}_1+x_2*\mathbf{i}_2+x_3*\mathbf{i}_3+x_4*\mathbf{i}_4+x_5*\mathbf{i}_5+x_6*\mathbf{i}_6+x_7*\mathbf{i}_7$, which have 4 quasi-real units $\mathbf{i}_0$, $\mathbf{i}_1$, $\mathbf{i}_2$, $\mathbf{i}_3$ and which have their own imaginary unit for each of these quasi-real units: $\mathbf{i}_4^2 = -\mathbf{i}_0$; $\mathbf{i}_5^2 = -\mathbf{i}_1$; $\mathbf{i}_6^2 = -\mathbf{i}_2$; $\mathbf{i}_7^2 = -\mathbf{i}_3$.

Bipolar algebras and multi-polar algebras, which have arisen in the field of matrix genetics and bioinformatics, possess many other interesting properties, which are described in special publications (Petoukhov, 2008a, 2008d). They allow developing new class of mathematical models of self-reproduction systems and new class of algorithm for information processing. They also allow investigating possible generalizations of known physical equations to find new results with a physical sense from there (it is the mathematical fact that known physical equations can be arrived at from appropriate bipolar equations by passage to the limit in values of appropriate bipolar coordinates). The idea of multi-dimensional numbers and multi-dimensional spaces works intensively for a long time in theoretical physics and other fields of science for modeling the phenomena of our physical world. This chapter adds this idea of multi-dimensional numbers and multi-dimensional spaces with appropriate mathematical formalisms into the fields of molecular genetics and bioinformatics.

The algebraic theory of the genetic code, which utilizes methods of symmetry and new genetic patterns, can say many useful and unexpected things about an origin of the genetic code and about laws of

Figure 22. Multiplication of two 3-polar numbers

$$\begin{vmatrix} x_0, x_1, x_3 \\ x_0, x_1, x_2 \\ x_0, x_1, x_2 \end{vmatrix} * \begin{vmatrix} y_0, y_1, y_3 \\ y_0, y_1, y_2 \\ y_0, y_1, y_2 \end{vmatrix} = (x_0+x_1+x_2)* \begin{vmatrix} y_0, y_1, y_3 \\ y_0, y_1, y_2 \\ y_0, y_1, y_2 \end{vmatrix}$$

Figure 23. The matrix form of presentation of 8-dimensional 4-polar numbers (the upper matrix) and the multiplication table of their basic elements (the lower table)

$$\begin{vmatrix} x_0 & x_1 & x_2 & x_3 & -x_4 & -x_5 & -x_6 & -x_7 \\ x_0 & x_1 & x_2 & x_3 & -x_4 & -x_5 & -x_6 & -x_7 \\ x_0 & x_1 & x_2 & x_3 & -x_4 & -x_5 & -x_6 & -x_7 \\ x_0 & x_1 & x_2 & x_3 & -x_4 & -x_5 & -x_6 & -x_7 \\ x_4 & x_5 & x_6 & x_7 & x_0 & x_1 & x_2 & x_3 \\ x_4 & x_5 & x_6 & x_7 & x_0 & x_1 & x_2 & x_3 \\ x_4 & x_5 & x_6 & x_7 & x_0 & x_1 & x_2 & x_3 \\ x_4 & x_5 & x_6 & x_7 & x_0 & x_1 & x_2 & x_3 \end{vmatrix}$$

	i_0	i_1	i_2	i_3	i_4	i_5	i_6	i_7
i_0	i_0	i_1	i_2	i_3	i_4	i_5	i_6	i_7
i_1	i_0	i_1	i_2	i_3	i_4	i_5	i_6	i_7
i_2	i_0	i_1	i_2	i_3	i_4	i_5	i_6	i_7
i_3	i_0	i_1	i_2	i_3	i_4	i_5	i_6	i_7
i_4	i_4	i_5	i_6	i_7	$-i_0$	$-i_1$	$-i_2$	$-i_3$
i_5	i_4	i_5	i_6	i_7	$-i_0$	$-i_1$	$-i_2$	$-i_3$
i_6	i_4	i_5	i_6	i_7	$-i_0$	$-i_1$	$-i_2$	$-i_3$
i_7	i_4	i_5	i_6	i_7	$-i_0$	$-i_1$	$-i_2$	$-i_3$

living matter. In particular we recommend investigations of the evolution of the dialects of the genetic code from the viewpoint of the genetic Yin-Yang-algebras. Some results of such initial investigation are described in the next chapter.

Degeneracy of the genetic code agrees with the multi-dimensional algebra, which is unknown in modern mathematical natural science. After the discovery of non-Euclidean geometries and of Hamilton quaternions, it is known that different natural systems can possess their own geometry and their own algebra. The genetic code is connected with its own multi-dimensional numerical systems or the multi-dimensional algebras. A bunch of these genetic algebras can be considered as a basis of an algebraic system of the pre-code or as the mathematical model of the genetic code. These algebras allow revealing hidden peculiarities of the structure of the genetic code and, perhaps, its evolution. The genetic code has its own forms of ordering. It seems that many difficulties of modern bioinformatics are connected with utilizing for its natural structures inadequate algebras, which were developed for completely different natural systems. Hamilton had similar difficulties in his attempts to describe 3D-space transformations by means of 3-dimensional numbers while this description needs quaternions. This chapter proposes a special algebraic system for bioinformatics and for mathematical biology. Revealed algebraic properties of the genetic code allow putting forward the problem of the algebraization of bioinformatics and of biology. They allow modeling not only molecular-genetic ensembles but also genetic inherited macro-physiological systems and phenomena.

REFERENCES

Crick, F. (1968). The origin of the genetic code. *Journal of Molecular Biology, 38*, 367–379. doi:10.1016/0022-2836(68)90392-6

Geodakian, V. A. (1999). The role of sex chromosomes in evolution: A new concept. *Journal of Mathematical Sciences*, *93*(4), 521–530. doi:10.1007/BF02365058

Ivanov, V. V. (1999). *The selected works on semiotics and history of culture*. Moscow: Yazyki russkoi kultury (in Russian).

Kline, M. (1980). *Mathematics. The loss of certainty*. New York: Oxford University Press.

Koestler, A. (1978). *Janus. A summing up*. New York: Random House.

Konopelchenko, B. G., & Rumer, Y. B. (1975). Classification of the codons in the genetic code. [in Russian]. *DAN SSSR*, *223*(2), 145–153.

Lochak, G. (1994). *La geometrisation de la physique*. France: Flammarion.

Pavlov, D. G. (2004). Leading article. *Hypercomplex numbers in geometry and in physics, 1*(1), 4-7 (in Russian).

Penrose, R. (1989). *The emperor's new mind*. Oxford: Oxford University Press.

Petoukhov, S. V. (1999). Genetic code and the ancient Chinese book of changes. *Symmetry: Culture and Science*, *10*, 211–226.

Petoukhov, S. V. (2001). The *bi-periodic table of genetic code and the number of protons*. Moscow: MKC (in Russian).

Petoukhov, S. V. (2001a). Genetic codes I: Binary sub-alphabets, bi-symmetric matrices, and the golden section; Genetic codes II: Numeric rules of degeneracy and the chronocyclic theory. *Symmetry: Culture and Science*, *12*(3-4), 255–306.

Petoukhov, S. V. (2003-2004). Attributive conception of genetic code, its bi-periodic tables, and problem of unification bases of biological languages. *Symmetry . Cultura e Scuola, 14-15*(part 1), 281–307.

Petoukhov, S. V. (2005). The rules of degeneracy and segregations in genetic codes. The chronocyclic conception and parallels with Mendel's laws. In *Advances in Bioinformatics and its Applications, Series in Mathematical Biology and Medicine, 8*, 512-532. World Scientific.

Petoukhov, S. V. (2006a). Bioinformatics: Matrix genetics, algebras of the genetic code, and biological harmony. *Symmetry: Culture and Science*, *17*(1-4), 251–290.

Petoukhov, S. V. (2008a). *Matrix genetics, algebras of the genetic code, noise-immunity*. Moscow: RCD (in Russian).

Petoukhov, S. V. (2008b). The degeneracy of the genetic code and Hadamard matrices (pp. 1-8).

Petoukhov, S. V. (2008c). Matrix genetics, part 1: Permutations of positions in triplets and symmetries of genetic matrices (pp. 1-12). Retrieved on March 8, 2008, from http://arXiv:0803.0888

Petoukhov, S. V. (2008d). Matrix genetics, part 2: The degeneracy of the genetic code and the octave algebra with two quasi-real units (the "yin-yang octave algebra") (pp. 1-27). Retrieved on March 8, 2008, from http://arXiv:0803.3330

Petoukhov, S. V. (2008e). Degeneracy of the genetic code and its own multi-dimensional algebra. In R. Balling & A. Nordheim (Eds.), *XX International Congress of Genetics* (p. 209). Berlin: German Genetic Society.

Petoukhov, S. V. (2008f). Matrix genetics, musical harmony, and the genetic matrices of hydrogen bonds. In R. Balling & A. Nordheim (Eds.), *XX International Congress of Genetics* (p. 209). Berlin: German Genetic Society.

Ratner, V. A. (2002). *Genetics, molecular cybernetics*. Novosibirsk: Nauka.

Retrieved on February 8, 2008, from http://arXiv:0802.3366

Schrodinger, E. (1944). *What is life? The physical aspect of the living cell.* Cambridge: Cambridge University Press.

Vernadsky, V. I. (1965). *Chemical structure of the earth and its surrounding*. Moscow: Nauka.

Ycas, M. (1969). *The biological code*. Amsterdam, London: North-Holland Publishing Company.

Chapter 8
The Evolution of the Genetic Code from the Viewpoint of the Genetic 8-Dimensional Yin-Yang-Algebra

ABSTRACT

The set of known dialects of the genetic code is analyzed from the viewpoint of the genetic 8-dimensional Yin-Yang-algebra. This algebra was described in Chapter 7. The octet Yin-Yang-algebra is considered as the model of the genetic code. From the viewpoint of this algebraic model, for example, the sets of 20 amino acids and of 64 triplets consist of sub-sets of "male," "female," and "androgynous" molecules, and so forth. This algebra allows one to reveal hidden peculiarities of the structure and evolution of the genetic code and to propose the conception of "sexual" relationships among genetic molecules. The first results of the analysis of the genetic code systems from such an algebraic viewpoint speak about the close connection between evolution of the genetic code and this algebra. They include 7 phenomenological rules of evolution of the dialects of the genetic code. The evolution of the genetic code appears as the struggle between male and female beginnings. The hypothesis about new biophysical factor of "sexual" interactions among genetic molecules is proposed. The matrix forms of presentation of elements of the genetic octet Yin-Yang-algebra are connected with Hadamard matrices by means of the simple U-algorithm. Hadamard matrices play a significant role in the theory of quantum computers, in particular. It leads to new opportunities for the possible understanding of genetic code systems as quantum computer systems. Revealed algebraic properties of the genetic code allow one to put forward the problem of algebraization of bioinformatics on the basis of the algebras of the genetic code. The described investigations are connected with the question: what is life from the viewpoint of algebra?

DOI: 10.4018/978-1-60566-124-7.ch008

INTRODUCTION AND BACKGROUND

This chapter is devoted to the first results of investigations of the evolution of the genetic code from the viewpoint of the genetic octet Yin-Yang-algebra, which was described in the previous Chapter 7. Owing to discovery of the connection of this algebra with the genetic code, new opportunities arise for algebraic systematizing and classification of binary-oppositional structures in molecular-genetic ensembles. Notions and formalisms of this algebra are used here to analyze ensembles of genetic molecules in connection with a traditional theme of male and female beginnings in living substance on various levels. This algebraic way leads to revealing a set of phenomenological rules of evolution of the genetic code and to new possibilities of understanding some interrelations between elements of molecular-genetic ensembles. New notions of "sexual" types of the triplets and amino acids can be proposed on a well-reasoned mathematical basis. The results of these investigations and of applications of such new notions are described and discussed.

The theme of male and female beginnings and biological reproduction, which is connected with them, is one of the main themes in human civilization. This binary opposition – man and woman - exists in different forms in many theories in the fields of psychology, biology and culture, etc. Existence of male and female types in psychology, of male and female chromosomes, of male and female gametal cells, etc. is known widely. This primeval theme presented in religions and myths of all times and people. For example, in Ancient China female and male beginnings (Yin and Yang) were considered as the main operating forces in the world, and the world has been created by them. The spiritual philosophical doctrines of the East, which are presented in many ancient books, asserted, that the soul at initial stage of its creation united both male and female beginnings and, in that way, the soul reflected the dual nature of the Creator.

Initial representations about bisexual nature of human being have been formulated in folklore and mythology of many nations of the world. In particular, these representations were developed by an ancient philosophy. For example, Plato's narration is known concerning androgynous beings from which modern people have been brought into the world. According to Plato, love is the instinctive aspiration of individuals, who love each other, to uniting them with their return to the initial state, which was before its division into two. Modern psychology considers bisexuality as the fundamental characteristic of constitutional nature of human being. And the notion of androgynous being is considered as a fixation of this duality which includes always the male and female beginnings but in the different proportions, which can be changed during a life. Many famous philosophers have presented to people a wide set of valuable thoughts about male and female beginnings of being and about nature of sexual relations of men and women. People have gotten accustomed to seeing mutual relations between men and women. A vast set of works in various fields of culture is devoted to these relations. It is considered ordinary that male and female beginnings in nature are necessary to continue life and its development (Bull, 1983; Geodakian, 1999; Karlin, & Lessard, 1986; Maynard Smith, 1978; Mooney, 1992; Williams, 1975).

The tendency of thinkers to reflect the natural fact of male and female beginnings on a formal language is known from the ancient time. For example, thoughts about fundamental meanings of male and female beginnings are reflected by thinkers of Ancient China and of the Pythagorean School into the thematic division of the series of natural numbers, where even numbers embody the female beginning (Yin) and odd numbers embody the male beginning (Yang).

Alternation of even and odd numbers in a series of natural numbers was considered as the form of an interpenetrating in the union of male and female beginnings. Especial value was given to the basic female number 2 (in Ancient China it was considered as the "number of the Earth") and to the basic male number 3 ("number of the Sky") (Schutskiy, 1997). These conceptions have penetrated into many countries and into the latest doctrines, including the Pythagorean School. Besides in the Ancient China, mutual relations in the world between female and male beginnings were expressed by means of square tables of "The book of changes" ("I Ching"). These tables contained combinations of female and male beginnings under their names Yin and Yang in various proportions and sequences.

Sexual attributes are inherited genetically in living substance. Biology has revealed long ago, that male and female sexual cells (gametes) exist, and that sexual chromosomes contain male and female chromosomes, etc. The results of matrix genetics, which are described in our book, allow one to make the following suppositions: 1) a set of binary-oppositional attributes relates directly to the problem of male and female beginnings in biology; 2) phenomenological peculiarities of these ensembles can be expressed by means of a language of generalized multi-dimensional Yin-Yang-numbers. These thoughts have led to continuations of the investigations, which were described in the previous chapter, about the connection of the genetic matrices with multi-dimensional Yin-Yang numbers. The genetic 8-dimensional Yin-Yang-algebra gives the new conceptual and formal instrument for analyzing and modeling many biological phenomena including phenomena of evolution of the genetic code, which has many dialects for unknown reasons.

This genetic algebra defines the system of 8-dimensional Yin-Yang numbers YY_8 (the matrix form of presentation of these numbers YY_8 is presented on Figure 3 and Figure 4 in Chapter 7.):

$$YY_8=x_0*\mathbf{f}_0+x_1*\mathbf{m}_1+x_2*\mathbf{f}_2+x_3*\mathbf{m}_3+x_4*\mathbf{f}_4+x_5*\mathbf{m}_5+x_6*\mathbf{f}_6+x_7*\mathbf{m}_7$$

Multiplication of any two members of such a set of octet numbers YY_8 generates a new octet number of the same set. Chapter 7 described that this numeric system has regular and sharp distinctions between the sub-set of the basic "female" (or Yin) elements $\mathbf{f}_0$, $\mathbf{f}_2$, $\mathbf{f}_4$, $\mathbf{f}_6$ and the sub-set of the basic male (or Yang) elements $\mathbf{m}_1$, $\mathbf{m}_3$, $\mathbf{m}_5$, $\mathbf{m}_7$. These distinctions are based on the features of the multiplication table of these Yin-Yang numbers YY_8. (Figure 6 in Chapter 7).

This genetic octet Yin-Yang-algebra is penetrated by the principle of binary opposition of elements with even and odd indexes. But one can note that the principle of binary opposition penetrates many systems of the genetic code as well. Really, DNA has the double spiral configuration; each letter of the genetic alphabet has its binary-oppositional partner in a complimentary pair; amino acids have amphoteric properties (they demonstrate acid properties and alkaline properties simultaneously; a non-dissociated form of amino acids is transformed into a dipolar form under conditions of neutral water solution); etc. It seems that many such facts of binary oppositions in genetic systems possess hidden connections with the genetic Yin-Yang-algebra, which exists not accidentally.

The five essential coincidences between structures of the Yin-Yang matrix YY_8 and the genetic matrix $[C\ A;\ G\ U]^{(3)}$ were described in Chapter 7. These structural coincidences allow one to consider the octet algebra YY_8 as the adequate model of the structure of the genetic code. One can postulate such an algebraic model and then deduce some peculiarities of the genetic code from this model.

Inheritance of sexual attributes exists in living nature. The results of investigations in the field of matrix genetics allows one to suppose the following: 1) ensembles of binary-oppositional attributes, which exist in molecular-genetic systems, are related to the problem of male and female beginnings; 2) many phenomenological features of these ensembles can be expressed in the language of multidimensional numbers, first of all, in the language of the genetic octet Yin-Yang-algebra which gives new possibilities to investigate genetic ensembles of such binary oppositions and interrelations inside them.

Taking these assumptions into account, let us analyze evolutionary interrelations among different dialects of the genetic code from the viewpoint of the genetic octet Yin-Yang-algebra (another name is the genetic bipolar algebra).

THE COMPARISON ANALYSIS AND PHENOMENOLOGICAL RULES OF DIALECTS OF THE GENETIC CODE

Chapter 3 described in details, that many dialects of the genetic code are known in modern science. For this book all initial data about these dialects were taken from the website of the National Center for Biotechnology Information http://www.ncbi.nlm.nih.gov/Taxonomy/Utils/wprintgc.cgi. These dialects differ one from another through their specifics of the degeneracy (through concrete relations between 20 amino acids and 64 triplets). One can find from the data of the mentioned website, that 17 dialects are known only which differ one from another by the numbers of degeneracy of the amino acids (see these 17 dialects in the table on Figure 1 in Chapter 3.). A small quantity of the dialects from the website differ one from another by their start-codons only but not by the of the amino acids; we consider these dialects as the same dialect in our investigation.

Only some triplets change their code meaning in the different dialects in comparison with the basic case of the vertebrate mitochondria genetic code in the sense that they begin to encode other amino acids or stop-signals. What are those limitations which are utilized by nature in its choice of such changeable (or evolutional) triplets? Has the matrix disposition of these variable triplets any relation to the *YY*-coordinates $x_0, x_1, ..., x_7$ of the matrix YY_8 (Figure 3 and Figure 4 in Chapter 7.) and to their disposition in the genomatrix? Or the *YY*-coordinates have no relation to evolution of the genetic code and to systemic disposition of the variable triplets in the genomatrix $[C\ A;\ U\ G]^{(3)}$?

If such a relation is discovered, it gives additional evidence that the genetic octet Yin-Yang-algebra can be utilized as the adequate model of the genetic code or as the algebraic basis of the genetic code (the algebraic pre-code). It can be useful in tasks of sorting, putting in order and in deeper understanding of the genetic language. It can help to create new effective methods of information processing for many applied tasks as well. The appropriate algebraic model of the genetic code should give opportunities to deduce some evolutional peculiarities of the genetic code from such a fundamental mathematical system.

The results of corresponding comparison analysis have discovered the expressed connection between the disposition of the variable triplets in the genomatrix $[C\ A;\ U\ G]^{(3)}$ and the disposition of the *YY*-coordinates $x_0, x_1,.., x_7$ together with their signs "+" and "-" in the matrix YY_8. The obtained results lead to a few phenomenological rules of evolution of the dialects of the genetic code on the basis of the genetic octet Yin-Yang-algebra. In other words the scheme, which is defined by this matrix algebra, holds true in the evolution of the genetic code in some significant aspects. These results give additional evidence of appropriateness of such algebraic an approach in bioinformatics.

The matrix form of presentation of members of the genetic octet Yin-Yang-algebra (Figure 3 and Figure 4 in Chapter 7) contains 32 components with the sign "+" and 32 components with the sign "-". The matrix disposition of the components with the sign "+" fits the disposition of the 32 black triplets (the notion of black triplets was introduced in Chapter 2). These black triplets encode 8 kinds of the high-degeneracy amino acids Ala, Arg, Gly, Leu, Pro, Ser, Thr, Val, each of which is encoded by 4 triplets or more in the vertebrate mitochondrial genetic code, which is considered as the basic dialect. Other 12 amino acids are encoded by the white triplets. These 12 acids Asn, Asp, Cys, Gln, Glu, His, Ile, Lys, Met, Phe, Trp, Tyr are the low-degeneracy ones because each of them is encoded by 3 triplets or less. So the set of 20 amino acids consists of the canonical sub-set of the 8 high-degeneracy amino acids and the canonical sub-set of the 12 low-degeneracy amino acids. In the case of the vertebrate mitochondrial genetic code, the matrix disposition of these two canonical sub-sets fits the matrix disposition of the *YY*-coordinates with the signs "+" and "-" correspondingly.

But do these two sub-sets, which fit the algebraic features of the matrix YY_8, play any role in many other dialects of the genetic code? The positive answer to this question was presented in Chapter 2 already: the two non-trivial phenomenological rules № 1 and № 2 of evolution of the genetic code were demonstrated there, which are connected closely with these canonical sub-sets and hence with the matrix YY_8. These results are one of the important evidences of the adequacy of the 8-dimensional octet algebra YY_8 for the genetic code and its evolutionary peculiarities. Below the phenomenological rules №№ 3-7 will be presented as well, which were discovered from the viewpoint of the genetic octet Yin-Yang-algebra YY_8.

Let us continue the comparative analysis. As we mentioned above, only some triplets change their code meaning in the different dialects in comparison with the case of the vertebrate mitochondria code. What are those formal attributes which are utilized by nature in its choice of these evolutional changeable triplets from the set of 64 triplets? How these triplets and their appropriate amino acids are disposed in the genomatrix $[C\ A;\ U\ G]^{(3)}$ (Figure 3 and Figure 4 in Chapter 7)? Has the matrix disposition of these variable triplets any relation to the *YY*-coordinates $x_0, x_1, ..., x_7$ and to their disposition in the genomatrix? Can these variable triplets be associated naturally with the groups of the male and female *YY*-coordinates and triplets? Or do the *YY*-coordinates have no relation to evolution of the genetic code and to a systemic disposition of the variable triplets in the genomatrix $[C\ A;\ U\ G]^{(3)}$? This section continues the comparison analysis to answer such questions.

The table on Table 1 gives data for analysing these questions. The vertebrate mitochondrial genetic code (the code № 1) is utilized as the standard for comparison of code meanings of triplets in different dialects. The second tabular column shows those changeable triplets, which possess another code meaning (relative to their meaning in the dialect № 1) in the dialect which is named in the first column. A name of encoded amino acid or stop-codon (Stop) is given near each triplet in the second column in connection with the appropriate dialect named in the first column. Brackets in the second column contain that amino acid or stop-codon, which is encoded by this triplet in the dialect № 1. Each row of the second column is finished by the *YY*-coordinate, which is disposed together with this triplet in the same cell of the genomatrix on Figure 4 of Chapter 7. At last, the third column demonstrates data about start-codons, which define the beginning of protein synthesis in the considered dialect. An appropriate YY-coordinate is shown for each start-codon as well.

About Triplets which Change their Code Meaning

Let us analyze the data from the second column of the table on Figure 1. This column shows 14 kinds of the changeable triplets which possess different code meanings in different dialects: AAA, AGA, AGG, AUA, CUA, CUC, CUG, CUG, CUU, UAA, UAG, UCA, UGA, UUA. Some of these triplets have several meanings. For example the triplet AGA encodes the stop-signal in the dialect № 1, the amino acid Arg in the dialect № 4; and the amino acid Gly in the dialect № 8. Or the triplet UAA encodes

Table 1. The table about changeable triplets and start-codons in the dialects of the genetic code. Initial data are taken from http://www.ncbi.nlm.nih.gov/Taxonomy/Utils/wprintgc.cgi

Dialects of the genetic code	Changeable triplets	Start-codons
1) The Vertebrate Mitochondrial Code		AUU, $-x_6$ AUC, $-x_6$ AUA, $-x_7$ AUG, $-x_7$ GUG, x_3
2) The Standart Code	UGA, Stop (Trp), $-x_5$ AGG, Arg (Stop), $-x_5$ AGA, Arg (Stop), $-x_5$ AUA, Ile (Met), $-x_7$	UUG, $-x_7$ CUG, x_3 AUG, $-x_7$
3) The Mold, Protozoan, and Coelenterate Mitochondrial Code and the Mycoplasma/Spiroplasma Code	AGG, Arg (Stop), $-x_5$ AGA, Arg (Stop), $-x_5$ AUA, Ile (Met), $-x_7$	UUG, $-x_7$ UUA, $-x_7$ CUG, x_3 AUC, $-x_6$ AUU, $-x_6$ AUG, $-x_7$ AUA, $-x_7$ GUG, x_3
4) The Invertebrate Mitochondrial Code	AGG, Ser (Stop), $-x_5$ AGA, Ser (Stop), $-x_5$	UUG, $-x_7$ AUU, $-x_6$ AUC, $-x_6$ AUA, $-x_7$ AUG, $-x_7$ GUG, x_3
5) The Echinoderm and Flatworm Mitochondrial Code	AGG, Ser (Stop), $-x_5$ AGA, Ser (Stop), $-x_5$ AUA, Ile (Met), $-x_7$ AAA, Asn (Lys), $-x_7$	AUG, $-x_7$ GUG, x_3
6) The Euplotid Nuclear Code	UGA, Cys (Trp), $-x_5$ AGG, Arg (Stop), $-x_5$ AGA, Arg (Stop), $-x_5$ AUA, Ile (Met), $-x_7$	AUG, $-x_7$
7) The Bacterial and Plant Plastid Code	UGA, Stop (Trp), $-x_5$ AGG, Arg (Stop), $-x_5$ AGA, Arg (Stop), $-x_5$ AUA, Ile (Met), $-x_7$	UUG, $-x_7$ CUG, x_3 AUC, $-x_6$ AUU, $-x_6$ AUA, $-x_7$ AUG, $-x_7$
8) The Ascidian Mitochondrial Code	AGG, Gly (Stop), $-x_5$ AGA, Gly (Stop), $-x_5$	UUG, $-x_7$ AUA, $-x_7$ AUG, $-x_7$ GUG, x_3
9) The Alternative Flatworm Mitochondrial Code	UAA, Tyr (Stop), $-x_7$ AGG, Ser (Stop), $-x_5$ AGA, Ser (Stop), $-x_5$ AUA, Ile (Met), $-x_7$ AAA, Asn (Lys), $-x_7$	AUG, $-x_7$
10) Blepharisma Nuclear Code	UGA, Stop (Trp), $-x_5$ UAG, Gln (Stop), $-x_7$ AGG, Arg (Stop), $-x_5$ AGA, Arg (Stop), $-x_5$ AUA, Ile (Met), $-x_7$	AUG, $-x_7$

continued on the following page

Table 1. continued

Dialects of the genetic code	Changeable triplets	Start-codons
11) Chlorophycean Mitochondrial Code	UGA, Stop (Trp), $-x_5$ UAG, Leu (Stop), $-x_7$ AGG, Arg (Stop), $-x_5$ AGA, Arg (Stop), $-x_5$ AUA, Ile (Met), $-x_7$	AUG, $-x_7$
12) Trematode Mitochondrial Code	AGG, Ser (Stop), $-x_5$ AGA, Ser (Stop), $-x_5$ AAA, Asn (Lys), $-x_7$	AUG, $-x_7$ GUG, x_3
13) Scenedesmus obliquus Mitochondrial Code	UGA, Stop (Trp), $-x_5$ UAG, Leu (Stop), $-x_7$ UCA, Stop (Ser), x_5 AGG, Arg (Stop), $-x_5$ AGA, Arg (Stop), $-x_5$ AUA, Ile (Met), $-x_7$	AUG, $-x_7$
14) Thraustochytrium Mitochondrial Code	UGA, Stop (Trp), $-x_5$ UUA, Stop (Leu), $-x_7$ AGG, Arg (Stop), $-x_5$ AGA, Arg (Stop), $-x_5$ AUA, Ile (Met), $-x_7$	AUU, $-x_6$ AUG, $-x_7$ GUG, x_3
15) The Alternative Yeast Nuclear Code	UGA, Stop (Trp), $-x_5$ AGG, Arg (Stop), $-x_5$ AGA, Arg (Stop), $-x_5$ AUA, Ile (Met), $-x_7$ CUG, Ser (Leu), x_3	CUG, x_3 AUG, $-x_7$
16) The Yeast Mitochondrial Code	AGG, Arg (Stop), $-x_5$ AGA, Arg (Stop), $-x_5$ CUG, Thr (Leu), x_3 CUU, Thr (Leu), x_2 CUA, Thr (Leu), x_3 CUC, Thr (Leu), x_2	AUA, $-x_7$ AUG, $-x_7$
17) The Ciliate, Dasycladacean and Hexamita Nuclear Code	UGA, Stop (Trp), $-x_5$ UAG, Gln (Stop), $-x_7$ UAA, Gln (Stop), $-x_7$ AGG, Arg (Stop), $-x_5$ AGA, Arg (Stop), $-x_5$ AUA, Ile (Met), $-x_7$	AUG, $-x_7$

the stop-signal in the dialect № 1, the amino acid Tyr in the dialect № 9, and the amino acid Gln in the dialect № 17.

All kinds of changeable triplets are met 69 times in the second column. But only two kinds of the male *YY*-coordinates "$-x_5$" and "$-x_7$" with the sign "-" correspond to these triplets in all dialects practically. Specifically the male coordinate "$-x_5$" is met 41 times (it is 59,4% of all cases), and the male coordinate "$-x_7$"is met 22 times (it is 31,9% of all cases). It composes in sum more than 90% of all cases. The male coordinate "$+x_5$" is met 1 time in the dialect № 13 but with the sign "+". One can name the male *YY*-coordinates "$-x_5$", "$-x_7$" and "$+x_5$" as canonical Yin-Yang-coordinates for the changeable triplets (Figure 4 of Chapter 7). The described statistics allows one to formulate the following rule (in addition to two phenomenological rules in Chapter 3).

The Phenomenological Rule № 3 Connected with the Octet Yin-Yang-Algebra

Those triplets possess different code meanings in the different dialects of the genetic code, which correspond to the canonical male coordinates "$-x_5$", "$-x_7$" and "$+x_5$" of the matrix YY_8.

This rule is held true precisely for all the dialects besides the case of yeast with its two dialects: the dialect № 15, where the non-canonical male coordinate "$+x_3$" appears (for the triplet CUG), and the dialect № 16, which has the following unique feature. In this dialect № 16 the four triplets CUA, CUG, CUC, CUU, which are begun with the same pair of the letters (CU), change their code meanings by the identical way: all of them encode the acid Thr instead of the acid Leu (it is the unusual case because, if any other four triplets are begun with the equal pair of any letters, they do not change jointly their code meanings in other dialects). These four triplets correspond to the non-canonical *YY*-coordinates "$+x_2$" and "$+x_3$".

Yeast is unicellular mushrooms, chemoorganoheterotrophs, which are possible to vegetative cloning (asexual reproduction). Probably, the genetic-code deviation of the yeast from the rule № 3 is connected with their asexual reproduction and heterotrophy. (We noted in Chapter 3 already, that the dialects of the genetic code of the heterotrophic organisms, which feed on ready living substance, can have some deviations from the canonical forms of the dialects of autotrophic organisms, which produce living substance by using solar energy). The additional evidence of molecular-genetic singularity of yeast is the fact that the histone H1 is not discovered in their genetic system at all (http://drosophila.narod.ru/Review/histone.html).

The Connection between Evolution of the Genetic Code and the Anisotropy of the YY_8-Space

Chapter 7 has described the anisotropy of the coordinate space of the YY_8-numbers (the YY_8-space). The 8-dimensional YY_8-numbers $YY_8 = x_0*\mathbf{f_0}+x_1*\mathbf{m_1}+x_2*\mathbf{f_2}+x_3*\mathbf{m_3}+x_4*\mathbf{f_4} +x_5*\mathbf{m_5}+x_6*\mathbf{f_6}+x_7*\mathbf{m_7}$ have been interpreted as the double genetic quaternion. If all female coordinates are equal to zero ($x_0 = x_2 = x_4 = x_6 = 0$), we have the male variant of YY_8 in the form $(YY_8)_{MALE}$:

$$(YY_8)_{MALE} = x_1*\mathbf{m_1} + x_3*\mathbf{m_3} + x_5*\mathbf{m_5} + x_7*\mathbf{m_7} \quad (1)$$

The multiplication table of the basic elements $\mathbf{m_1}$, $\mathbf{m_3}$, $\mathbf{m_5}$, $\mathbf{m_7}$ of $(YY_8)_{MALE}$ coincides with the multiplication table of genetic quaternions $g = y_0*\mathbf{1} + y_1*\mathbf{i_1} + y_2*\mathbf{i_2} + y_3*\mathbf{i_3}$ on Figure 10 of Chapter 7. By analogy with Hamilton's quaternion, the first item $y0*\mathbf{1}$ (or $x_1*\mathbf{m_1}$ in the expression (1)) of genetic quaternions is called as their scalar part, and the sum of other three items is called as the vector part of genetic quaternions.

In accordance with Figure 11 of Chapter 7, these genoquaternions "*g*" possess the norm

$$x_1^2 + x_3^2 - x_5^2 - x_7^2 \quad (2)$$

The signature (+, +, -, -) of the norm (equation 2) of genoquaternions differs from the signature (+, +, +, +) of the norm of quaternions by Hamilton. This difference is very significant because it defines the following fundamental circumstance. The vector part $x_3*\mathbf{m_3} + x_5*\mathbf{m_5} + x_7*\mathbf{m_7}$ of genetic quaternions corresponds to the case of some anisotropic space in contrast to quaternions by Hamilton, the vector part of which corresponds to the case of the isotropic space. In the expression (2), this difference in the signatures of the norms is connected with the *YY*-coordinates x_5 and x_7, which can be named "anisotropic coordinates" for this reason. But these coordinates x_5 and x_7 are those, which correspond to the changeable triplets of the genetic code in accordance with the rule № 3. It is a very interesting fact that all evolution of code meanings of genetic triplets occurs practically in connection only with these anisotropic coordinates of the model space. Consequently the close connection between evolution of the genetic code and the anisotropy of this YY_8-space exists. For this reason, one can formulate the following rule № 4, which is a continuation of the rule № 3.

The Phenomenological Rule № 4, which is Connected with the Octet Yin-Yang Numbers and with the Anisotropy of YY_8-Space

In evolution of dialects of the genetic code, all changeable triplets correspond to the anisotropic male coordinates of genetic YY_8-numbers.

Similarly to rule № 3, rule № 4 has one exception: the case of yeast, which is characterized by asexual reproduction and heterotrophy and which changes the code meanings of the coordinates x_2 and x_3 additionally. It is obvious that the following **prediction** can be made. If new dialects of the genetic code are discovered in the future for organisms with bisexual reproduction, changeable triplets will correspond to the anisotropic male coordinates of genetic YY_8-numbers as well.

One can make one more remark about the male coordinates "$-x_5$", "$-x_7$", which are connected with more than 90% of all changeable triplets, as was mentioned above. All triplets, which correspond to these coordinates, change their code meanings besides the four invariable triplets: UGG with the coordinate "$-x_5$", and AAG, AUG, UUG with the coordinate "$-x_7$". Perhaps new dialects of the genetic code will be discovered in the future, where these triplets change their code meanings as well.

The Phenomenological Rule № 5 Connected with the Genetic Octet Yin-Yang-Numbers

All those 16 triplets, which correspond to the *YY*-coordinate x_0 and x_1 of the scalar part of genetic YY_8-numbers, never change their code meanings in the dialects of the genetic code (including the case of yeast).

Really, one can see that these coordinates x_0 and x_1 of the scalar part of YY_8-numbers are absent in the table on Figure 1 together with their 16 triplets CCC, CCA, CCU, CCG, CGC, CGA, CGU, CGG, GCC, GCA, GCU, GCG, GGC, GGA, GGU, GGG. So, the coordinates of the scalar part of the genetic YY_8-numbers define the absolute invariable part of the set of the genetic triplets.

About Stop-Codons

Encoding of stop-signals of protein synthesis turns on a special interest. Stop-signals are encoded by different triplets (stop-codons) in different dialects of the genetic code. The 7 kinds of triplets play the role of stop-codons in these dialects. Three of them (UUU, UAG, UUA) fit the *YY*-coordinate "$-x_7$". The other three triplets (AGA, AGG, UGA) fit the coordinate "$-x_5$". The seventh triplet (UCA) fits the coordinate "$+x_5$". All these coordinates are the anisotropic male *YY*-coordinates. Consequently the function of stop-codons is closely connected with the anisotropy of YY_8-space. The results of the investigation of stop-codons in the genetic dialects from the viewpoint of YY_8-algebra allow one to formulate the following rule.

The Phenomenological Rule № 6 Connected with the Octet Yin-Yang-Algebra and with the Anisotropy of the YY_8-Space

Those triplets serve as stop-codons in the dialects of the genetic code, which correspond to the anisotropic male *YY*-coordinates "$-x_5$", "$-x_7$" and "$+x_5$".

This rule is held true for all 17 dialects without exceptions. It draws attention to the fact that the function of stop-codons is the "male function" always from the viewpoint of YY_8-algebra because stop-codons are connected with the male coordinates. A few triplets exist (for example UUA and UGG), which correspond to the same coordinates "$-x_5$", "$-x_7$" and "$+x_5$" but which are not stop-codons in known dialects of the genetic code. Will such a dialect of the genetic code be discovered in the future, where these triplets play the role of stop-codons? Time will tell.

About Start-Codons

Till now we did not analyze start-codons (function of start-codons is the additional function of some triplets which they execute besides their basic function of coding of amino acids). The third column of the table on Figure 1 demonstrates those start-codons of the 17 dialects of the genetic code, which are presented in basic sets of code meanings of 64 triplets of the considered 17 dialects on the website http://www.ncbi.nlm.nih.gov/Taxonomy/Utils/wprintgc.cgi. Eight triplets play the role of start-codons in these 17 cases. The four of them (AUA, AUG, UUA, UUG) correspond to the *YY*-coordinate "$-x_7$". The two triplets (AUC, AUU) correspond to the coordinate "$-x_6$". The other two triplets (CUG, GUG) correspond to the coordinate "$+x_3$". The set of start-codons of the dialect № 1 corresponds to all these coordinates "$-x_7$", "$-x_6$" and "$+x_3$". These data allow one to formulate the additional rule about start-codons.

The Phenomenological Rule № 7 Connected with the Octet Yin-Yang-Algebra

All start-codons in the dialects of the genetic code correspond to *YY*-coordinates "$-x_7$", "$-x_6$" and "$+x_3$".

This rule is held true for all 17 dialects of the genetic code without exceptions. One can add that the start-codon AUG, which corresponds to the *YY*-coordinate "$-x_7$", is included in all the 17 dialects. All start-codons, which are presented in the table on Figure 1, have the letter U on their second position that reminds one about the U-algorithm of connection between genomatrices and Hadamard matrices (see Chapter 6).

THE MOLECULAR-SEXUAL APPROACH IN MOLECULAR GENETICS

Let us name each low-degeneracy amino acid, which is encoded by one of the female *YY*-coordinates x_0, x_2, x_4, x_6 only (see Figure 4 of Chapter 7), as the female amino acid conditionally. Such female amino acids are Asn_F, Asp_F, Cys_F, His_F, Ile_F, Phe_F, Tyr_F (we mark the female amino acids by means of the lower index F). Each low-degeneracy amino acid, which is encoded by one of the male *YY*-coordinates x_1, x_3,

x_5, x_7 only, is named the male amino acid correspondingly. Such male amino acids are Gln_M, Glu_M, Lys_M, Met_M, Trp_M (we mark the male acids by means of the lower index M).

The case of the high-degeneracy amino acids is more complex because such acids correspond to male and female *YY*-coordinates simultaneously (Figure 4 of Chapter 7). For example the acid Arg corresponds to x_0 and x_1. For this reason we name each of the high-degeneracy amino acids as the androgynous acid conditionally by analogy with androgynous individuals which possess male and female attributes simultaneously. The pure androgynous acids, each of which corresponds to the male and female *YY*-coordinates in equal degree, are Ala_A, Arg_A, Gly_A, Pro_A, Thr_A, Val_A (we mark androgynous acids by means of the lower index A). The amino acid Ser_A is disposed in 6 cells in the octet genomatrix (Figure 2 in Chapter 2), which correspond to the unequal quantities of the female and male *YY*-coordinates: the quantity of the female coordinates is equal to 4 and the quantity of the male coordinates is equal to 2. For this reason Ser_A is named the "androgynous acid of the female type". The amino acid Leu_A possess the symmetric-oppositional character relative to Ser_A because Leu_A is disposed in 6 matrix cells also but these cells correspond to 4 male coordinates and to 2 female coordinates (Figure 2 of Chapter 2). For this reason Leu_A is named the "androgynous acid of the male type". Leu_A and Ser_A form the sub-set of the quasi-androgynous acids.

All these additions to the names of amino acids are introduced on the basis of the vertebrate mitochondrial code (the code № 1 on Figure 1 in Chapter 3, and Table 1 in this chapter). But do what we have for the other 16 dialects of the genetic code, where some amino acids receive new correspondence to triplets and to *YY*-coordinates? What do such changes mean from the viewpoint of the notions of the genetic YY_8-algebra?

In accordance with the rule № 3, some male triplets of the dialect № 1 change their code meanings in the course of evolution of the genetic code (the case of yeast is the exceptional one). These male triplets encode the male and androgynous acids in the dialect № 1. The additional rule is that, if such changeable male triplet encodes another amino acid in another dialect of the genetic code, this new amino acid is one of the female acids necessarily (the case of yeast is the exceptional one). In other words, the expansion of the female acids (Asn_F, Cys_F, Ile_F, Tyr_F) into the male columns of the genomatrix $[C\ A;\ U\ G]^{(3)}$ (Figure 4 in Chapter 7) take place in the course of evolution. But the male amino acids never come to the female columns. As a result the genomatrix $[C\ A;\ U\ G]^{(3)}$, which possesses the equal qualities of the male and female amino acids in the dialect № 1, becomes the more female matrix in other dialects due to the prevalence of the female amino acids in the matrix cells there.

One can add that the androgynous acids Arg_A, Gly_A, Leu_A force out the male stop-codons in some dialects and take their places in the male columns of the genomatrix $[C\ A;\ U\ G]^{(3)}$. Figuratively speaking, the female beginning forces out the male beginning in the set of amino acids. On the other hand, the category of the male triplets increases its positions in the set of start-codons and stop-codons to guide punctuations of protein synthesis.

The male triplets encode not only all stop-codons in all dialects, but the set of start-codons becomes the more male set in the course of evolution: the single female coordinate "$-x_6$", which exists in the dialect № 1, is eliminated in most dialects. Really the dialects №№ 2, 5, 6, 8-13, 15-17 have no start-codons with the female *YY*-coordinates (Table 1). On the whole the evolution of the genetic code is the struggle between the male and female beginnings on the molecular-genetic level from the viewpoint of this algebraic model. It reminds one of the struggles between matriarchy and patriarchy in the history of human civilization. It reminds one of many other famous confrontations between the male and female beginnings as well. The creator of analytic psychology C. Yung subdivides the soul into the male and female beginnings, ratios of which can be changed in different periods of human life. The described data allow to formulate the following generalized rule of the struggle between male and female beginnings in evolution of the genetic code from the viewpoint of the genetic YY_8-algebra: in evolution of dialects of the genetic code, an increase of the set of triplets, which encode the female and androgynous amino acids, exists concerning the analogical set in vertebrate mitochondrial genetic code. The set of triplets, which encode the start-codons, becomes the more male set in this process.

The revelation of the structural division of the set of 20 amino acids into the sub-sets of the male, female and androgynous amino acids can be useful for modeling many astonishing phenomena in molecular genetics. The discussion is about the phenomena of mutual disclosure and of mutual attraction between two molecular one-specific partners, which lead to a formation of new molecular pairs; they take place in a medium of huge number of other molecules (molecular bouillon). Let us consider the following example.

THE EXAMPLE OF THE PAIRS OF HISTONES

It is known that nucleosome histones are important protein components of chromosomes. Below we utilize the well-known data about histones from the website of the National Center for Biotechnology Information http://www.ncbi.nlm.nih.gov/books/bv.fcgi?rid=mboc4.figgrp.632.

In eukaryote cells, filaments of DNA are coiled around nucleosomes, each of which is a shank consisting of the histones of the four types: H2A, H2B, H3 and H4. This set of four types is divided by nature into the pairs of one-specific histones. The histones H2A and H2B possess the important possibility to create the pair just one with another on the basis of their mutual revealing and mutual "attraction" in a molecular bouillon (by analogy with a male and a female individuals of one species among macroscopic biological organisms). Another pair consists of the histones H3 and H4, which possess the similar possibility to create the pairs just one with another on the analogical basis of their mutual revealing and mutual "attraction" in molecular bouillon.

A single nucleosome contains the ensemble of eight histones, where two histones of each of the four types H2A, H2B, H3 and H4 are included. The DNA molecule is reeled up on this octamer shank in the form of the left spiral. The structure of nucleosome plays the main role in the packing of DNA on all levels. Each nucleosome is formed in accordance with the principle of the multi-level recognition defined by the structures of the histones. Each histone molecule contains a central structured 3-spiral domain and non-structured N- and C-"tails". The one-specific histones identify one another and create their pairs. All creation of the octamer shank is based on the consecutive creation of pairs of the two one-specific molecular objects (Figure 1).

In the first step, the spiral domains cooperate among themselves. As a result, pairs (dimers) arise: one pair H3-H4 and two pairs H2A-H2B. In the second step, two first dimers form the pair association of the following level of complexity: the tetramer arises with two pairs H3-H4. In the third step, this tetramer forms a pair association of the higher level with two pairs H2A-H2B. As a result, the octamer of the histones arises. All these searches and copulations of one-specific histones into pairs, and then into new pairs from previous pairs occur in a molecular bouillon with a huge bedlam of biological molecules of other kinds and their splinters. It occurs despite of effects of electric shielding and other noise circumstances there.

Figure 1. The multistage association of pairs of the one-specific histones H3-H4 and H2A-H2B into the dimers, the tetramers and the octamers (this figure is taken from http://www.ncbi.nlm.nih.gov/books/bv.fcgi?rid=mboc4.figgrp.636)

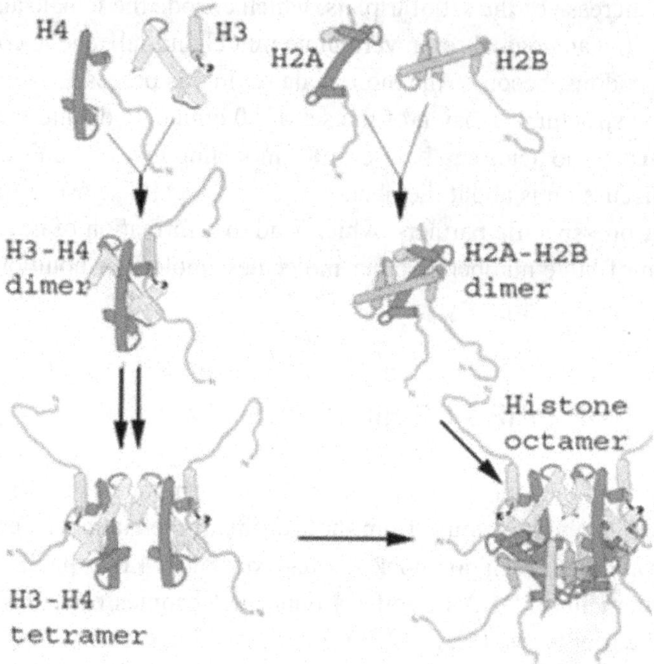

These phenomena of the micro world of molecules should be subordinated to principles of quantum mechanics. But their conclusion from these known principles is an excessive problem for modern science. These paired associations of the histones in molecular genetics carry the art name "hand shakes of molecules" traditionally (the multistage association of pairs of the one-specific histones H3-H4 and H2A-H2B into the dimers, the tetramers and the octamers can be found at http://www.ncbi.nlm.nih.gov/books/bv.fcgi?rid=mboc4.figgrp.636). But from the viewpoint of the our bisex theory, which is based on the genetic Yin-Yang algebras and which speaks about "sexual" interactions of genetic molecules, one can propose another art name for such pair search and association: "marriage", "love-crossing" or "love-copulation".

WHETHER AN UNKNOWN QUANTUM MECHANICAL FACTOR OF A "SEXUAL ATTRACTION" AMONG GENETIC MOLECULES EXISTS?

Such molecular-genetic facts form the basis to suspect, that phenomena of love or a love search of a sexual partner, which exist at a level of animal organisms, have arisen not on an empty place. But they are the continuation of those quantum mechanical phenomena of search of the one-specific partner into a mutual pair, which exist already at the level of genetic molecules, at least. The interrelations among genetic molecules in their search of each other can be interpreted as sexual relations to some extent.

Plato had formulated the famous statement about a congenital aspiration of each person to look for the second half. From the viewpoint of our bisex conception, which is based on the genetic Yin-Yang algebra, Plato's statement can be transferred into the world of those congenital properties of genetic molecules which are reflected in their search of their second halves.

In our opinion, taking into account the described facts, one can put forward the working hypothesis about existence of "a sexual intermolecular attraction" (or a "bipolar attraction") between genetic one-specific elements as a new biophysical factor of a quantum mechanical sense. This new hypothetical factor or principle is presented, first of all, as an explanation of molecular-genetic phenomena of search of the one-specific pair partner by multi-atomic bio-molecules to create a specific pair in complex conditions of multi-component bullion. The genetic Yin-Yang-algebra can be useful to model and investigate such a factor. This factor can have a force character and/or information character. It does not reject the existence of other known factors (for example, interactions of electric charges and so forth), but it is additional to them. Of course, it would be wrong to extend an action of this factor of "a sexual intermolecular attraction", which is proposed in connection with phenomena of assembly of pairs of one-specific multi-atomic molecular elements (multi-atomic quantum mechanical "modules"), into the field of all aspects of molecular-genetic organization.

Another example of the possible display of the hypothetical factor of "a sexual intermolecular attraction" (or a genetic bipolar attraction) in genetic systems gives the phenomenon which was discovered by Mirzabekov in studying a transfer RNA (Mirzabekov, 1997). The claim is that halves and quarters of these molecules can find each other in molecular bouillon and can gather in one molecule, which possesses a typical function of a transfer RNA.

THE ANALYSIS OF THE INSULIN STRUCTURE AS THE SIMPLEST EXAMPLE

The discovered connection between the genetic code and the Yin-Yang-algebra gives the opportunity of classification of many molecular elements in accordance with their "sexual" characteristics from the table on Table 2. The knowledge about such hidden structure of the set of genetic molecules can be useful for a study, an explanation and a prediction of features of interactions of molecular elements with different sexual characteristics.

For example, all sets of proteins (and their genes) can be divided conditionally into sub-sets of male, female and androgynous proteins in accordance with the sexual type of those amino acids, which form the majority in compositions of these proteins. As a result new information arises about structural characteristics of proteins. Let us make the first attempt to analyze a protein from the viewpoint of the genetic Yin-Yang-algebra. For this first attempt we take the simplest protein – insulin, which contains

Table 2. The classification of the 20 amino acids as female acids, male acids and androgynous acids in accordance with the kinds of their connections with the female and male YY-coordinates on Figure 4 in Chapter 7.

Female acids	*Male acids*	*Androgynous acids*	*Quasi-androgynous acids*
Asn_P Asp_P Cys_P His_P Ile_P Phe_P Tyr_F	Gln_M, Glu_M, Lys_M, Met_M, Trp_M	Ala_A, Arg_A, Gly_A, Pro_A, Thr_A, Val_A	Leu_A, Ser_A

51 amino acids. The set of these acids is encoded by 51 triplets in the gene of insulin. Does this genetic sequence of triplets have any regularity from the viewpoint of the genetic Yin-Yang algebra?

The insulin consists of two chains: the α-chain and the β-chain. The α-chain contains 21 amino acids and the β-chain contains 30 amino acids. These chains are encoded by the genetic sequences (Table 3), which are taken from the text-book (Inge-Vechtomov, 1983, p. 321-323). Taking into account the table on Table 2, the following intrinsic "sexual" structure is revealed for each sequence with its male, female and androgynous amino acids (Table 3).

The first results of this analysis are the following. The genetic sequences of both chains contain equal quantities of the female triplets (10 female triplets) and of the male triplets (4 male triplets). All other triplets are androgynous ones. From the described algebraic viewpoint, insulin and its gene are a female protein and a female gene because female acids and female triplets dominate there.

The α-chain has the following set of the female links: 2ATC(Ile,$-x_6$)$_F$, 6TGT(Cys,$-x_4$)$_F$, 7TGC(Cys,$-x_4$)$_F$, 10ATC(Ile,$-x_6$)$_F$, 11TGC(Cys,$-x_4$)$_F$, 14TAC(Tyr,$-x_6$)$_F$, 18AAC(Asn,$-x_6$)$_F$, 19TAC(Tyr,$-x_6$)$_F$, 20TGT(Cys,$-x_4$)$_F$, 21AAC(Asn,$-x_6$)$_F$. This set of female links shows unexpectedly the regularity of double composition in its structure (the phenomenon of doubling).

The essence of this phenomenon is that each kind of female triplet and female amino acid is met twice in the α-chain exactly. (Below we note that the analogical phenomenon of doubling takes place for the set of male links as well). Really, the female triplet ATC exists there in the links №№ 2 and 10; the triplet TGT – in the links №№ 6 and 20; the triplet TGC – in the links №№ 7 and 11; the triplet TAC – in the links №№ 14 and 19; the triplet AAC – in the links №№ 18 and 21.

Table 3. The genetic sequences of triplets for the α-chain and the β-chain of insulin. The following data are shown for each triplet: its current number in the sequence; the encoded amino acid; its YY-coordinate from Figure 4 in Chapter 7. Female links, which contain female amino acids, are marked by the index F; male links are marked by the index M; androgynous links are marked by the index A. The genetic letter T (thymine) is used instead of the letter U, but it does not matter

α-chain:
1GGC(Gly, x_0)$_A$ → 2ATC(Ile, $-x_6$)$_F$ → 3GTT(Val, x_2)$_A$ → 4GAA(Glu, $-x_3$)$_M$ → 5CAG(Gln, $-x_3$)$_M$ → 6TGT(Cys, $-x_4$)$_F$ → 7TGC(Cys, $-x_4$)$_F$ → 8ACT(Thr, x_4)$_A$ → 9TCT(Ser, x_4)$_A$ → 10ATC(Ile, $-x_6$)$_F$ → 11TGC(Cys, $-x_4$)$_F$ → 12TCT(Ser, x_4)$_A$ → 13CTT(Leu, x_2)$_A$ → 14TAC(Tyr, $-x_6$)$_F$ → 15CAG(Gln, $-x_3$)$_M$ → 16CTT(Leu, x_2)$_A$ → 17GAG(Glu, $-x_3$)$_M$ → 18AAC(Asn, $-x_6$)$_F$ → 9TAC(Tyr, $-x_6$)$_F$ → 20TGT(Cys, $-x_4$)$_F$ →21AAC(Asn, $-x_6$)$_F$
β-chain:
1TTC(Phe, $-x_6$)$_F$ → 2GTC(Val, x_2)$_A$ → 3AAT(Asn, $-x_6$)$_F$ → 4CAG(Gln, $-x_3$)$_M$ → 5CAC(His, $-x_2$)$_F$ → 6CTT(Leu, x_2)$_A$ → 7TGT(Cys, $-x_4$)$_F$ → 8GGT(Gly, x_0)$_A$ → 9TCT(Ser, x_4)$_A$ → 10CAC(His, $-x_2$)$_F$ → 11CTC(Leu, x_2)$_A$ → 12GTT(Val, x_2)$_A$ → 13GAA(Glu, $-x_3$)$_M$ → 14GCT(Ala, x_0)$_A$ → 15TTG(Leu, $-x_7$)$_A$ → 16TAC(Tyr, $-x_6$)$_F$ → 17CTT(Leu, x_2)$_A$ → 18GTT(Val, x_2)$_A$ → 19TGC(Cys, $-x_4$)$_F$ → 20GGT(Gly, x_0)$_A$ → 21GAA(Glu, $-x_3$)$_M$ → 22CGT(Arg, x_0)$_A$ → 23GGT(Gly, x_0)$_A$ → 24TTC(Phe, $-x_6$)$_F$ → 25TTC(Phe, $-x_6$)$_F$ → 26TAC(Tyr, $-x_6$)$_F$ → 27ACT(Thr, x_4)$_A$ → 28CCT(Pro, x_0)$_A$ → 29AAG(Lys, $-x_7$)$_M$ → 30ACT(Thr, x_4)$_A$

Let us make a small deviation into the field of linguistics. It is well-known in molecular genetics that "*the more we understand laws of coding of genetic information, the more their similarity to principles of linguistics amazes*" (Ratner, 2002, p.203).

Does the described phenomenon of doubling in the case of the simplest protein (insulin) have a similarity with the structural principles of the simplest words of human languages? Yes, it has. Really the simplest words of different human languages demonstrate the same phenomenon of doubling or of their construction on the basis of doubling syllables. Such words are utilized by babies, when they start to speak; they are used intuitively by mothers in dialogue with babies; they are most digestible and exploitable at training speech in the case of deaf-and-dumb people; they are utilized by different nations for speech imitation of the sounds of world around: "mama", "papa", "baba", "wee-wee", etc. (the Russian language, which is native for one of the authors of this book, has a lot of examples of such simple words with doubling syllables). In process of development of speech, this primitive principle of construction of the simplest words with doubling letters is overcome gradually. These data are the addition to the famous conception that linguistic languages have arisen not in an empty space but they are a continuation of the genetic language (Baily, 1982; Jacob, 1974, 1977; Makovskiy, 1992; Petoukhov, 2003, 2003-2004; Jacobson, 1987, 1999; Yam, 1995; etc).

One small addition can be made also to this concept. The theory of artificial and computer languages demonstrates that there is no necessity at all to include in languages a division of the whole set of nouns (and some other language elements) into sub-sets of nouns of masculine gender, of feminine gender and of neuter gender. But the natural human languages possess such division of the set of all nouns into sub-sets of nouns of such three genders. If the human languages are a continuation of the genetic language (Jacobson, 1987, 1999), then the genetic language should possess such division of the whole set of its elements into sub-sets of masculine, feminine and neuter genders. Our algebraic investigation of the genetic code confirms this conception by means of the discovery of the algebraic division of the sets of elements of the genetic languages into sub-sets of elements of masculine, feminine and neuter (androgynous) genders. In our opinion, these three genders of elements in genetic systems exist due to intrinsic features of the genetic Yin-Yang-algebra in connection with the fundamental task of noise- immunity of genetic coding.

Let us return to insulin. The set of the female links of the α-chain contains 5 different triplets ATC, TGT, TGC, TAC, AAC, but 4 different amino acids exist there because the acid Cys exists in two pairs of links. Each of these pairs is encoded by its own triplet – TGT or TGC, but these triplets do not differ by their coding meanings because they encode the same amino acid Cys in all dialects of the genetic code. This whole set of the female links corresponds to two *YY*-coordinates "$-x_6$" and "$-x_4$" only. The number of repetitions of each of these coordinates is an even number: "$-x_6$" is repeated 6 times, and "$-x_4$" is repeated 4 times. One can recall that all *YY*-coordinates have the one-to-one relation with the letter composition of triplets (Figure 4 of Chapter 7). In accordance with this connection, each abstract *YY*-coordinate (for example "$-x_6$") presents an algorithmically defined number ("$-\beta\beta\gamma$" in this example), which is based on real parameters of the molecules of the genetic alphabet.

The male links of the α-chain are 4GAA(Glu,$-x_3$)$_M$, 5CAG(Gln,$-x_3$)$_M$, 15CAG(Gln,$-x_3$)$_M$, 17GAG(Glu,$-x_3$)$_M$. Each of their amino acids Glu and Gln is repeated twice again. All these links correspond to the same *YY*-coordinate "$-x_3$". The triplet CAG is repeated twice. The triplets GAA and GAG do not differ in their coding meanings because they encode the same amino acid Gln in all dialects of the genetic code. So the phenomenon of doubling exists for the male links of the α-chain as well.

Each of the quasi-androgynous acids Ser_A and Leu_A exists in the α-chain twice as well: 9TCT(Ser, $x_4)_A$, 12TCT(Ser, $x_4)_A$, 13CTT(Leu, $x_2)_A$, 16CTT(Leu, $x_2)_A$. These acids and their triplets TCT and CTT correspond to the female YY-coordinates "x_2" and "x_4".

Now let us consider the β-chain of insulin (Table 3) with the following female links: 1TTC(Phe,-$x_6)_F$, 3AAT(Asn,-$x_6)_F$, 5CAC(His,-$x_2)_F$, 7TGT(Cys,-$x_4)_F$, 10CAC(His,-$x_2)_F$, 16TAC(Tyr,-$x_6)_F$, 19TGC(Cys,-$x_4)_F$, 24TTC(Phe,-$x_6)_F$, 25TTC(Phe,-$x_6)_F$, 26TAC(Tyr,-$x_6)_F$. The phenomenon of doubling exists for the whole β-chain except for its first three links Phe-Val-Asn (which correspond to the tripeptide). Really the main part of the β-chain contains the following female amino acids twice: His (encoded by the triplet CAC); Tyr (encoded by the triplet TAC); Phe (encoded by the triplet TTC); Cys (encoded by the triplets TGT and TGC, which do not differ in their code meanings in all dialects of the genetic code because they encode the same acid Cys always). The female links in the β-chain correspond to the *YY*-coordinates "-x_6", "-x_4" and "-x_2".

The set of the male links in the β-chain contains 4CAG(Gln, -$x_3)_M$, 13GAA(Glu, -$x_3)_M$, 21GAA(Glu, -$x_3)_M$, 29AAG(Lys, -$x_7)_M$. This set coincides with the set of the male links of the α-chain with the exception of the link № 29, which is next to last in the β-chain.

In contrast to the α-chain, where all 4 male triplets and acids correspond to the *YY*-coordinate "-x_3", the coordinate "-x_3" of the last male link in the β-chain is replaced by the coordinate "-x_7". It breaks one of the male pairs: the triplet AAG, which encodes Lys, exists here instead of the triplet CAG, which differs by the first letter only and which encodes the acid Gln. The second male pair of the links №№ 13 and 21 submits to the phenomenon of doubling because the triplet GAA encodes the acid Glu in both these links.

Is this phenomenon of doubling connected with the 3D-construction of insulin (and of those proteins, which possess the same phenomenon) by means of any regular metric or vector relations in space dispositions of pairs of such links? How widely and precisely is the phenomenon of doubling in male and female sub-sets carried out for different proteins? Many such questions arise as a result of the analysis of objects of molecular genetics from the viewpoint of the genetic octet Yin-Yang-algebra. They should be investigated in the future.

SOME APPLICATIONS OF RESULTS OF MATRIX GENETICS IN BIOINFORMATICS AND ALGEBRAIC BIOLOGY

Symmetries of genetic systems in their matrix forms of presentation are connected closely to algebraic biology which is a branch of theoretical and mathematical biology. Algebraic biology uses tools from symbolic computation, algebra, algebraic geometry, and discrete mathematics for the modeling and analysis of biological systems. Examples of application areas include all aspects of systems biology and -omics data analysis, functional genomics, evolutionary biology, synthetic biology, and cell biology. Algebraic analysis of matrix forms of presentation of genetic systems described in this book can be useful for many of these application areas.

Each of these possible applications needs special historical, biological and mathematical introductions. Taking into account a limited volume of this book we describe here one example of such applications. The example is connected to a method of Chaos Game Representation (CGR) of genetic sequences which is well-known in bioinformatics and molecular biology. This iterative mapping method allows one to convert a nucleotide sequence into a scale-independent and unique visual image. This was introduced in genome analysis in the work (Jeffrey, 1990). In this work J. Jeffrey proposed a visualization of a non-randomness character of DNA sequences by means of a chaos game algorithm for four points. Such an approach permits the representation and investigation of patterns in sequences, visually revealing previously unknown structures. Based on a technique from chaotic dynamics, the method produces an image of a gene sequence which displays both local and global patterns. The images have a complex structure which varies depending on the sequence. CGR raises a new set of questions about the structure of DNA sequences, and is a new tool for investigating gene structure. Here we should reproduce some materials about CGR from the work by Jeffrey because we will describe a new variant of this method for investigations of amino acids sequences of proteins on the basis of the genetic Yin-Yang algebra.

During the past 15 years a new field of physics known as 'non-linear dynamics', 'chaotic dynamical systems', or simply 'chaos' (Barnsley, 1988; Devaney, 1989) has been developed. Central parts of the field are questions of the structure of certain complex curves known as 'fractals'. The Chaos Game is an algorithm which allows one to produce pictures of fractal structures. In simplest form, it proceeds as follows:

1. Locate three dots on a piece of paper. They can be anywhere, as long as they are not all on a line. We will call these dots vertices.
2. Label one vertex with the numerals 1 and 2, one of the others with the numerals 3 and 4, and the third with the numerals 5 and 6.
3. Pick a point anywhere on the paper, and mark it. This is the initial point.
4. Roll a 6-sided die. Since in Step 2 the vertices were labeled, the number that comes up on the die is a label on a vertex. Thus, the number rolled on the die picks out a vertex. On the paper, place a mark half way between the previous point and the indicated vertex. (The first time the die is rolled, the 'previous point' is the initial point picked in Step 3.) For example, if 3 is rolled, place a mark on the paper half way between the previous point and the vertex labelled '3'.
5. Continue to roll the die, on each roll marking the paper at the point half way between the previous point and the indicated vertex.

One might expect that this procedure, if repeated many times, would yield a paper covered with random dots or, perhaps, a triangle filled with random dots. It turns out this is not the case. In fact, if the Chaos Game is run for several thousand points, the result is a beautiful fractal figure which is known in mathematics for many years under the name 'Sierpinski triangle', after the mathematician who first defined it. For the cases of five points, six, or seven initial points the chaos game produces a figure with visible patterns (pentagons within pentagons, a striated hexagon, or heptagons within heptagons), but for eight or more point the game yields essentially a filled-in polygon, except that the center is empty.

With four initial points, however, the result is different. It is not squares within squares, as one might expect; in fact there is no pattern at all. The chaos game on four points produces a square uniformly and randomly filled with dots in the case of random rolling of a 6-sided die. The picture produced by the chaos game is known as the attractor. Mathematically, the chaos game is described by an iterated function system (IFS).

If a sequence of numbers is used to produce an attractor for an IFS code and that attractor has visually observable then we have, intuitively, revealed some underlying structure in the sequence of numbers. Experiments (Jeffrey, 1990) had shown that the Chaos Game can be used to display certain kinds of non-randomness visually. This led to the following question. Since a genetic sequence can be treated formally as a string composed from the four letters 'a', 'c', 'g', and 't' (or 'u'), what patterns will it arise if we use Chaos Game not a series of random numbers but DNA sequences in their relevant numeric form? Instead of 'rolling a 4-sided die', use the next base (a, c, g, t/u) to pick the next point. Each of the four corners of the square is labeled by symbols 'a', 'c', 'g', or 'u'. If a 'c', for example, is the next base, then a point is plotted half way between the previous point and the 'c' corner. In such a way Jeffrey has obtained a set of CGR patterns of different DNA sequences. These patterns had fractal characters and they were used for many tasks of comparative analysis in bioinformatics and for formulating new scientific questions.

After Jeffrey's work, many scientists studied CGR patterns of nucleotide sequences (Goldman, 1993; Gutierrez, Rodriguez & Abramson, 2001; Joseph & Sasikumar, 2006; Oliver al., 1993; Tavassoly et al., 2007a, 2007b; Wang et al., 2005). As the result, "*alignment free methods based on Chaos Game Representation (CGR), also known as sequence signature approaches, have proven of great interest for DNA sequence analysis. Indeed, they have been successfully applied for sequence comparison, phylogeny, detection of horizontal transfers or extraction of representative motifs in regulation sequences*" (Deschavanne, Tuffey, 2008, p. 615). The most of these works were devoted to CGR patterns of nucleotide sequences. It is interesting to use the analogical CGR method for studying of amino acids sequences in proteins. This using can lead to a disclosure of hidden symmetries and other regularities in sets of protein for problems of bioinformatics, evolutionary biology, etc. In contrast to a quantity of works about CGR of nucleotides sequences, much lesser quantity of works about CGR of amino acids sequences of proteins exist because twenty (not four) different amino acids exist (Basu, et al., 1997; Fiser, Tusnady & Simon, 1994; Yu, Anh, & Lau, 2004). But the discovery of the genetic Yin-Yang-algebra described in this book allows getting round this difficulty. One can say that a new type of CGR of proteins exists now which is based on the notions from the genetic algebra.

The genetic Yin-Yang-algebra is considered as the model of the genetic code. From the viewpoint of this algebraic model, the set of 20 amino acids contains the following 4 subsets in the case of the most symmetrical dialect of the genetic code (the vertebrate mitochondrial genetic code). The first subset contains amino acids Gln, Glu, Lys, Met, Trp which are coded by triplets possessing purine suffixes (purine A or G is on the third positions in these triplets). The acids from this subset were named as "male" acids or "purine acids" conditionally. The second subset contains amino acids Asn, Asp, Cys, His, Ile, Phe, Tyr which are coded by triplets possessing pyrimidine suffixes (pyrimidine C or U/T is on the third positions in these triplets). The acids from this subset were named as "female" acids or "pyrimidine acids". The third subset contains amino acids Ala, Arg, Gly, Pro, Thr, Val which are coded by triplets possessing purine suffixes and pyrimidine suffixes in equal quantities. The acids from this subset were named as "pure androgynous". The fourth subset contains amino acids Ser and Leu which are coded by triplets with purine suffixes and pyrimidine suffixes in unequal quantities. The acids from this subset were named as "androgynous of the female and male types" (see more details in this Chapter above).

These four subsets form a special kind of a 4-letters alphabet for the types of amino acids as considered. Such an alphabet can be used to obtain CGR patterns in the case of amino acids sequences of proteins by analogy with the described case of CGR patterns for nucleotide sequences. O. Tavassoly was the first researcher who proposed using this new "sexual" alphabet of the types of amino acids for a construction of CGR patterns of proteins. Initial results of such a construction are presented in details in the work (Tavassoly, Petoukhov, & Vahedi, 2009). In these works each of mentioned 4 subsets were symbolized by a number: the first subset was symbolized by number 0, the second subset – by number

2, the third – by number 1, the fourth – by number 3. These numbers are used as vertexes of a square for CGR patterns. Each of amino acids in a considered sequence is replaced by a relevant number 0, 1, 2 or 3. In this way, for instance, a symbolic sequence Asp-Glu-Ser-Arg-Cys-Leu-... is transformed into a numeric sequence 2-0-3-1-2-3-....The first point of a future CGR pattern is placed half way between the center of the square and the vertex corresponding to the first letter of a considered numeric sequence; its i-th point is then placed half way between the (i-1)-th point and the vertex corresponding to the i-th letter.

Here we present an example of a CGR pattern obtained in such way (as in Tavassoly, Petoukhov & Vahedi, 2009) for the longest protein - Human Titin which contains 34,350 acids. Its sequence used for producing the CGR image was obtained from GeneBank (http://www.ncbi.nlm.nih.gov), where this protein has Accession Number: CAD12456. The algorithms of chaos game for four vertexes from this genetic-algebraic viewpoint were coded in MATLAB 7.6 (http://www.mathworks.com).

Titin is the longest polypeptide yet described and an abundant protein of striated muscle. Titin also contains binding sites for muscle associated proteins so it serves as an adhesion template for the assembly of contractile machinery in muscle cells (Labeit et al., 1990). Mutations in the titin gene are associated with familial hypertrophic cardiomyopathy (Siu et al.,1999) and tibial muscular dystrophy (Hackman et al.,2002). Autoantibodies to titin are produced in patients with the autoimmune disease scleroderma (Machado et al., 1998).

The produced CGR pattern has a fractal structure or a self-similarity character (Figure 2). One can see that the CGR pattern of the whole square is reproduced in each quadrant of the square, and in each subquadrant, etc. It has a certain analogy with CGS patterns in the work (Jeffrey, 1990). In addition the whole CGR pattern and its small pieces in the Titin case possess some kinds of symmetries along diagonals, etc. Identifying chaos in experimental data such as biological sequences, one can search for a

Figure 2. The CGR pattern of the titin protein (from Tavassoly, Petoukhov & Vahedi, 2009)

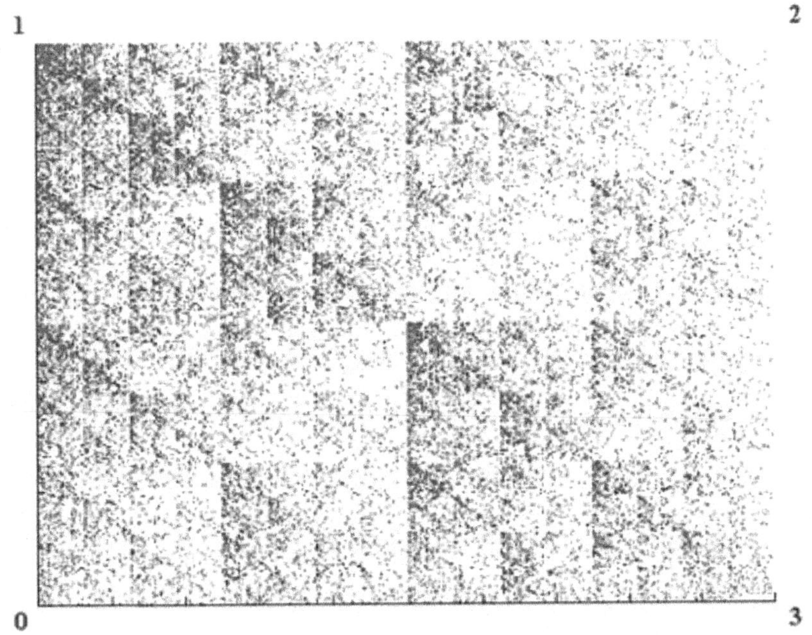

strange attractor in the special dynamics, identified by its fractal structure. Having found such an attractor, one can try to estimate its dimension, which is a measure of the number of active variables and hence the complexity of the equations required to model the dynamics. Fractals are useful geometric manifestation of the chaotic dynamics. They are called "the fingerprints of chaos" sometimes. The revealing the fractal structure of such CGR patterns of titin and many other proteins shows hidden connections of protein structures with non-linear dynamics or chaotic dynamical systems as the new field of physics. It reveals also some relations between the genetic Yin-Yang-algebras and non-linear dynamics.

The disclosure of fractals which are connected with proteomic sequences in the described variant of the CGR method, allows using this new knowledge in different branches of bioinformatics, molecular and evolutionary biology. For example, fractal properties of CGR images such as fractal dimension and multifractal spectra are the tools for genomic and proteomic sequences comparison and phylogeny studies including protein classifications where many new symmetrical patterns arise in bioinformatics.

SOLUTIONS AND RECOMMENDATIONS

The results described in this chapter about connections of the evolution of the genetic code with the features of the 8-dimensional Yin-Yang algebra generate many thoughts, including thoughts about possible modeling of biological evolutionary processes on the basis of this algebra. But evolutionary processes last during time; they depend on time. Till now in relation to biological applications of the Yin-Yang-algebra we spoke about static parameters of the genetic code only. Does this algebra give any opportunities to model kinematic processes, which depend on time? For example, does this algebra give opportunities to model cyclic processes, which are so typical for biologic organisms and which were discussed in Chapter 3 in the chronocyclic conception already? Yes, the system of Yin-Yang numbers allows one to work with kinematic processes as well.

Really, each of *YY*-coordinates can be presented as a variable function of time. In this case we have expression (1) in the form:

$$YY_8 = x_0(t)*\mathbf{f}_0+x_1(t)*\mathbf{m}_1+x_2(t)*\mathbf{f}_2+x_3(t)*\mathbf{m}_3+x_4(t)*\mathbf{f}_4+x_5(t)*\mathbf{m}_5+x_6(t)*\mathbf{f}_6+x_7(t)*\mathbf{m}_7 \quad (3)$$

Multiplication or addition of any two Yin-Yang numbers, coordinates of which are functions of time, give a new Yin-Yang number, coordinates of which are new functions of time as well. One of many possible kinds of such functions is the trigonometric functions: *sin*(*wt*), *cos*(*wt*), etc. These functions are used in Fourier spectral analysis of signals, and this Fourier analysis can be used in the case of the 8-dimensional functions YY_8 (3) as well. The other interesting possible kind is Walsh-Hadamard functions. We recommend using these mathematical opportunities, first of all, for the description of many kinds of hierarchical cyclic processes in biologic systems.

One should note the interesting problem of the left-right symmetry in biological objects, which was mentioned in Chapter 7 already. The binary opposition "male-female" is connected with the binary opposition "the left side and the right side" in the history of various nations. It is known also, that the opposition "left-right" was one of the kinds of interpretation of the opposition "Yin-Yang" in Ancient China. Phenomena of curling of some genetic molecules into the left side or into the right side are well known. One of the examples is the helix structure of DNA.

The described data about connections of molecular-genetic systems with the Yin-Yang algebra put forward some questions about "the left" aspects and "the right" aspects in the organization of ensembles of genetic molecules including proteins, which can be twisted into the left side or into the right side on their various parts. For example, are there some differences between proteins of the male kind and of the female kind relative to a direction of their preferred (left or right) twisting on the whole? Do some differences of this twist category exist between male and female regions of separate proteins, which include various regions with male and female amino acids in their composition? One can recommend special investigations to answer such questions.

All history of science shows the significance of such investigation of natural systems, which discover their hidden sub-structures with different characteristics. The union of these sub-structures defines the specific properties of the whole system. Modern science gives many examples of the importance of discoveries of hidden sub-structures and of their interactions in natural systems. One of them is the important theoretical discovery of quarks in the physics of elementary particles. The theory of quarks speaks about internal compositions of elementary particles and suggests a classification of a set of such particles. Nobody saw quarks in experiments, but this fact does not prevent the use of the theory of quarks widely. The described Yin-Yang-algebraic theory of the genetic code implies a classification and investigations of molecular-genetic systems by means of well-reasoned mathematical instruments.

It is difficult to list all possible consequences and all new possible investigations, which are brought by the discovery of the algebraic structure of the genetic code. We recommend the further wide development of this Pythagorean approach to genetic systems and to the problem of their evolution. Such a Pythagorean approach can give very useful results in the fields of bioinformatics, theoretical biology and applied sciences.

FUTURE TRENDS AND CONCLUSION

The main task of the mathematical natural sciences is the creation of mathematical models of natural systems, which can describe these systems in adequate manner. One can see that the algebraic model YY_8, which is described in this book for the genetic code, fits this task.

It seems that the results of these algebraic-genetic investigations can be useful in some practical tasks as well, for example, in the task of selection of the appropriate sex-partner for an individual by means of personal analysis of molecular-genetic structures of different persons. Similar tasks are not fantasy but they exist already on the world market of genetic services (see for example the website https://www.23andme.com/).

Are the genetic Yin-Yang-algebras suitable for the description of the phenomena in the non-living substance and of evolution of these phenomena? Or does a line of demarcation exist between living substance and non-living substance? It is one of the important questions for investigations in the future. Other important questions concern new understanding of Yin-Yang oppositions in genetic systems on the basis of Yin-Yang-algebraic models. In our opinion, this new understanding and new algebraic models will give many useful results.

Probably along a way of introduction of systems of Yin-Yang numbers into mathematical natural sciences, the conceptual and formal instruments of natural sciences will come closer to a structure of living substance. And science will get anthropomorphic features and will become a continuation of the person with its male and female beginnings. Briefly speaking, it will be realized "an anthropomorphic principle of development of natural sciences". The given principle is associated with the famous opinion (Teilhard de Chardin, 1959) that future synthetic science will take for its basis of the person, and that the person, as a subject of knowledge, is a key to all science about nature.

The results of this chapter give additional evidences that the order and the evolution of the genetic code are connected with the special 8-dimensional Yin-Yang numeric system. One can think that this numeric system is the algebraic key to many secrets of the genetic code, nature of which has system-numeric character. This approach coincides with the Pythagorean tradition to explain all things by means of systems of numbers (see Chapter 7). For modeling of the genetic code, the systems of Yin-Yang numbers are prominent in the infinite set of other possible numeric systems.

The structure of the vertebrate mitochondrial genetic code, which is the most symmetric dialect of the genetic code, is a mould (or cast, or copy) of the structure of the Yin-Yang numeric system in significant aspects. In the course of evolution, small deviations from this mould arise in connection with new nascent forms of nutrition, reproduction, etc. In our opinion, the Pythagorean approach to such evolutionary overcoming the initial dialect of the genetic code can be useful in future investigations as well.

New molecular-sexual approach has arisen to study interactions between biological molecules, which belong to different "sexual" kinds. New research mathematical instruments are proposed, which connected closely with methods of symmetries and with new patterns in molecular genetics and bioinformatics.

REFERENCES

Almeida, J. S., Carrico, J. A., Maretzek, A. M., Noble, P. A., & Fletcher, M. (2001). Analysis of genomic sequences by chaos game representation. *Bioinformatics (Oxford, England)*, *17*, 429–437. doi:10.1093/bioinformatics/17.5.429

Baily, C. J. L. (1982). *On the yin and yang nature of language*. London: Ann Arbor.

Barnsley, M. F. (1988). *Fractals everywhere*. New York: Springer-Verlag.

Basu, S., Pan, A., Dutta, C., & Das, J. (1997). Chaos game representation of proteins. *Journal of Molecular Graphics & Modelling*, (October): 279–289. doi:10.1016/S1093-3263(97)00106-X

Bull, J. J. (1983). *Evolution of sex determining mechanisms*. Menlo Park, CA: WA Benjamin/Cummings.

Deschavanne, P., Giron, A., Vilain, J., Fagot, G., & Fertil, B. (1999). Genomic signature: Characterization and classification of species assessed by chaos game representation of sequences. *Molecular Biology and Evolution*, *16*(October), 1391–1399.

Deschavanne, P., & Tuffey, P. (2008). Exploring an alignment free approach for protein classification and structural class prediction. *Biochimie*, *90*, 615–625. doi:10.1016/j.biochi.2007.11.004

Devaney, R. L. (1989). *An introduction to chaotic dynamical systems*. Redwood City, CA: Addison Wesley.

Dutta, C., & Das, J. (1992). Mathematical characterization of chaos game representation: New algorithms for nucleotide sequence analysis. *Journal of Molecular Biology*, *228*, 715–729. doi:10.1016/0022-2836(92)90857-G

Fiser, A., Tusnady, G. E., & Simon, I. (1994). Chaos game representation of protein structures. *Journal of Molecular Graphics, 12*, 302–304. doi:10.1016/0263-7855(94)80109-6

Geodakian, V. A. (1999). The role of sex chromosomes in evolution: A new concept. *Journal of Mathematical Sciences, 93*(4), 521–530. doi:10.1007/BF02365058

Goldman, N. (1993). Nucleotide, dinucleotide, and trinucleotide frequencies explain patterns observed in chaos game representations of DNA sequences. *Nucleic Acids Research*, (May): 2487–2491. doi:10.1093/nar/21.10.2487

Gutierrez, J. M., Rodriguez, M. A., & Abramson, G. (2001). Multifractal analysis of DNA sequences using novel chaos-game representation. *Physica A, 300*, 271–284. doi:10.1016/S0378-4371(01)00333-8

Inge-Vechtomov, S. G. (1983). *Introduction into molecular genetics*. Moscow: Vysshaya Shkola (in Russian).

Jacob, F. (1974). Le modele linguistique en biologie. *Critique, Mars, XXX*(322), 197-205.

Jacob, F. (1977). The linguistic model in biology. In D. Armstrong & C. H. van Schooneveld (Eds.), *Roman Jakobson: Echoes of his scholarship*. Lisse: Peter de Ridder (pp. 185-192).

Jacobson, R. (1987). *Language in literature*. R. Pomorska & S. Rudy (Eds.). Cambridge, MA.

Jacobson, R. (1999). *Texts, documents, studies*. Moscow, RGGU (in Russian).

Joseph, J., & Sasikumar, R. (2006). Chaos game representation for comparison of whole genomes. *BMC Bioinformatics*, 7(May), 243-246. Hackman, P., Vihola, A., Haravuori, H., Marchand, S., Sarparanta, J., De Seze, J., Labeit, S., Witt, C., Peltonen, L., Richard, I., & Udd, B. (2002). Tibial muscular dystrophy is a titinopathy caused by mutations in TTN, the gene encoding the giant skeletal-muscle protein titin. *American Journal of Human Genetics, 71*(September), 492–500.

Karlin, S., & Lessard, S. (1986). *Sex ratio evolution*. New Jersey: Princeton University Press.

Labeit, S., Barlow, D. P., Gautel, M., Gibson, T., Holt, J., & Hsieh, C. L. (1990). A regular pattern of two types of 100-residue motif in the sequence of titin. *Nature, 345*(May), 273–276. doi:10.1038/345273a0

Machado, C., Sunkel, C. E., & Andrew, D. J. (1998). Human autoantibodies reveal titin as a chromosomal protein. *The Journal of Cell Biology, 141*(April), 321–333. doi:10.1083/jcb.141.2.321

Makovskiy, M. M. (1992). *Linquistic genetics*. Moscow: Nauka (in Russian).

Maynard Smith, J. (1978). *The evolution of sex*. Cambridge: Cambridge Univ. Press.

Mirzabekov, A. D. (1997). The teacher and the mentor. In A. D. Mirzabekov (Ed.), *Aleksandr Aleksandrovich Baev* (pp. 325-336). Moscow: Nauka (in Russian).

Mooney, S. M. (1992). *The evolution of sex: A historical and philosophical analysis*. Unpublished doctoral dissertation, Boston University, Boston.

Oliver, J. L., Bernaola-Galvan, P., Guerrero-Garcia, J., & Roman-Roldan, R. (1993). Entropic profiles of DNA sequences through chaos-game-derived images. *Journal of Theoretical Biology*, (February): 21, 457–470.

Petoukhov, S. V. (2003). The biperiodic table and attributive conception of genetic code. A problem of unification bases of biological languages. In *Proceedings of "The 2003 International Conference on Mathematics and Engineering Techniques in Medicine and Biological Sciences,"* session "Bioinformatics 2003," Las Vegas, June 23-26.

Petoukhov, S. V. (2003-2004). Attributive conception of genetic code, its bi-periodic tables, and problem of unification bases of biological languages. *Symmetry: Culture and Science, 14-15*(part 1), 281–307.

Ratner, V. A. (2002) *Genetics, molecular cybernetics*. Novosibirsk, Nauka (in Russian).

Schutskiy, Y. K. (1997). *The Chinese classical "the book of changes."* Moscow: Vostochnaya literatura (in Russian).

Siu, B. L., Niimura, H., Osborne, J. A., Fatkin, D., MacRae, C., & Solomon, S. (1999). Familial dilated cardiomyopathy locus maps to chromosome 2q31. *Circulation, 99*(March), 1022–10266.

Tavassoly, I., Tavassoly, O., Rad, M., & Dastjerdi, N. (2007a). Multifractal analysis of chaos game representation images of mitochondrial DNA. In D. Howard (Ed.), *IEEE Conference: Frontiers in the Convergence of Bioscience and Information Technologies* (pp. 224-229). Jeju City, South Korea: IEEE Press.

Tavassoly, I., Tavassoly, O., Rad, M., & Dastjerdi, N. (2007b). Three dimensional chaos game representation of genomic sequences. In D. Howard (Ed.), *IEEE Conference: Frontiers in the Convergence of Bioscience and Information Technologies*. (pp. 219-223). Jeju City, South Korea: IEEE Press.

Tavassoly, O., Petoukhov, S., & Vahedi, S. (2009). Visualization of hidden symmetries of protein sequences. *Symmetry: Culture and Science, 20*, (in print).

Teilhard de Chardin, P. (1959). *The phenomenon of man.* New York: Harper and Row.

Wang, Y., Hill, K., Singh, S., & Kari, L. (2005). The spectrum of genomic signatures: From dinucleotides to chaos game representation. *Gene*, (February): 173–185. doi:10.1016/j.gene.2004.10.021

Williams, G. C. (1975). *Sex and evolution.* Princeton.

Yam, P. (1995). Talking trash (linguistic patterns show up in junk DNA). *Scientific American, 272*(3), 12–15.

Yu, Z. G., Anh, V., & Lau, K. S. (2004). Chaos game representation of protein sequences based on the detailed HP model and their multifractal and correlation analyses. *Journal of Theoretical Biology, 226*(February), 341–348. doi:10.1016/j.jtbi.2003.09.009

Chapter 9
Multidimensional Numbers and the Genomatrices of Hydrogen Bonds

ABSTRACT

This chapter returns to the kind of numeric genetic matrices, which were considered in Chapter 4-6. This kind of genomatrices is not connected with the degeneracy of the genetic code directly, but it is related to some other structural features of the genetic code systems. The connection of the Kronecker families of such genomatrices with special categories of hypercomplex numbers and with their algebras is demonstrated. Hypercomplex numbers of these two categories are named "matrions of a hyperbolic type" and "matrions of a circular type." These hypercomplex numbers are a generalization of complex numbers and double numbers. Mathematical properties of these additional categories of algebras are presented. A possible meaning and possible applications of these hypercomplex numbers are discussed. The investigation of these hyperbolic numbers in their connection with the parameters of molecular systems of the genetic code can be considered as a continuation of the Pythagorean approach to understanding natural systems.

INTRODUCTION AND BACKGROUND

The discovery of quaternions by Hamilton in 1843 has led to many important consequences for all mathematical natural sciences (Kline, 1980). Hamilton believed that this discovery was the most significant achievement of his life though his name is connected in modern science with many other essential notions and formalisms: the famous equation by Hamilton, which is called "the canonical equation of mechanics" and which underlies the whole of theoretical physics; "functions by Hamilton"; "Hamiltonians", etc. After the discovery of quaternions, Hamilton devoted to their study last 20 years of his life,

DOI: 10.4018/978-1-60566-124-7.ch009

specially having refused a post as the president of academy of sciences of Ireland. Within these years he published 109 scientific works devoted to quaternions, including two fundamental monographs. Only in the XIX-th century almost 600 scientific works were published and which were devoted to the theory of these hypercomplex numbers and to their successful applications to various problems in physics, geometry, the theory of numbers, etc. One of the main consequences was the understanding that various natural systems can possess their own algebra or that they can be connected closely with various systems of multi-dimensional numbers.

It seems an important task to investigate from different viewpoints is what systems of multi-dimensional numbers (or what types of multi-dimensional algebras) are connected or can be connected with ensembles of parameters of the genetic code and with relevant bioinformation spaces. And what symmetries and patterns are typical for these numerical systems and for their matrix forms of presentation. This chapter describes some results of such investigations.

A development of the notion of "number" has a long history. This history is described in many monographs and it abounds with interesting discoveries and generalizations, many of which are known widely. An appearance of two-dimensional numbers, which were named "complex numbers" by "the king of mathematics" C. Gauss at the end of the XVIII century, was one of them. This kind of numbers utilized the notion of "imaginary unit", which was proposed in the middle of XVI century by the Italian doctor, the mathematician and the designer of mechanisms (in particular, a cardan shaft) G. Cardano. Similarly to real numbers, complex numbers possess the commutative property on multiplication: the result of multiplication of two complex numbers does not depend on the order of factors. This important property allowed developing the theory of functions of complex variables, which plays a significant role in modern science. These two-dimensional numbers have appeared very useful not only in the sphere of pure mathematics, but in many applied fields as well. These complex numbers are the mathematical basis of quantum mechanics and of many other branches of mathematical natural sciences.

There is no doubt that development of formalized theories depends highly on those mathematical notions and instruments, on which they are based. Two-dimensional complex numbers, which are a sum of real item and imaginary item, have appeared as magic instruments for the development of theories and calculations in the field of problems of heat, light, sounds, vibrations, elasticity, gravitation, magnetism, electricity, liquid streams, and phenomena of a micro-world. Therefore attempts were repeatedly undertaken to construct the generalized numbers with greater dimension by means of inclusion of additional items into the structure of complex numbers. Those "numbers", which can be constructed by means of the addition of imaginary units $\mathbf{i}_k = -1$ to real unit, were named "hypercomplex numbers". As a result of such attempts, W.Hamilton has created 4-dimensional numbers, which contain one real unit and three imaginary units and which were named quaternions. Quaternions by Hamilton do not possess the commutative property.

Hypercomplex numbers in their modern comprehension can contain not only real and imaginary units but also a special unit $\mathbf{e}_k$, which possesses the property $\mathbf{e}_k^2 = +1$ but which differs from real unit. In this chapter we shall name these units e_k as "semi-imaginary units" because in the multiplication tables of matrions (Figures 5 and 10) they possess some properties of real unit and of imaginary unit simultaneously. A review of many known kinds of multi-dimensional numbers is presented on the websites http://hypercomplex.xpsweb.com. An important place among hypercomplex numbers belong to those numbers which possess the commutative property and the associative property simultaneously. Matrions, which are described in this chapter, belong to this category of hypercomplex numbers.

Table 1.

$x_0 x_1$ $x_1 x_0$	and	$x_0 x_1$ $-x_1 x_0$

Chapters 4-6 considered the Kronecker families of numerical genomatrices, the kernel of which were the (2x2)-matrices of the following kinds (see Table 1).where x_0 and x_1 are real numbers. These kinds of matrices are well-known in mathematics and physics.

The matrix on the left side in the expression (1) is the matrix form of presentation of the so called "double numbers", which were introduced by Clifford in 1872 year: $x_0*\mathbf{1}+x_1*\mathbf{e}_1$, where $\mathbf{e}_1^2 = +1$ (e_1 is the semi-imaginary unit in the mentioned terminology). From a geometric viewpoint, this matrix defines a transformation of scaling with a hyperbolic turn. Upon its normalization, the determinant of which is equal to 1, it coincides with the matrix of hyperbolic turn. This transformation of hyperbolic turn is connected with hyperbolic functions, the geometric theory of logarithms, the special theory of relativity and the theory of Sine-Gordon solitons. These double numbers are connected with Lorentz transformations and for this reason they are named Lorentz numbers sometimes. A few geometrical applications of double numbers are described in the book (Rosenfeld, 1966). One can mention the known synonymous names of the same double numbers: hyperbolic complex numbers, split-complex numbers, countercomplex numbers, anormal-complex numbers, motors, perplex numbers, tessarines.

The matrix on the right side in the expression (1) is the matrix form of presentation of complex numbers. This matrix form of complex numbers is used in computer informatics usually. One of the kinds of Hadamard matrices [1 1; -1 1] is a particular case of this matrix. Upon its normalization, the determinant of which is equal to 1, this matrix from (1) coincides with the matrix of a classical circular turn.

The algorithm of doubling by Grassmann-Clifford is known, which produces a huge set of 2^n-dimensional generalizations of complex and double numbers (Silvestrov, 1998). Only minor part of this set is investigated till now. Most of these investigated kinds of hypercomplex numbers do not possess the commutative property. This chapter describes those two kinds of hypercomplex numbers, which are connected with the Kronecker families of the genetic matrices and which possess the commutative property and the associative property simultaneously. We did not find a deep investigation of such kinds of hypercomplex numbers in literature sources in the past. Hypercomplex numbers of these two kinds were named "matrions of hyperbolic kind" and "matrions of circular kind". These kinds of matrions were constructed as generalizations of double numbers and of complex numbers correspondingly by means of a special block-fractal algorithm (Petoukhov, 2008a). This algorithm has advantages of obviousness and simplicity, but it can be interpreted as a particular case of the algorithm of doubling by Grassmann-Clifford.

THE HYPERBOLIC MATRIONS AS A SPECIAL KIND OF HYPERCOMPLEX NUMBERS

Let us return to the Kronecker families of numerical genomatrices from Chapters 4-6. This chapter demonstrates that, from the algebraic viewpoint, these Kronecker families of genomatrices are special

Figure 1. The creating of 2^n-dimensional hyperbolic matrions by means of the block-fractal algorithm

$$\begin{vmatrix} x_0 & x_1 \\ x_1 & x_0 \end{vmatrix} \Rightarrow \begin{vmatrix} y_0 & y_1 & y_2 & y_3 \\ y_1 & y_0 & y_3 & y_2 \\ y_2 & y_3 & y_0 & y_1 \\ y_3 & y_2 & y_1 & y_0 \end{vmatrix} \Rightarrow \begin{vmatrix} z_0 & z_1 & z_2 & z_3 & z_4 & z_5 & z_6 & z_7 \\ z_1 & z_0 & z_3 & z_2 & z_5 & z_4 & z_7 & z_6 \\ z_2 & z_3 & z_0 & z_1 & z_6 & z_7 & z_4 & z_5 \\ z_3 & z_2 & z_1 & z_0 & z_7 & z_6 & z_5 & z_4 \\ z_4 & z_5 & z_6 & z_7 & z_0 & z_1 & z_2 & z_3 \\ z_5 & z_4 & z_7 & z_6 & z_1 & z_0 & z_3 & z_2 \\ z_6 & z_7 & z_4 & z_5 & z_2 & z_3 & z_0 & z_1 \\ z_7 & z_6 & z_5 & z_4 & z_3 & z_2 & z_1 & z_0 \end{vmatrix} \Rightarrow \ldots$$

families of 2^n-dimensional hypercomplex numbers, which are named "matrions" because they have arisen due to a consideration of block matrices. Sets of these 2^n-dimensional numbers contain unitary matrices as their own units.

We begin with the generalization of double numbers, the matrix form of presentation of which is shown in the expression (1, on the left side). The proposed algorithm of such generalization includes a series of steps (Figure 1). The first step creates the generalized (2^2x2^2)-matrix with four components y_0, y_1, y_2, y_3 from the initial (2x2)-matrix with the two components x_0, x_1. On this step, each real component x_0 and x_1 of the initial (2x2)-matrix (1) is considered as a (2x2)-matrix of some double number with its components y_s: $x_0 = [y_0y_1; y_1y_0]$, $x_1 = [y_2y_3; y_3y_2]$. When these new (2x2)-matrices are put into the initial (2x2)-matrix, the last matrix becomes such (2^2x2^2)-matrix, which is a matrix form of presentation of 4-dimensional hypercomplex numbers. On the second step, we act analogically: each component y_s of the (2^2x2^2)-matrix is interpreted as a (2x2)-matrix of some double number with its components z_p: $y_0=[z_0z_1; z_1z_0]$, $y_1=[z_2z_3; z_3z_2]$, $y_2=[z_4z_5; z_5z_6]$, $y_3=[z_6z_7; z_7z_6]$. When these new (2x2)-matrices are put into the (2^2x2^2)-matrix, the last matrix becomes such (2^3x2^3)-matrix, which is a matrix form of presentation of 8-dimensional hypercomplex numbers. A quantity of such steps is not limited.

The 2^n-dimensional hypercomplex numbers, which arise as a result of such generalization of double numbers, are named matrions of hyperbolic type or hyperbolic matrions.

The (2^nx2^n)-matrices, which are constructed by such local replacement of each matrix component by a matrix block with four components, possess a global block character of a fractal type. Really, their (2^{n-1}x2^{n-1})-quadrants, which are disposed along each diagonal, are identical to each other. If such identical quadrants along the main diagonal and along the second diagonal are replaced by components A_1 and A_2 correspondingly, the whole of the block matrix becomes a matrix form of presentation of double numbers: $[A_1A_2; A_2A_1]$. This property of identity along diagonals relates to quadrants, sub-quadrants, sub-sub-quadrants, etc. In this sense we have fractals of matrix blocks. In view of this, the described algorithm is named a block-fractal algorithm. All (2^nx2^n)-matrices of hyperbolic matrions (Figure 1) are bisymmetrical: they are symmetrical relative to both diagonals. The cross-wise character of the blocks of these matrices reminds one of the cross-wise character of the blocks of many morpho-functional and information systems, which were mentioned in Figure 6 of Chapter 1.

Those readers, who are familiar with the Kronecker product (its second name is the tensor product) of matrices, will see easily that the block-fractal algorithm coincides with the Kronecker product of appropriate (2x2)-matrices from the viewpoint of the final matrix. Figure 2 illustrates this application of the Kronecker product for the case of arriving at an 8-dimensional hyperbolic matrion.

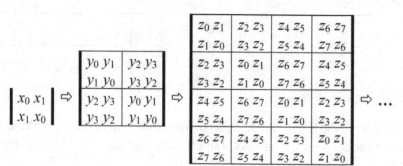

Figure 2. The Kronecker form of the block-fractal algorithm

$$\begin{Vmatrix} x_0 & x_1 \\ x_1 & x_0 \end{Vmatrix} \otimes \begin{Vmatrix} x_2 & x_3 \\ x_3 & x_2 \end{Vmatrix} \otimes \begin{Vmatrix} x_4 & x_5 \\ x_5 & x_4 \end{Vmatrix} = \begin{Vmatrix} z_0 & z_1 & z_2 & z_3 & z_4 & z_5 & z_6 & z_7 \\ z_1 & z_0 & z_3 & z_2 & z_5 & z_4 & z_7 & z_6 \\ z_2 & z_3 & z_0 & z_1 & z_6 & z_7 & z_4 & z_5 \\ z_3 & z_2 & z_1 & z_0 & z_7 & z_6 & z_5 & z_4 \\ z_4 & z_5 & z_6 & z_7 & z_0 & z_1 & z_2 & z_3 \\ z_5 & z_4 & z_7 & z_6 & z_1 & z_0 & z_3 & z_2 \\ z_6 & z_7 & z_4 & z_5 & z_2 & z_3 & z_0 & z_1 \\ z_7 & z_6 & z_5 & z_4 & z_3 & z_2 & z_1 & z_0 \end{Vmatrix}$$

In view of this, the described algorithm of construction of generalized numbers can be named Kronecker algorithm. However, the majority of potential readers of this book, who are interested in mathematical and theoretical biology, do not know the notion of the Kronecker product, but they know the notion of fractals. In addition this notion stimulates many heuristic associations. These reasons explain the used name: "the block-fractal algorithm".

Each matrion (2^nx2^n)-matrix can be presented in its decomposition form as a sum of its basic matrices. For example, 4-dimensional hyperbolic matrion G_2 can be presented in the following decomposition form (Figure 3).

In the right part of the equation on Figure 3, the first basic matrix is the unitary matrix, which can be marked by the symbol **1**. We mark the next three basic matrices by symbols $\mathbf{e}_1$, $\mathbf{e}_2$, $\mathbf{e}_3$ correspondingly. Each of these basic matrices differs from the unitary matrix and possesses the following mutual property: $\mathbf{e}_\kappa^2 = +\mathbf{1}$, where $k = 1, 2, 3$. We call such basic matrices semi-imaginary units, as it was mentioned above.

It can be checked easily, that in the case of replacement of basic matrices by such symbols, any 2^n-dimensional hyperbolic matrion receives a poly-linear or vector form of its representation (Figure 4), which is more habitual for hypercomplex numbers.

The following general "rule of enclosure" exists for 2^n-dimensional hyperbolic matrions. It concerns an enclosure of a poly-linear form of a matrion of a less dimension into a poly-linear form of a matrion of higher dimension. According to this rule, the first half of a poly-linear form of presentation of 2^n-dimensional hyperbolic matrion repeats the whole of poly-linear form of presentation of 2^{n-1}-dimensional

Figure 3. The decomposition of the matrix form of presentation of a 4-dimensional hyperbolic matrion G_2. The right part of the equation is a sum of basic matrices of the matrion

$$G_2 = \begin{Vmatrix} x_0 & x_1 & x_2 & x_3 \\ x_1 & x_0 & x_3 & x_2 \\ x_2 & x_3 & x_0 & x_1 \\ x_3 & x_2 & x_1 & x_0 \end{Vmatrix} = x_0 * \begin{Vmatrix} 1 & 0 & 0 & 0 \\ 0 & 1 & 0 & 0 \\ 0 & 0 & 1 & 0 \\ 0 & 0 & 0 & 1 \end{Vmatrix} + x_1 * \begin{Vmatrix} 0 & 1 & 0 & 0 \\ 1 & 0 & 0 & 0 \\ 0 & 0 & 0 & 1 \\ 0 & 0 & 1 & 0 \end{Vmatrix} +$$

$$+ x_2 * \begin{Vmatrix} 0 & 0 & 1 & 0 \\ 0 & 0 & 0 & 1 \\ 1 & 0 & 0 & 0 \\ 0 & 1 & 0 & 0 \end{Vmatrix} + x_3 * \begin{Vmatrix} 0 & 0 & 0 & 1 \\ 0 & 0 & 1 & 0 \\ 0 & 1 & 0 & 0 \\ 1 & 0 & 0 & 0 \end{Vmatrix}$$

Figure 4. Examples of poly-linear forms of presentations of hyperbolic matrions in the cases of their 2-, 4-, 8-dimensions

Dimension	Hyperbolic matrion in its poly-linear form
2^1	$x_0*\mathbf{1} + x_1*\mathbf{e_1}$
2^2	$x_0*\mathbf{1} + x_1*\mathbf{e_1} + x_2*\mathbf{e_2} + x_3*\mathbf{e_3}$
2^3	$x_0*\mathbf{1}+x_1*\mathbf{e_1}+x_2*\mathbf{e_2}+x_3*\mathbf{e_3}+x_4*\mathbf{e_4}+x_5*\mathbf{e_5}+x_6*\mathbf{e_6}+x_7*\mathbf{e_7}$

hyperbolic matrion. In other words, a matrion of less dimension is enclosed into the first half of a matrion of doubled dimension (see Figure 4). In view of this, any hyperbolic matrion of high dimension unites a set of all hyperbolic matrions of fewer dimensions in a form of hierarchical ensemble.

An analogical rule of enclosure is true for multiplication tables of basic matrices (basic elements) of 2^n-dimensional hyperbolic matrions (see an example on Figure 5). Concerning such multiplication tables, one should note that the set of basic matrices of 2^n-dimensional hyperbolic matrions forms the closed set of basic elements: multiplication of any two members of the set of the basic matrices produces the matrix from this set again. It means that algebras of 2^n-dimensional hyperbolic matrions exist with appropriate multiplication tables. Figure 5 shows an example of such a multiplication table for the case of 8-dimensional hyperbolic matrions. Each cell of this multiplication table demonstrates the result of multiplication of appropriate basic elements from the left column and the upper row of the table. This multiplication table defines the appropriate algebra completely.

The (2x2)-square and the (4x4)-square are marked in the left upper corner of the multiplication table on Figure 5. These squares are the multiplication tables for the cases of 2-dimensional and 4-dimensional hyperbolic matrions correspondingly. By analogy this multiplication table of 8-dimensional hyperbolic matrions is enclosed in the appropriate multiplication table of 16-dimensional hyperbolic matrions, etc. It means that the hierarchical ensemble of appropriate algebras exists.

A unique feature of hyperbolic matrions is that the structure of their multiplication table coincides with their own matrix structure always. Really, if the symbols in the multiplication table on Figure 5 are interpreted as real numbers, this multiplication table becomes the matrix form of 8-dimensional hyperbolic matrions (see Figures 1 or 2). The multiplication tables of hyperbolic matrions are symmetrical relative to their main diagonals. This fact reflects the existence of the commutative property of hyperbolic ma-

Figure 5. The multiplication table of the basic elements of 8-dimensional hyperbolic matrions

	1	$\mathbf{e_1}$	$\mathbf{e_2}$	$\mathbf{e_3}$	$\mathbf{e_4}$	$\mathbf{e_5}$	$\mathbf{e_6}$	$\mathbf{e_7}$
1	**1**	$\mathbf{e_1}$	$\mathbf{e_2}$	$\mathbf{e_3}$	$\mathbf{e_4}$	$\mathbf{e_5}$	$\mathbf{e_6}$	$\mathbf{e_7}$
$\mathbf{e_1}$	$\mathbf{e_1}$	**1**	$\mathbf{e_3}$	$\mathbf{e_2}$	$\mathbf{e_5}$	$\mathbf{e_4}$	$\mathbf{e_7}$	$\mathbf{e_6}$
$\mathbf{e_2}$	$\mathbf{e_2}$	$\mathbf{e_3}$	**1**	$\mathbf{e_1}$	$\mathbf{e_6}$	$\mathbf{e_7}$	$\mathbf{e_4}$	$\mathbf{e_5}$
$\mathbf{e_3}$	$\mathbf{e_3}$	$\mathbf{e_2}$	$\mathbf{e_1}$	**1**	$\mathbf{e_7}$	$\mathbf{e_6}$	$\mathbf{e_5}$	$\mathbf{e_4}$
$\mathbf{e_4}$	$\mathbf{e_4}$	$\mathbf{e_5}$	$\mathbf{e_6}$	$\mathbf{e_7}$	**1**	$\mathbf{e_1}$	$\mathbf{e_2}$	$\mathbf{e_3}$
$\mathbf{e_5}$	$\mathbf{e_5}$	$\mathbf{e_4}$	$\mathbf{e_7}$	$\mathbf{e_6}$	$\mathbf{e_1}$	**1**	$\mathbf{e_3}$	$\mathbf{e_2}$
$\mathbf{e_6}$	$\mathbf{e_6}$	$\mathbf{e_7}$	$\mathbf{e_4}$	$\mathbf{e_5}$	$\mathbf{e_2}$	$\mathbf{e_3}$	**1**	$\mathbf{e_1}$
$\mathbf{e_7}$	$\mathbf{e_7}$	$\mathbf{e_6}$	$\mathbf{e_5}$	$\mathbf{e_4}$	$\mathbf{e_3}$	$\mathbf{e_2}$	$\mathbf{e_1}$	**1**

trions. The existence of the commutative property and the associative property in the case of hyperbolic matrions can be proved directly as well (see the book (Petoukhov, 2008a)).

Any row of a matrix of a hyperbolic matrion contains the same set of matrix components, which are permuted among them in comparison with the case of the first row. These permutations of components from one row to another row are connected with the algorithm of diadic shifts which was described in Chapter 1. A disposition of all kinds of components in a (2^nx2^n)-matrix of hyperbolic matrion produces a certain global mosaic or pattern, which is a typical visual attribute of 2^n-dimensional hyperbolic matrions. At transition from the case of 2^n-dimensional hyperbolic matrions to the case of 2^{n+1}-dimensional hyperbolic matrions, this typical global mosaic is transmitted "by right of succession" into its (2^nx2^n)-quadrants as the local mosaic of these quadrants. The global mosaic of 2^{n+1}-dimensional hyperbolic matrions differs from the global mosaic of 2^n-dimensional hyperbolic matrions always. Since hyperbolic matrions are connected with structures of the genetic code, these mosaics can possess some physiological meaning from the viewpoint of the doctrine by C. Jung about archetypes. In view of this, these genetic mosaics or the patterns are interesting things for their application in design, culture, art therapy and some other fields.

A matrix of a 2^n-dimensional hyperbolic matrion with non-zero components, all of which are equal to each other, is a singular matrix. Such matrices have no inverse matrices, and the operation of division is not defined for them. Nonsingular matrices of hyperbolic matrions satisfy a definition of metric tensors of Riemanian geometry and they can be used for modeling internal geometry of surfaces of biological bodies.

The quint and golden matrices, which were considered in Chapter 4 and which are connected with the Pythagorean musical scale and other presented materials, are hyperbolic matrions from the algebraic viewpoint. This algebraic investigation reveals algebraic roots of some characteristics of ensembles of molecular parameters of the genetic code additionally.

THE CIRCULAR MATRIONS

Chapter 6 has considered questions about connections of the genetic code with Hadamard matrices, which lead to relations of molecular-genetic systems with unitary operators of quantum mechanics and with Hadamard gates of quantum computers. Any Hadamard (2x2)-matrix in a Kronecker power is a Hadamard matrix again. But the Hadamard matrix [1 1; -1 1] coincides with the matrix form of presentation of the complex number (1***1** + 1***i**). The general matrix form $[x_0 x_1; -x_1 x_0]$ of presentation of complex numbers $z = x_0$***1** + x_1***i** was shown in the expression (1, on the right side). The matrix $[1\ 1; -1\ 1]^{(n)}$, where (n) means the Kronecker power n =1, 2, 3, …, is a Hadamard matrix again. But what can one say about the Kronecker family of the matrices $[x_0 x_1; -x_1 x_0]^{(n)}$, which is a Kronecker generalization of the complex numbers?

Now we will demonstrate that the Kronecker family of matrices $[x_0 x_1; -x_1 x_0]^{(n)}$ is the family of matrix forms of presentation of 2^n-dimensional hypercomplex numbers, which are named matrions of circular type (or circular matrions). This type of hypercomplex number, which possesses interesting mathematical properties, can be useful in future algebraic investigations of molecular-genetic structures in connection with Hadamard matrices, etc.

In the case of the generalization of complex numbers, one can apply the same block-fractal algorithm, which was used above for the described generalization of double numbers (Figure 1). The first step of

Figure 6. The creating of 2^n-dimensional circular matrions by means of the block-fractal algorithm

$$\begin{vmatrix} x_0 & x_1 \\ -x_1 & x_0 \end{vmatrix} \Rightarrow \left[\begin{array}{cc|cc} y_0 & y_1 & y_2 & y_3 \\ -y_1 & y_0 & -y_3 & y_2 \\ \hline -y_2 & -y_3 & y_0 & y_1 \\ y_3 & -y_2 & -y_1 & y_0 \end{array}\right] \Rightarrow \left[\begin{array}{cc|cc|cc|cc} z_0 & z_1 & z_2 & z_3 & z_4 & z_5 & z_6 & z_7 \\ -z_1 & z_0 & -z_3 & z_2 & -z_5 & z_4 & -z_7 & z_6 \\ \hline -z_2 & -z_3 & z_0 & z_1 & -z_6 & -z_7 & z_4 & z_5 \\ z_3 & -z_2 & -z_1 & z_0 & z_7 & -z_6 & -z_5 & z_4 \\ \hline -z_4 & -z_5 & -z_6 & -z_7 & z_0 & z_1 & z_2 & z_3 \\ z_5 & -z_4 & z_7 & -z_6 & -z_1 & z_0 & -z_3 & z_2 \\ \hline z_6 & z_7 & -z_4 & -z_5 & -z_2 & -z_3 & z_0 & z_1 \\ -z_7 & z_6 & z_5 & -z_4 & z_3 & -z_2 & -z_1 & z_0 \end{array}\right]$$

the algorithm of the generalization of complex numbers creates the generalized (2^2x2^2)-matrix with four components y_0, y_1, y_2, y_3 from the initial (2x2)-matrix with the two components x_0, x_1. On this step, each real component x_0 and x_1 of the initial (2x2)-matrix in the expression (1) is considered as a (2x2)-matrix of some complex number with its components y_s: $x_0 = [y_0y_1; -y_1y_0]$, $x_1 = [y_2y_3; -y_3y_2]$. When these new (2x2)-matrices are put into the initial (2x2)-matrix, the last matrix becomes such a (2^2x2^2)-matrix, which is a matrix form of presentation of 4-dimensional hypercomplex numbers. On the second step, we act analogically: each component y_s of the (2^2x2^2)-matrix is interpreted as a (2x2)-matrix of some complex number with its components z_p: $y_0=[z_0z_1; -z_1z_0]$, $y_1=[z_2z_3; -z_3z_2]$, $y_2=[z_4z_5; -z_5z_6]$, $y_3=[z_6z_7; -z_7z_6]$. When these new (2x2)-matrices are put into the (2^2x2^2)-matrix, the last matrix becomes such a (2^3x2^3)-matrix, which is a matrix form of presentation of 8-dimensional hypercomplex numbers (Figure 6). The quantity of such steps is not limited. The 2^n-dimensional hypercomplex numbers, which arise as a result of such generalization of complex numbers, are named matrions of circular type or circular matrions.

The (2^nx2^n)-matrices of circular matrions are nonsingular as a rule. But for the cases $n \geq 2$ a set of circular matrions has so called zero divisor. It means that non-null members of such a set exist, multiplication of which produces a null matrion. Existence of zero divisor does not allow utilizing such operation of division, which can work on the whole of the set of circular matrions. In the case of 4-dimensional circular matrions, an example of zero divisor is shown on Figure 7.

The decomposition of 2^n-dimensional circular matrions into a sum of their basic matrices is realized by the usual way (see an example on Figure 8).

The first basic matrix on the right side of the expression on Figure 8 is the unitary matrix marked by the symbol **1**. The next three basic matrices will be marked by symbols $\mathbf{i}_1$, $\mathbf{i}_2$, $\mathbf{e}_3$ correspondingly. One can check easily that the basic matrices i_1 and i_2 are imaginary units because $\mathbf{i}_1^2 = \mathbf{i}_2^2 = \mathbf{-1}$. The fourth basic matrix is a semi-imaginary unit because $\mathbf{e}_3^2 = \mathbf{+1}$. In view of this, 2^n-dimensional circular matrions, beginning with $n = 2$, are based on three kinds of mathematical units: real unit, imaginary units and

Figure 7. An example of divisors of zero in the case of 4-dimensional circular matrions: multiplication of two non-null circular matrions gives null matrix. Here "a" and "b" are real numbers

$$\begin{vmatrix} 0 & a & a & 0 \\ -a & 0 & 0 & a \\ -a & 0 & 0 & a \\ 0 & -a & -a & 0 \end{vmatrix} * \begin{vmatrix} b & 0 & 0 & b \\ 0 & b & -b & 0 \\ 0 & -b & b & 0 \\ b & 0 & 0 & b \end{vmatrix} = \begin{vmatrix} 0 & 0 & 0 & 0 \\ 0 & 0 & 0 & 0 \\ 0 & 0 & 0 & 0 \\ 0 & 0 & 0 & 0 \end{vmatrix}$$

Figure 8. An Example of the decomposition of a 4-dimensional circular matrion

$$\begin{Vmatrix} x_0 & x_1 & x_2 & x_3 \\ -x_1 & x_0 & -x_3 & x_2 \\ -x_2 & -x_3 & x_0 & x_1 \\ x_3 & -x_2 & -x_1 & x_0 \end{Vmatrix} = x_0 * \begin{Vmatrix} 1&0&0&0 \\ 0&1&0&0 \\ 0&0&1&0 \\ 0&0&0&1 \end{Vmatrix} + x_1 * \begin{Vmatrix} 0&1&0&0 \\ -1&0&0&0 \\ 0&0&0&1 \\ 0&0&-1&0 \end{Vmatrix} +$$

$$+ x_2 * \begin{Vmatrix} 0&0&1&0 \\ 0&0&0&1 \\ -1&0&0&0 \\ 0&-1&0&0 \end{Vmatrix} + x_3 * \begin{Vmatrix} 0&0&0&1 \\ 0&0&-1&0 \\ 0&-1&0&0 \\ 1&0&0&0 \end{Vmatrix}$$

semi-imaginary units. Any 2^n-dimensional circular matrion can be expressed in its poly-linear form, if its basic matrices are replaced by their symbols (Figure 9).

By analogy with the case of hyperbolic matrions, the general "rule of enclosure" exists for 2^n-dimensional circular matrions. It concerns an enclosure of a poly-linear form of a matrion of a less dimension into a poly-linear form of a matrion of higher dimension. According to this rule, the first half of a poly-linear form of presentation of 2^n-dimensional circular matrion repeats the whole of poly-linear form of presentation of 2^{n-1}-dimensional circular matrion. In other words, a matrion of less dimension is enclosed into the first half of a matrion of doubled dimension (see Figure 9). In view of this, any circular matrion of high dimension unites a set of all circular matrions of fewer dimensions in a form of hierarchical ensemble.

The analogical rule of enclosure holds true for the multiplication table of basic elements of circular matrions. Figure 10 shows the example of the multiplication table for the 8-dimensional case. Each cell of this multiplication table demonstrates the result of multiplication of appropriate basic elements from the left column and the upper row of the table.

The (2x2)-square and the (4x4)-square are marked in the left upper corner of the multiplication table on Figure 10. These squares are the multiplication tables for the cases of 2-dimensional and 4-dimensional circular matrions correspondingly. By analogy this multiplication table of 8-dimensional circular matrions is enclosed in the appropriate multiplication table of 16-dimensional circular matrions, etc. It means that the hierarchical ensemble of appropriate algebras exists. This multiplication table is symmetrical relative to the main diagonal. This feature reflects the commutative property of the algebras of circular matrions.

The real unit and semi-imaginary units of circular matrions will be named "anti-imaginary units" in contrast to imaginary units. The set of basic elements of 2^n-dimensional circular matrions is divided

Figure 9. Poly-linear forms of presentations of circular matrions for the cases of 2-, 4-, and 8-dimensional cases

Dimension	A kind of circular matrion
2^1	$x_0*\mathbf{1} + x_1*\mathbf{i_1}$
2^2	$x_0*\mathbf{1} + x_1*\mathbf{i_1} + x_2*\mathbf{i_2} + x_3*\mathbf{e_3}$
2^3	$x_0*\mathbf{1} + x_1*\mathbf{i_1} + x_2*\mathbf{i_2} + x_3*\mathbf{e_3} + x_4*\mathbf{i_4} + x_5*\mathbf{e_5} + x_6*\mathbf{e_6} + x_7*\mathbf{i_7}$

Figure 10. The multiplication table of the basic elements of 8-dimensional circular matrions

	1	i_1	i_2	e_3	i_4	e_5	e_6	i_7
1	1	i_1	i_2	e_3	i_4	e_5	e_6	i_7
i_1	i_1	-1	e_3	$-i_2$	e_5	$-i_4$	i_7	$-e_6$
i_2	i_2	e_3	-1	$-i_1$	e_6	i_7	$-i_4$	$-e_5$
e_3	e_3	$-i_2$	$-i_1$	+1	i_7	$-e_6$	$-e_5$	i_4
i_4	i_4	e_5	e_6	i_7	-1	$-i_1$	$-i_2$	$-e_3$
e_5	e_5	$-i_4$	i_7	$-e_6$	$-i_1$	+1	$-e_3$	i_2
e_6	e_6	i_7	$-i_4$	$-e_5$	$-i_2$	$-e_3$	+1	i_1
i_7	i_7	$-e_6$	$-e_5$	i_4	$-e_3$	i_2	i_1	-1

into two sub-sets always: one sub-set consists of anti-imaginary units only and another sub-set consists of imaginary units only. The sequence of anti-imaginary and imaginary units in the poly-linear form of presentation of 2^n-dimensional circular matrions possesses a regular character: this sequence coincides with the sequence of elements "+1" and "-1" on the main diagonal of Hadamard matrix $[1\ 1;\ 1\ -1]^{(n)}$, which plays an important role in the theory of discrete signals and of quantum computers (see Chapter 6). Moreover the disposition of all signs "+" and "-" in multiplications tables of 2^n-dimensional circular matrions is identical to the disposition of the same signs "+" and "-" in the appropriate Hadamard matrices. But a disposition of the signs "+" and "-" in matrix forms of presentation of 2^n-dimensional circular matrions coincides with the disposition of the same signs "+" and "-" in the Hadamard matrices of another kind: $[1\ 1;\ -1\ 1]^{(n)}$. If magnitudes of all coordinates $x_0, x_1, \ldots$ of a circular matrions are equal to 1, then the matrix of this circular matrion is identical to this kind of Hadamard matrix.

RULES OF EIGENVALUES OF MATRICES OF CIRCULAR MATRIONS

Eigenvalues of matrices of circular matrions (Figure 6) are complex numbers. A construction of these eigenvalues reflects a division of the whole set of coordinates of 2^n-dimensional circular matrions into the two sub-sets of anti-imaginary and imaginary units. Really, coordinates $x_0, x_1, \ldots$ at imaginary and anti-imaginary basic elements form isolated sub-sets, which are collected in the real part and the imaginary part of these eigenvalues. For example, a 4-dimensional matrion $x_0*\mathbf{1} + x_1*\mathbf{i}_1 + x_2*\mathbf{i}_2 + x_3*\mathbf{e}_3$ possess the following four eigenvalues in the form of complex numbers: $x_0 + x_3 + j*(-x_1 + x_2)$; $x_0 + x_3 - j*(-x_1 + x_2)$; $-x_3 + x_0 + j*(x_1 + x_2)$; $-x_3 + x_0 - j*(x_1 + x_2)$ (here j means the imaginary unit of the complex numbers). One can see that all anti-imaginary coordinates x_0 and x_3 are collected in the real part of these eigenvalues only, and all imaginary coordinates x_1 and x_2 are collected in the imaginary part of these eigenvalues only. Another example is the case of 8-dimensional circular matrions $x_0*\mathbf{1} + x_1*\mathbf{i}_1 + x_2*\mathbf{i}_2 + x_3*\mathbf{e}_3 + x_4*\mathbf{i}_4 + x_5*\mathbf{e}_5 + x_6*\mathbf{e}_6 + x_7*\mathbf{i}_7$, which possesses the following eight eigenvalues:

$x_5 - x_3 + x_6 + x_0 + j*(- x_1 - x_2 - x_7 + x_4)$;

$x_5 - x_3 + x_6 + x_0 - j*(- x_1 - x_2 - x_7 + x_4)$;

$- x_5 - x_3 - x_6 + x_0 + j*(x_1 + x_2 - x_7 + x_4);$

$- x_5 - x_3 - x_6 + x_0 - j*(x_1 + x_2 - x_7 + x_4);$

$x_5 + x_3 - x_6 + x_0 + j*(- x_1 + x_2 + x_7 + x_4);$

$x_5 + x_3 - x_6 + x_0 - j*(- x_1 + x_2 + x_7 + x_4);$

$- x_5 + x_3 + x_6 + x_0 + j*(x_1 - x_2 + x_7 + x_4);$

$- x_5 + x_3 + x_6 + x_0 - j*(x_1 - x_2 + x_7 + x_4)$

One can see again that all anti-imaginary coordinates x_0, x_3, x_5, x_6 are collected in the real part of these eigenvalues only, and all imaginary coordinates x_1, x_2, x_4, x_7 are collected in the imaginary part of these eigenvalues only.

Sets of eigenvalues of 2^n-dimensional circular matrions are characterized by an additional regularity: a sum of all eigenvalues of such matrion is equal to the real number 2^n*x_0, that is the magnitude of the dimensionality of the matrion with a factor, which is equal to the coordinate x_0 at real basic element of the matrion. For the two considered examples, sums of 4-dimensional and 8-dimensional circular matrions are equal to $4*x_0$ and $8*x_0$ correspondingly, as one can check easily.

Investigations of connections of the matrix structures of molecular-genetic systems with circular matrions and Hadamard matrices should be continued.

CIRCULAR MATRIONS AND THE SERIES OF NATURAL NUMBERS

This paragraph describes some new symmetrical patterns which were arrived at in matrix genetics in the course of investigations of circular matrions. In accordance with an ancient tradition, the whole series of natural numbers is considered as the union of odd and even numbers. The theory of 2^n-dimensional circular matrions suggests that the series of natural numbers can be considered as the union of two sub-sets of another kind: one sub-set consists of those natural numbers, which coincide with indexes of anti-imaginary coordinates, and another sub-set consists of those natural numbers, which coincide with indexes of imaginary coordinates. The described "rule of enclosure" plays a useful role here. Each natural number "*k*" is an index of anti-imaginary basic element of 2^n-dimensional circular matrions ($n > k$) or it is an index of imaginary basic element of such a circular matrion. Thus the whole of the set of natural numbers is divided into two equal sub-sets. Members of one sub-set including numbers 0, 3, 5, 6, 9, 10, 12, 15, 17, 18, 20, 23, 24, 27, 29, 30, 33, 34, 36, 39, 40, 43, 45, 46, 48, 51, 53, 54, 57, 58, 60, 63,... correspond to the indexes of anti-imaginary basic units of circular matrions. One can name these numbers as "anti-imaginary natural numbers" conditionally (or "black natural numbers", because cells with such numbers are marked by black color in matrices on Figure 11). Another sub-set including numbers 1, 2, 4, 7, 8, 11, 13, 14, 16, 19, 21, 22, 25, 26, 28, 31, 32, 35, 37, 38, 41, 42, 44, 47, 49, 50, 52, 55, 56, 59, 61, 62,... correspond to the indexes of imaginary basic units of these matrions. Such numbers can be named "sub-imaginary natural numbers" conditionally (or "white natural numbers", because cells with such numbers are marked by white color in matrices on Figure 11).

Figure 11. (2^nx2^n)-matrices with enumerated cells in sequence, cells with anti-imaginary natural numbers are marked by black color and cells with sub-imaginary natural numbers are marked by white color

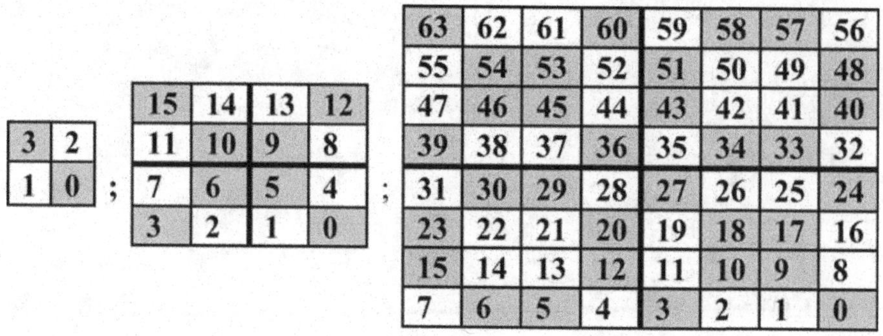
3 2
1 0 ;
15 14 13 12
11 10 9 8
7 6 5 4
3 2 1 0 ;
63 62 61 60 59 58 57 56
55 54 53 52 51 50 49 48
47 46 45 44 43 42 41 40
39 38 37 36 35 34 33 32
31 30 29 28 27 26 25 24
23 22 21 20 19 18 17 16
15 14 13 12 11 10 9 8
7 6 5 4 3 2 1 0

The algorithm of alternation of anti-imaginary and sub-imaginary numbers in the series of natural numbers is connected not only with Hadamard matrices but can be expressed graphically in connection with square matrices. Really, let us draw initial matrices from a family of (2^nx2^n)-matrices with numerated cells in sequence (Figure 11). Cells with anti-imaginary (sub-imaginary) numbers are marked by black (white) colors. Mosaics of these matrices are symmetrical relative to both diagonals. Their left side and their right side are anti-symmetrical to each other. Their quadrants and sub-quadrants possess the analogical properties. One can see that the black-and-white mosaic of the (2x2)-matrix is repeated in those quadrants of the next (4x4)-matrix of this family, which are disposed along the main diagonal. The same mosaic is repeated in the anti-symmetrical kind in those quadrants which are disposed along the second diagonal. By analogy the whole of mosaic of (4x4)-matrix is repeated in the next (8x8)-matrix of this family in its quadrants which are disposed along the main diagonal; and its quadrants along the second diagonal reproduce the same mosaic in anti-symmetrical kind. In other words, in this matrix family, a global mosaic of a matrix of the previous generation is inherited as a local mosaic of quadrants of a matrix of the next generation.

The described graphical algorithm allows one to define what kinds of natural numbers are anti-imaginary or sub-imaginary natural numbers without knowledge of multiplication rules at all. For this aim it is enough to have a family of (2^nx2^n)-matrices with cells enumerated in sequence and to remember the mosaic pattern of the simplest (2x2)-matrix on Figure 11. Then this mosaic pattern should be reproduced in quadrants of the next (4x4)-matrix in accordance with this algorithm. Repeating this graphic procedure along the sequence of matrices of this family gives the division of the series of natural numbers into the two named sub-sets. (2^nx2^n)-matrices with such binary-mosaic character possess one interesting property as well. If all their black cells contain elements "+1" and all white cells contain elements "-1", an exponentiation of these matrices into an integer positive power reproduce these matrices with regular factors. For example, the result of exponentiation of such (2x2)-matrix S in the power "*k*" is the following: $S^k = 2^{k-1} * S$.

As we can judge, the "rule of closing of the sub-set" holds true for the sub-sets of anti-imaginary and sub-imaginary natural numbers: a product of two anti-imaginary (sub-imaginary) natural numbers gives new anti-imaginary (sub-imaginary) natural numbers always. It means that each of these sub-sets is the closed sub-set relative to multiplication. Such division of the series of natural numbers into the sub-sets of anti-imaginary and sub-imaginary numbers puts forward many interesting questions for further investigations. For example, a problem arises about a distribution of prime anti-imaginary numbers and of prime sub-imaginary numbers. The claim is about interrelations of numbers inside each of these sub-sets. A prime sub-imaginary number is any sub-imaginary number which cannot be factorized with using other sub-imaginary natural numbers, which differ from the real unit and this number itself. A prime anti-imaginary number is any anti-imaginary number which cannot be factorized using other anti-imaginary natural numbers. It is known that the classical problem about a distribution of prime numbers inside the series of natural numbers has generated many important investigations, methods and theorems. But this classical problem is not solved till now. In our opinion, the new problem about distributions of prime anti-imaginary numbers and of prime sub-imaginary numbers can generate new interesting investigations as well. Can this new problem, which has arisen in the theory of circular matrions, help in completely solving the classical problem of the distribution of prime numbers? It is an open question.

Another example of interesting questions, which has arisen from the theory of circular matrions, is a question of rational numbers. The set of all rational numbers is divided into several sub-sets by criteria of their numerator and denominator belonging to the sub-set of anti-imaginary numbers or to the sub-set of sub-imaginary numbers. These sub-sets of rational numbers possess a special character of distribution in the set of all rational numbers and they possess a specificity of transfer of members of one sub-set into another sub-set under actions of various mathematical operations.

We can mention briefly here only that questions of anti-imaginary and sub-imaginary natural numbers are connected with problems of modeling the parametric ensembles of the genetic code in the field of matrix genetics as well.

SOLUTIONS AND RECOMMENDATIONS

The 2^n-dimensional hypercomplex numbers described in this chapter are connected with matrix presentations of ensembles of parameters of the genetic code. They give new symmetric graphical patterns and allow one to construct some mathematical models of the mentioned ensembles. Simultaneously they introduce interesting materials into the science of hypercomplex numbers. Applications of the theory of matrions allow revealing such features of biological information processing, disclosures of which are difficult by other ways.

In view of the block character of their matrix structures, 2^n-dimensional matrions can be applied in many fields where block matrices have been utilized successfully for a long time. One can recall that properties of block matrices provide an obviousness of complex interrelations for comfortable logic analysis of complex structures. In the case of industrial plants, for example, block matrices, which reflect a movement of documents and indexes, have served for a long time in analysis of rationality of industrial structures, in work organization and utilization of administrative departments.

Mathematics of matrions proposes new interesting possibilities for various branches of mathematical biology, for example, for biomechanics. Special the variants of theory of locomotion can be created for a comparative analysis and of a classification of gaits of various animals. In this case the claim is about an interpretation of a multi-linked locomotor apparatus in a form of such multi-dimensional functional matrion, coordinates of which are variable in regular manner for simulating movements of links of locomotors apparatus. A control of such matrion simulating systems is based on specific algorithms, which reflect integral essence of a matrion as an integral multi-dimensional number. Such algorithms will be useful not only for biomechanics of movements, but for robotics as well.

The theory of matrions which has arisen in matrix genetics is in the very beginnings of its development. In this book we have no ability to present many interesting materials on this topic. For example, introducing appropriate systems of matrion coordinates into 2^n-dimensional spaces allows one to work with multi-dimensional vectors of these matrion spaces by means of operations of addition, subtraction and multiplication of these vectors.

Since matrions reflect structural features of genetic code systems, complex numbers and double numbers, which have been known in mathematics for a long time, receive new aspects of their importance in fields of biology, physiology, psychophysis, etc. For example, double numbers prove to be connected with the Pythagorean musical scale and with the famous tables of the Ancient Chinese book "I Ching" (see Chapter 12). One can recommend the development of mathematical aspects of the theory of matrions. One can recommend using numeric systems of multi-dimensional matrions in structural investigations of the molecular-genetic systems as well.

FUTURE TRENDS AND CONCLUSION

Hypercomplex numbers had many essential applications in mathematical natural sciences in the past. Matrix genetics proposes additional kinds of hypercomplex numbers in the form of matrions for utilization in the field of bioinformatics and mathematical biology. A further development of the theory of matrions of hyperbolic type and of circular type in their matrix forms of presentation can give many new interesting results including new symmetrical patterns connected with biological and other structures.

Of course, the notions of matrions can be generalized as well, for example, by means of their consideration not over a field of real numbers but over a field of complex numbers or over rings of other kinds of numbers including matrions themselves. A development of a theory of functions of matrion variables for both – hyperbolic and circular - categories of matrions is interesting because of their commutative property.

The described results demonstrate that the double numbers, which were introduced by Clifford in 1872 year, are connected with the Pythagorean musical scale, with the system of molecular parameters of the genetic code, with a matrix-genetic definition of the golden section, etc. The theory of proposed kinds of hypercomplex numbers in the forms of hyperbolic matrions and of circular matrions leads to new symmetric patterns in the field of bioinformatics and to some new ideas about mathematical simulation of ensembles of molecular-genetic structures.

REFERENCES

Kline, M. (1980). *Mathematics. The loss of certainty*. New York: Oxford University Press.

Petoukhov, S. V. (2008a). *Matrix genetics, algebras of the genetic code, noise-immunity*. Moscow: RCD (in Russian).

Rosenfeld, B. A. (1966). *Multi-dimensional spaces*. Moscow: Nauka (in Russian).

Silvestrov, V. V. (1998). Systems of numbers. [in Russian]. *Soros Educational Journal*, *8*, 121–127.

Section 4
Connections of Matrix Genetics with Other Fields of Science and Culture

Section 4 is devoted to many connections of the genetic code systems in their matrix forms of presentation with various fields of science and culture. Some concrete examples are presented including inherited phyllotaxis laws of biological morphogenesis, physiological peculiarities of color perception, and morphogenetic invariants of projective geometry.

Chapter 10
Genetic System, Fibonacci Numbers, and Phyllotaxis Laws

ABSTRACT

This chapter describes data suggesting a connection between matrix genetics and one of the most famous branches of mathematical biology: phyllotaxis laws of morphogenesis. Thousands of scientific works are devoted to this morphogenetic phenomenon, which relates with Fibonacci numbers, the golden section, and beautiful symmetrical patterns. These typical patterns are realized by nature in a huge number of biological bodies on various branches and levels of biological evolution. Some matrix methods are known for a long time to simulate in mathematical forms these phyllotaxis phenomena. This chapter describes connections of the famous Fibonacci (2x2)-matrices with genetic matrices. Some generalizations of the Fibonacci matrices for cases of (2^nx2^n)-matrices are proposed. Special geometrical invariants, which are connected with the golden section and Fibonacci numbers and which characterize some proportions of human and animal bodies, are described. All these data are related to matrices of the genetic code in some aspects.

INTRODUCTION BACKGROUND

The complexity of biological objects complicates the creation of appropriate mathematical simulators of their morphogenetic and other features. But one phenomenon has been famous in the field of biological morphology for a long time, which allows one to create non-trivial mathematical simulators. It is a morphogenetic phenomenon of phyllotaxis or a phenomenon of regular dispositions of leaves and some other parts in configurations of biological bodies. This inherited phenomenon is connected with Fibonacci numbers and with the golden section and it is observed at very various levels and branches

DOI: 10.4018/978-1-60566-124-7.ch010

of biological evolution (Adler, 1974, 1990; Bowman, Eshed, & Baum, 2002; Clark, 2001; Douady, & Couder, 1992; Jean, 1995; Lee, & Levitov, 1998; Stieger, Reinhardt, & Kuhlemeier, 2002; Thompson, 1942; Waites, & Hudson, 1995). The task of understanding the questions of how and why genetic systems provide such regular dispositions of biological parts is one of the important ones in the field of mathematical and theoretical biology.

Matrix genetics reveals some new ways to study such questions because it involves a matrix approach and a matrix presentation of the golden section and of Fibonacci numbers. Simultaneously new forms of manifestations of phyllotaxis or quasi-phyllotaxis laws can be revealed on the bases of data of matrix genetics, because it deals, for example, with Hadamard matrices and with genetic multi-dimensional algebras, which can lead to new facts of manifestations of Fibonacci numbers and the golden section in biological morphogenesis. Some mathematical aspects arise in this field, which need their accurate investigations from different sides. Initial results of such investigations are presented in this chapter.

The main objectives of this chapter are investigations of possibilities of generalizations of Fibonacci matrices and of golden genomatrices, which are connected with inherited phyllotaxis laws and with many manifestations of the golden section in physiological phenomena. Mathematical properties of these generalizations should be investigated as well. The hypothesis is put forward that invariants of projective geometry, which are observed as ontogenetic invariants of the kinematic scheme of the inherited human body, are connected with peculiarities of the genetic code and with its Hadamard matrices.

A possible connection of the genetic matrices with invariants of projective geometry, which are observed as ontogenetic invariants of the kinematic scheme of the human body, is presented.

Biological phenomena of phyllotaxis are one of the most famous and popular phenomena in the field of mathematical biology. Thousands of scientific works are devoted to this morphogenetic phenomenon, which relates with Fibonacci numbers and the golden section (see for example the review in the book (Jean, 1995) with references including 1000 thematic sources approximately). It is important that these inherited phenomena are illustrated by a huge number of regular helical patterns of biological morphogenesis and they are described by means of simple mathematical laws. These phenomena and their patterns interest not only biologists, but also mathematicians and physicists (Douady, & Couder, 1992; Koch, & Meinhardt, 1994; Levitov, 1991a,b; Mandelbrot, 1983).

According to these phyllotaxis laws, pairs of Fibonacci numbers are realized at once at helical arrangement of leaves (or seeds) in plant bodies (shoots). These pairs belong to the sequences of the two following types:

$$\{Q_n' = F_{n+1}/F_n\}: 2/1, 3/2, 5/3, 8/5, 13/8, \ldots \Rightarrow \varphi = (1+5^{0.5})/2 = 1.618\ldots,$$

$$\{Q_n'' = F_{n+2}/F_n\}: 2/1, 3/1, 5/2, 8/3, 13/5, \ldots \Rightarrow \varphi^2 = 2.618\ldots \quad (1)$$

These sequences of fractions (equation 1) are named the parastichous sequence and the orthostichous sequence correspondingly, and the magnitude "φ" is the golden section here. The regular helical arrangement of leaves on plant shoots is characterized by the orthostichous sequence. Such arrangement at various plants can differ by a magnitude of a fraction among the orthostichous sequence of fractions Q_n. For example, leave arrangements in the cases of a lime tree, a elm tree, a beech are characterized by the fraction 2/1; in cases of an alder tree, a nut-tree, a vine, a sedge – by the fraction 3/1; in the cases of a raspberry, a pear tree, a poplar, a barberry – by the fraction 8/3; in the cases of an almond tree, a sea-

Figure 1. The Fibonacci series

n	0	1	2	3	4	5	6	7	8	9	10	11	...
F_n	0	1	1	2	3	5	8	13	21	34	55	89	...

buckthorn – by the fraction 13/5. Cones of coniferous trees correspond to fractions 21/8, 34/13, 55/21 in various cases.

Fractions from the parastichous sequences Q_n' (equation 1) are realized in many other cases as well. They are demonstrated in leave arrangements on a transverse section of a bud, in arrangements of sunflower seeds, scales of cones, etc. These arrangements possess left helixes and right helixes, quantities of which are equal to adjacent Fibonacci numbers as a rule. For example, the ratio of quantities of such left and right helixes in the case of cones of spruces is equal to 5/3; in the case of cones of larches the ratio is equal to 8/5; in the case of cones of pines the ratio is equal to 13/8, 21/13, 34/21; in the case of cones of sunflower heads the ratio is equal to 13/8, 21/13, 34/21, as a rule. Phyllotaxis laws were formulated in botany initially, but then they were revealed on very different branches and levels of biological evolution. The same phyllotaxis ratios exist in helical arrangements of scales in cases of fishes and of mammals, buds of hydras, organs of medusas, chambers of foraminifera shells, seams on shells of mollusca, separate components of muscles, etc. The phyllotaxis laws are demonstrated not only on a level of the whole organisms, but on a level of biological molecules as well in arrangements of alpha-helixes of polypeptides (see a review in (Petoukhov, 1981, § 1.1)). In his studying of the problem of consciousness and its connection with quantum coherence and with tubulin (one of kinds of proteins), R. Penrose (1989) has met the phyllotaxis phenomenon in microtubules of a neuron cytoskeleton: 13 rows of dimers exist, which form such microtubule, and a hexagonal pattern of microtubules consists of the 5 right helixes and of the 8 left helixes. An external layer of double microtubules contains 21 rows of dimers of tubulin (numbers 5, 8, 13, 21 are Fibonacci numbers).

As we have considered in Chapter 4, the series of Fibonacci numbers F_n (where $n = 0, 1, 2, 3,...$) begins with the numbers 0 and 1. Each next member of this series is equal to the sum of two previous members: $F_{n+2} = F_n + F_{n+1}$. Fibonacci numbers are used widely in the theory of optimization and in many other fields. One can find a rich collection of data about the golden section and the Fibonacci numbers on the web-site of "The museum of harmony and the golden section" by A. Stakhov (www.goldenmuseum.com) and in works (Jean, 1995; Kappraff, 1990, 1992). Fibonacci sequences and their generalizations provide fast algorithms of discrete orthogonal transformations in numeration systems with irrational basis. Figure 1 shows the first numbers F_n of the Fibonacci series.

This series of Fibonacci numbers possesses an interesting connection with the so called Fibonacci matrices $Q_{left} = [0\ 1;\ 1\ 1]$ or $Q_{right} = [1\ 1;\ 1\ 0]$, which are known for a long time.

Figure 2. Exponentiation of Fibonacci matrices of two kinds produces Fibonacci numbers

$$Q_{left}^{\,n} = \begin{vmatrix} 0 & 1 \\ 1 & 1 \end{vmatrix}^n = \begin{vmatrix} F_{n-1} & F_n \\ F_n & F_{n+1} \end{vmatrix}; \quad Q_{right}^{\,n} = \begin{vmatrix} 1 & 1 \\ 1 & 0 \end{vmatrix}^n = \begin{vmatrix} F_{n+1} & F_n \\ F_n & F_{n-1} \end{vmatrix}$$

Exponentiation of such (2x2)-matrices in the power "n" produces new (2x2)-matrices, the components of which are equal to three adjacent Fibonacci numbers F_{n-1}, F_n, F_{n+1} (Figure 2).

Consequently in this matrix approach each three adjacent Fibonacci numbers F_{n-1}, F_n, F_{n+1} are defined by one number "n", which is a power of a corresponding Fibonacci matrix. In view of this, one can define the series of Fibonacci numbers by means of these Fibonacci matrices without using the traditional algorithm of addition of two adjacent Fibonacci numbers at all. For example, the Fibonacci number F_n (beginning with $n = 3$) can be defined as a middle component among all components of the matrix Q_{left}^n or of the matrix Q_{right}^n. Or the Fibonacci number F_n can be defined as that number which is a component of each of the three matrices Q_{left}^{n-1}, Q_{left}^n and Q_{left}^{n+1} simultaneously. Below we will show that the Fibonacci matrices allow one to provide a non-classical definition of the golden section as well.

Euclidean symmetries exist in biological bodies together with symmetries of highest geometries, first of all, with symmetries of projective geometry. According to the famous Erlangen program by F. Klein, Euclidean geometry and many kinds of non-Euclidean geometries are defined as the sciences of invariants of relevant groups of transformation (Yaglom, 1988). For example, Euclidean geometry is the science of invariants of the group of motions and the projective geometry is the science of invariants of the group of projective transformations. Each of these geometries has its own set of invariants. The main invariant of the projective geometry is the so called "double ratio" (or the "wurf"). This invariant relates to the inherited proportions of kinematic scheme of human and animal bodies (Petoukhov, 1981, 1989) and it will be described in the last paragraph of this chapter in connection with the genetic matrices.

FIBONACCI'S MATRICES AND BIOLOGICAL LAWS OF PHYLLOTAXIS

Chapter 6 has described the two molecular features of the genetic letter U, which distinguish it from the other three letters A, C, G of the genetic alphabet: the absence of the amino group NH_2 and the replacement of the letter U by the kindred letter T at transfer from RNA to DNA. In the genetic kernel matrix [C A; U G] this opposition of the letter U to the other three genetic letters can be shown by expressing each of the letters C, A, G by means of the symbol "+1" and the opposite letter U by means of the symbol "-1". Such an operation leads to the transformation of the symbolic genomatrix [C A; U G] into the numeric Hadamard matrix [1 1; -1 1], as it was considered in Chapter 6. Since such binary opposition exists on the molecular level really, a computer system of organism allows the possibility to work with Hadamard genomatrices due to this opposition.

But the same binary opposition can be expressed by means of another pair of opposite numeric elements: number "0" can express the letter U and number "1" can express each of the other three letters in the genomatrix [C A; U G]. Both pairs of these binary symbols together with mathematical operations with them are realized easily in usual decisions of digital technique.

Besides using Hadamard matrices in digital technique and in spectral analysis, sometimes the method of replacements of both their elements "+1" and "-1" by another pair of elements "0" and "1" is utilized in some steps of information processing (Tolmachev, 1976; Solovieva, 2006). By analogy one can suppose that both these pairs of binary-oppositional symbols together with mathematical operations with them are utilized in a computer of organism as well. In this case, the same binary-oppositional traits of molecular-genetic systems can be expressed in genomatrices by means of the pair of binary elements "+1" and "-1" or by means of the pair of binary elements "0" and "1". It increases possibilities of bioinformation technologies significantly.

Figure 3. On the upper row: three kinds of generalized Fibonacci (8x8)-matrices G1, G2, G3. On the lower row: two examples of exponentiation of the matrices G1 and G3

$$G_1=\begin{vmatrix} 1&1&1&1&1&1&1&1\\ 1&1&1&1&1&1&1&1\\ 1&1&1&1&1&1&1&1\\ 1&1&1&1&1&1&1&1\\ 1&1&1&1&0&0&0&0\\ 1&1&1&1&0&0&0&0\\ 1&1&1&1&0&0&0&0\\ 1&1&1&1&0&0&0&0 \end{vmatrix};\ G_2=\begin{vmatrix} 1&1&1&1&1&1&1&1\\ 1&1&1&1&1&1&1&1\\ 1&1&0&0&1&1&0&0\\ 1&1&0&0&1&1&0&0\\ 1&1&1&1&1&1&1&1\\ 1&1&1&1&1&1&1&1\\ 1&1&0&0&1&1&0&0\\ 1&1&0&0&1&1&0&0 \end{vmatrix};\ G_3=\begin{vmatrix} 1&1&1&1&1&1&1&1\\ 1&0&1&0&1&0&1&0\\ 1&1&1&1&1&1&1&1\\ 1&0&1&0&1&0&1&0\\ 1&1&1&1&1&1&1&1\\ 1&0&1&0&1&0&1&0\\ 1&1&1&1&1&1&1&1\\ 1&0&1&0&1&0&1&0 \end{vmatrix}$$

$$G_1^7=4^6*\begin{vmatrix} 21&21&21&21&13&13&13&13\\ 21&21&21&21&13&13&13&13\\ 21&21&21&21&13&13&13&13\\ 21&21&21&21&13&13&13&13\\ 13&13&13&13&8&8&8&8\\ 13&13&13&13&8&8&8&8\\ 13&13&13&13&8&8&8&8\\ 13&13&13&13&8&8&8&8 \end{vmatrix};\ G_3^7=4^6*\begin{vmatrix} 21&13&21&13&21&13&21&13\\ 13&8&13&8&13&8&13&8\\ 21&13&21&13&21&13&21&13\\ 13&8&13&8&13&8&13&8\\ 21&13&21&13&21&13&21&13\\ 13&8&13&8&13&8&13&8\\ 21&13&21&13&21&13&21&13\\ 13&8&13&8&13&8&13&8 \end{vmatrix}$$

The possibility of a similar different interpretation of the same material depending on surrounding conditions is known in genetics for a long time: for example, the trait of alopecia is dominating for men, but this trait is recessive for women. It means that this trait depends on the internal environment of an organism.

As mentioned earlier, using the symbols "0" and "1" many genomatrices, for example, [C A; G U] or [U A; C G], become the famous Fibonacci matrices [1 1; 1 0] and [0 1; 1 1] correspondingly (see Figure 2 and the description of Fibonacci matrices above). The inherited phyllotaxis laws are connected with Fibonacci numbers and Fibonacci matrices. In such a way matrix genetics leads to new variants of mathematical simulations and an understanding of genetic bases of phyllotaxis phenomena.

It is interesting that eigenvalues of the Fibonacci matrices [0 1; 1 1] or [1 1; 1 0] are equal to the golden section "1.618…" and its reverse magnitude "-0.618…" with the sign "-". This feature allows one to propose new definition of the golden section as well. The golden section is the positive eigenvalue of the Fibonacci matrix. Kronecker exponentiation of the Fibonacci matrix (for example $[0\ 1;\ 1\ 1]^{(n)}$, where n = 2, 3, 4,..) gives the $(2^n\text{x}2^n)$-matrices, all eigenvalues of which are equal to the golden section in different integer powers only (magnitudes of the eigenvalues can be positive or negative, of course).

Generalizations of the Fibonacci (2x2)-matrices in a form of families of $(2^n\text{x}2^n)$-matrices are possible (Petoukhov, 2003-2004). Figure 3 (on the upper row) shows an example of such generalizations in a form of the family of $(2^3\text{x}2^3)$-matrices G_1, G_2 and G_3. Exponentiation of these matrices gives the matrices G_1^n, G_2^n, G_3^n, all components of which are three adjacent Fibonacci numbers F_{n-1}, F_n, F_{n+1} with a factor 4^{n-1}. Figure 3 on the lower row demonstrates an example of exponentiation of these generalized Fibonacci matrices into 7 power.

Two non-zero eigenvalues of the generalized Fibonacci matrices G_1, G_2, G_3 are equal to "$4*\varphi$" and "$-4*\varphi^{-1}$", where φ is the golden section. They are distinguished from appropriate eigenvalues of the classical Fibonacci (2x2)-matrices by the factor 4 only.

The following equation holds true for these matrices (by analogy with the recurrent equation $F_{n+2} = F_{n+1} + F_n$ for Fibonacci numbers):

$(G_i)^{n+2} = (G_i)^{n+1} + (G_i)^n$, where $i = 1, 2, 3$ (2)

One should note that the generalized Fibonacci matrices G_1, G_2, G_3 arise, for example, from the genomatrix [A C; G U]$^{(3)}$ of 64 triplets in the following algorithmic cases. The matrix G_1 arises if each triplet in the genomatrix [A C; G U]$^{(3)}$ is replaced by the element "0" in the case, when the letter U occupies the first position of the triplet, and by the element "1" in all other cases. The matrix G_2 arises if each triplet in the same genomatrix is replaced by the element ``0" in the case, when the letter U occupies the second position of the triplet, and by the element "1" in all other cases. The matrix G_2 arises if each triplet in the same genomatrix is replaced by the element "0" in the case, when the letter U occupies the third position of the triplet, and by the element "1" in all other cases.

One can add that the interesting formula exists for a connection of the Fibonacci matrices Q_{left} and Q_{right} (Figure 2) with the golden (2x2)-genomatrix Φ from Figure 3 in Chapter 4:

$F^2 = Q_{left}^{\ 2} + Q_{right}^{\ 2}$. (3)

This square formula is similar to a classical formula of length in a 2-dimensional vector space. This formula was revealed in a process of studying numerical genetic matrices as metric tensors of Riemannian geometry, which are used for researches of a problem of inherited morphological surfaces with their special internal geometry (Petoukhov, 2003-2004, 2008a).

ADDITIONAL FACTS ABOUT THE GOLDEN SECTION AND MATRICES

The previous paragraph was devoted to Fibonacci matrices and their generalizations. Now we will consider the golden matrices and their generalizations. The notion of golden matrices arose in Chapter 4 as a result of studying the genetic matrices. Such a name was given to a matrix, all components of which are equal to the golden section in its various integer powers (Figure 3 in Chapter 4). Figure 4 shows new kinds of golden matrices Φ_1, Φ_2, Φ_3.

The revelation of these additional kinds Φ_1, Φ_2, Φ_3 was made in the course of studying the genetic matrices of hydrogen bonds (some initial investigation was described in Chapter 4). These golden matrices Φ_1, Φ_2, Φ_3 are connected with matrices B_1, B_2, B_3 by means of the following simple expression:

$4*B_k = \Phi_k^{\ 2}$, where $k = 1, 2, 3$. (4)

One should note that these matrices B_1, B_2, B_3 arise from the genomatrix [C A; U G]$^{(3)}$ of 64 triplets in the following algorithmic cases. The matrix B_1 arises if each triplet in the genomatrix [C A; U G]$^{(3)}$ is replaced by the number of hydrogen bonds of the first letter of this triplet (C = G = 3 and A = U = 2). The matrix B_2 arises if each triplet in the same genomatrix is replaced by number of hydrogen bonds of the second letter of this triplet. The matrix B_3 arises if each triplet in the same genomatrix is replaced by the number of hydrogen bonds of the third letter of this triplet.

A turn of the golden matrices Φ_1, Φ_2, Φ_3 on 90^0 together with the same turn of the matrices B_1, B_2, B_3 leads to new matrices, which satisfy the equation (4) as well. Such a turn can be presented as a consequence of a cyclic alphabetic permutation of the genetic letters (for example, C→U→G→A→C) in the initial genomatrix [C A; U G]$^{(3)}$. We will return to such cyclic permutations in Chapter 11.

Figure 4. On the upper row: variants B_1, B_2, B_3 of numeric presentation of the numeric genomatrix [C A; U G]$^{(3)}$. On the lower row: three kinds of appropriate golden matrices Φ_1, Φ_2, Φ_3. Here φ is the golden section; $\tau = \varphi^{-1}$

B_1 =
3 3 3 3 2 2 2 2
3 3 3 3 2 2 2 2
3 3 3 3 2 2 2 2
3 3 3 3 2 2 2 2
2 2 2 2 3 3 3 3
2 2 2 2 3 3 3 3
2 2 2 2 3 3 3 3
2 2 2 2 3 3 3 3

B_2 =
3 3 2 2 3 3 2 2
3 3 2 2 3 3 2 2
2 2 3 3 2 2 3 3
2 2 3 3 2 2 3 3
3 3 2 2 3 3 2 2
3 3 2 2 3 3 2 2
2 2 3 3 2 2 3 3
2 2 3 3 2 2 3 3

B_3 =
3 2 3 2 3 2 3 2
2 3 2 3 2 3 2 3
3 2 3 2 3 2 3 2
2 3 2 3 2 3 2 3
3 2 3 2 3 2 3 2
2 3 2 3 2 3 2 3
3 2 3 2 3 2 3 2
2 3 2 3 2 3 2 3

Φ_1 =
φ φ φ φ τ τ τ τ
φ φ φ φ τ τ τ τ
φ φ φ φ τ τ τ τ
φ φ φ φ τ τ τ τ
τ τ τ τ φ φ φ φ
τ τ τ τ φ φ φ φ
τ τ τ τ φ φ φ φ
τ τ τ τ φ φ φ φ

Φ_2 =
φ φ τ τ φ φ τ τ
φ φ τ τ φ φ τ τ
τ τ φ φ τ τ φ φ
τ τ φ φ τ τ φ φ
φ φ τ τ φ φ τ τ
φ φ τ τ φ φ τ τ
τ τ φ φ τ τ φ φ
τ τ φ φ τ τ φ φ

Φ_3 =
φ τ φ τ φ τ φ τ
τ φ τ φ τ φ τ φ
φ τ φ τ φ τ φ τ
τ φ τ φ τ φ τ φ
φ τ φ τ φ τ φ τ
τ φ τ φ τ φ τ φ
φ τ φ τ φ τ φ τ
τ φ τ φ τ φ τ φ

The product of any two matrices from the set Φ_1, Φ_2, Φ_3 possesses the commutative property and gives such matrix in all cases, all components of which are equal to 10:

$$\Phi_1 * \Phi_2 = \Phi_2 * \Phi_1 = \Phi_2 * \Phi_3 = \Phi_3 * \Phi_2 = \Phi_1 * \Phi_3 = \Phi_3 * \Phi_1 \quad (5)$$

The analogical generalization of the described golden (2x2)-matrices exists for cases of (2^nx2^n)-matrices, where n = 2, 4, 5, 6,… . These cases correspond to genetic matrices [C A; U G]$^{(n)}$ of 2-plets, 4-plets, 5-plets, etc.

HADAMARD MATRICES, PROJECTIVE GEOMETRY AND THE GOLDEN WURF

It is known that Hadamard matrices of order ($4*n+4$) have connections with a projective plane and with projective geometry (Craigen, 1996; Dinitz, & Stinson, 1992; Lindner, & Rodger, 1997; Sachkov, 2004; Seberry, & Yamada, 1992). This fact can relate to the known fact about the existence of ontogenetic invariants of projective geometry in kinematic schemes of human and animal bodies at their ontogenetic growth (Petoukhov, 1981, 1989).

The main invariant of the projective geometry is the double ratio, which is defined by the following way (Klein, 1928). If one has four points *A*, *B*, *C*, and *D* on a straight line, the following magnitude W is named as "the double ratio":

$$W = [(C-A)*(D-B)] / [(B-C)*(D-A)] \quad (6)$$

The second name of the double ratio is the "wurf". This name has been introduced by German mathematician G. von Staudt; "wurf" means "throw" in a translation from German. Projective transformations can change Euclidean lengths of separate segments of a line and they can change simple ratios (or affine proportions) of lengths of any two segments, but the magnitude of the wurf W (equation 6) is never changed at any projective transformations. In other words, the wurf W is an invariant of projective transformations. (Note that the founder of the contemporary projective geometry, J. Desargues, French architect and engineer (1593-1662 years), widely utilized biological terminology, probably assuming a kinship between the projective structures and nature). It is interesting that an analogical situation is observed at ontogenetic changes of proportions of kinematic scheme of human and animal bodies. Let us describe this situation.

Numerous biological bodies are constructed on the bases of a principle of hierarchy of symmetrical blocks: a body is constructed from symmetric blocks of the first level, which are combined into symmetric blocks of the second level, etc. In particular, this principle manifests itself in the kinematic scheme of the human body, where the mirror symmetry of the two halves of the body, which act as second-level blocks, is added by an approximate projective symmetry of the longitudinal proportions of the three-part kinematic blocks, ensemble of which forms the kinematic scheme of the whole body. These discussions

Figure 5. Changes of the human body with age (according to (Petten, 1959)). Here "a)" and "b)" – antenatal and postnatal stages: "a)" – in lunar months; "b)" – in years (the first on the left is a newborn). "c)" – Examples of various variants of proportions in three-part segments, whose wurfs are equal to 1.31 and whose end points are A, B, C, D.

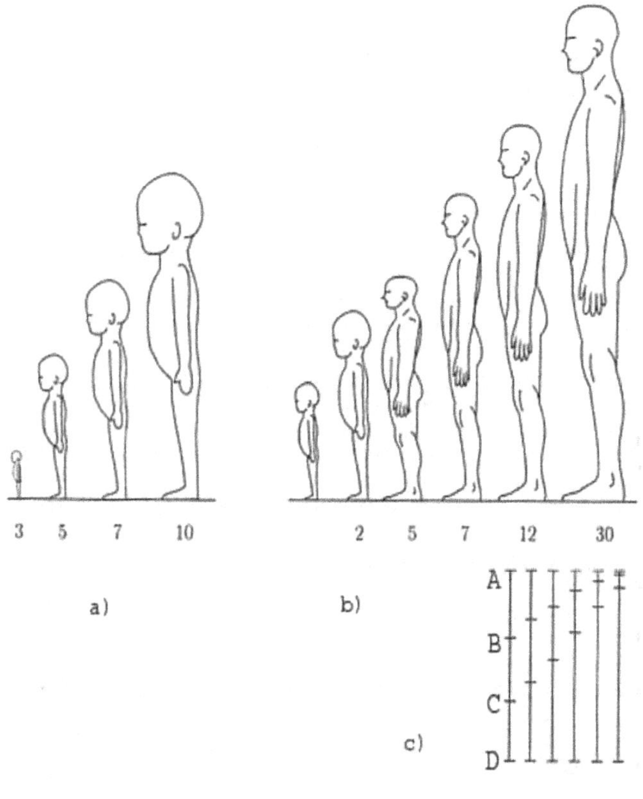

are about the following three-part kinematic blocks: the three phalanxes of fingers, the three-part extremities ("shoulder-forearm-wrist" and "hip-shin-foot") and the whole three-part body (in anthropology the body is subdivided into the upper part, the trunk part and the bottom part). We will designate the four end points of each of these three-part blocks by means of letters "*A*", "*B*", "*C*", "*D*". For example, in the case of the three-part block of the whole human body, the letter "*A*" marks the crown of the head; the letter "*B*" marks the base of the neck; the letter "*C*" marks the hip joint; the letter "*D*" marks the end of fingers of the straightened feet.

Lengths of separate segments of this ensemble of three-part blocks can differ from each other; ontogenetic changes of their lengths can differ significantly without conservation of their affine ratios (or of simple proportions) between two segments (Figure5 - 9). For example, throughout a life from a birth the upper part of the kinematic scheme of human body grows in 2.4 times, the trunk part- in 2.8 times, the bottom part - in 3.8 times. But the magnitude of the wurf *W*, which is calculated by means of the expression (6) for different ages, is not changed practically during all development of an individual after his birth. Moreover the magnitudes of the wurfs *W* for all mentioned three-part blocks of the kinematic scheme of the human body are equal to each other practically (Figure 6, 7, 8 and 9). Their general or reference magnitude is equal to 1.31 approximately. These calculations of wurfs (or projective proportions) in tables on Figure 6 - 9 were made on the bases of the known anthropological data (Bunak, 1957; Rokhlin, 1936).

The analysis of these wurf proportions in relation to the construction and in relation to the ontogenetic growth of the human body reveals that all mentioned three-part blocks of the human kinematic scheme are identical to each other from the viewpoint of their projective geometrical proportions (or from the viewpoint of projective geometry). It is true not only for normally developed people, but also for dwarfs and giants in many respects. It is true also for a wide set of highly organized animals (see details in (Petoukhov, 1981)).

Figure 6. The lengths of three segments in the kinematic block "shoulder(A_1B_1)-forearm(B_1C_1)-wrist(C_1D_1)" of human body at different ages. Initial data are taken from (Bunak, 1957). Relevant magnitudes of the wurfs W are shown in the right column

embrio age (lunar months)	schoulder - forearm - wrist in centimeters	*W*
4	2.37 – 1.90 – 1.55	1.33
6	4.40 – 3.65 – 3.23	1.34
8	6.17 – 5.05 – 4.53	1.35
9	6.95 – 5.70 – 5.14	1.35
newborn	8.80 – 7.40 – 6.30	1.33
age (years)		
1	12.5 – 10.4 – 8.7	1.33
4	16.7 – 13.7 – 11.2	1.33
7	20.7 – 16.5 – 13.1	1.33
10	23.2 – 18.4 – 14.4	1.32
13	26.3 – 20.7 – 16.0	1.32
17	30.5 – 23.9 – 18.3	1.32
20	32.3 – 24.5 – 18.8	1.33

Figure 7. The lengths of three segments in the kinematic block "hip(A_2B_2) - shin(B_2C_2) - foot(C_2D_2)" of human body at different ages. Initial data are taken from (Bunak, 1957). Relevant magnitudes of the wurfs W are shown in the right column

embrio age (lunar months)	hip – shin – foot in centimeters	W
4	2,76 – 2.66 – 1.84	1.26
6	5.45 – 5.45 – 4.30	1.27
8	7.40 – 7.60 – 5.80	1.27
9	8.40 – 8.60 – 6.61	1.27
newborn	10.90 – 9.90 – 7.80	1.30
age (years)		
1	15.8 – 14.4 – 11.5	1.30
4	22.4 – 20.2 – 15.8	1.30
7	28.0 – 24.8 – 18.9	1.30
10	32.1- 27.8 – 20.7	1.30
13	37.3 – 32.2 – 23.6	1.30
17	44.4 – 37.3 – 26.9	1.29
20	45.4 – 37.5 – 27.0	1.30

Figure 8. The lengths of three segments in the kinematic block "upper part(A_3B_3) - trunk(B_3C_3) – bottom part(C_3D_3)" of human body at different ages. Initial data are taken from (Bunak, 1957). Relevant magnitudes of the wurfs W are shown in the right column

age (years)	upper part-trunk-bottom part in centimeters	W
newborn	10.6 – 18.3 – 28.6	1.29
1	15.6 – 25.0 – 41.7	1.32
4	18.9 – 31.4 – 58.4	1.32
7	21.0 – 35.3 – 71.7	1.33
10	22.5 – 38.0 – 80.6	1.34
13	24.5 – 42.2 – 93.5	1.34
17	26.9 – 47.8 – 108.6	1.33
20	25.3 -51.8 – 109.9	1.29

Relevant Magnitudes of the Wurfs W are Shown in the Right Column

The reference magnitude of these wurfs $W \approx 1.31$ for these three-part kinematic blocks is connected with the phyllotaxis laws of morphogenesis, Fibonacci numbers and the golden section φ. This connection is revealed, first of all, by a consideration of magnitudes of wurfs of adjacent numbers F_n, F_{n+1}, F_{n+2} of the Fibonacci series (Figure 1). One can interpret these numbers as lengths of three adjacent segments of a straight line. In this case the relevant magnitude of the wurf W_n of such three segments is equal to the following:

$$W_n = [(F_n + F_{n+1}) * (F_{n+1} + F_{n+2})] / [F_{n+1} * (F_n + F_{n+1} + F_{n+2})] \quad (7)$$

The magnitudes of the wurfs W_n of the adjacent groups of three Fibonacci numbers form a new sequence:

Figure 9. The lengths of phalanxes of the middle finger of human hand (data are taken from (Rokhlin, 1936)). AB – the basal phalanx; BC – the medium phalanx; CD – the end phalanx

Age (years)	AB – BC - CD	*W*
4	2,42 – 1.43 – 0.86	1.31
6	2.64 – 1.65 – 1.02	1.31
8	3.00 – 1.88 – 1.19	1.31
10	3.10 – 1.96 – 1.25	1.31
12	3.34 – 2.13 – 1.37	1.31
14	3.56 – 2.27 – 1.46	1.31
16	4.08 – 2.57 -1.64	1.31
18	4.19 – 2.65 – 1.69	1.31
21	4.41 – 2.78 – 1.76	1.31

$$\{W_n\}: 1;\ 3/2;\ 5/4;\ 8/6;\ 13/10;\ \ldots;\ F_{n+2}/(2^*F_n);\ \ldots \rightarrow \mathrm{P} \ldots \quad (8)$$

The limit value P of the wurf sequence (8) at $n \rightarrow \infty$ is equal to the following:

$$\mathrm{P} = \varphi^2/2 = 1.309\ldots, \quad (9)$$

where $\varphi = (1+5^{0,5})/2 = 1.618\ldots$ is the golden section. This limit value P (equation 9) is equal to the magnitude of the wurf of three adjacent segments with relative lengths 1: φ: φ^2. One should note that the lengths of three phalanxes of the middle finger of the human hand produce this series of ratios 1: φ: φ^2 obviously (see Figure 9). This fact is additional evidence that the approximate magnitude 1.31 of the wurf of the three-part kinematic blocks of human and animal body is connected with the phyllotaxis laws of morphogenesis and with the golden section. To emphasize this connection with the golden section, this wurf with the magnitude $\varphi^2/2$ (equation 9) was named 'the golden wurf'' (Petoukhov, 1981, 1989). This invariant of projective geometry has a biological meaning and is utilized in aesthetics of proportions and in the theory of music harmony (see Chapter 4).

In accordance to the Erlangen program by F. Klein, much non-Euclidean geometries are particular cases of projective geometry (Yaglom, 1988). But physiology knows long ago that inherited physiological spaces of visual perception relate with non-Euclidean geometry. This kind of research was pioneered by R. Luneburg (1950), who has discovered essential distinctions of geometry of visual spatial perception from Euclidean geometry. These findings were followed by scores of papers in various countries where the idea of a non-Euclidean geometry of visual spatial perception was extended and refined. A brief review of these works is in (Petoukhov, 1981, 1989). The mentioned connection of the genetic code systems with projective geometry by means of genetic Hadamard matrices can be utilized for explanation of such inherited non-Euclidean geometric phenomena to some extent.

FUTURE TRENDS AND CONCLUSION

Matrix genetics proposes new mathematical objects in the form of the generalized golden matrices and of generalized Fibonacci matrices. These kinds of matrices possess interesting mathematical properties which can be utilized for creating of new mathematical simulators to describe inherited phyllotaxis phenomena in living matter. In particular they form series of matrices with some recurrent properties and with other interesting mathematical properties which should be investigated specially for their possible applications not only in mathematical and theoretical biology, but also in other scientific fields including theory of signal processing, mathematical theory of games, etc.

The results of the matrix genetics gives new possibilities for understanding the phenomena of genetic inheritance of kinematic schemes of human and animal bodies, which are constructed utilizing the invariants of projective geometry.

The mentioned connection of genetic Hadamard matrices with a projective plane allows one to research many questions of application of formalisms of projective geometries and of some other non-Euclidean geometries in biological morphogenesis, where initial facts of inherited realizations of projective invariants are known long ago. The adjacent questions of non-Euclidean geometries of visual spatial perception can be considered from this new viewpoint as well.

The theme of the golden section and Fibonacci numbers in inherited biological phenomena is a classical theme, which exists for a long time. One may expect that this theme get new impulses due to described data of matrix genetics (see Chapters 4 and 10).

Matrix analysis and matrix forms of presentation of natural systems possess many advantages including a possibility of a wide set of analogies because they are utilized in many fields of science. It concerns the matrix forms of presentation of the genetic code ensembles in high extent as well. We think that many inherited biological phenomena will get new theoretical considerations and mathematical simulators in the near future on the bases of data of matrix genetics. Beautiful symmetrical patterns of biological morphology will be added by beautiful patterns and symmetries of abstract mathematical theories provoked by results of matrix genetics.

Special studies of various variants of the golden matrices and of the Fibonacci matrices together with their generalizations should be made in connection with the genetic matrices.

The theme of the golden section and Fibonacci numbers in matrix genetics is important because many physiological systems and processes are connected with the golden section. Chapter 4 has emphasized already, that proportions of the golden section characterize cardio-vascular processes, respiratory processes, electric activity of the brain, locomotion activity, aesthetic phenomena, etc. The additional data presented in this chapter give new opportunities for deeper understanding genetic bases of phenomena of the golden section and of Fibonacci numbers in biology. One can hope that described results of matrix genetics will be useful for the explanation and numeric forecast of separate parameters in a set of different physiological sub-systems of organisms with their cooperative essence and golden section phenomena. In our opinion, many realizations of the golden section in nature on the whole are connected with its matrix essence. One can conjecture the same situation is true for Fibonacci numbers relates to special Fibonacci matrices.

An additional direction for future study in the field of matrix genetics is a presentation of genetic matrices (for example $[C\ A;\ U\ G]^{(n)}$) in a numeric form of Hermitian matrices, which play an important role in quantum mechanics. Particularly it can be made by means of a test replacement of the genetic letters by their numbers of hydrogen bonds in complex numeric form: for example, one can test a variant C = G = 3, A = 2*i, U = -2*i, where "i" is imaginary unit. In this case we obtain a Hermitian matrix [C A; U G] = [3 2*i; -2*i 3], eigenvalues of which are equal to 1 and 5. A question about a benefit of such test approach on a base of Hermitian forms of genetic matrices is under investigation now.

REFERENCES

Adler, I. (1974). A model of contact pressure in phyllotaxis. *Journal of Theoretical Biology*, *45*, 1–79. doi:10.1016/0022-5193(74)90043-5

Adler, I. (1990). Symmetry in phyllotaxis. *Symmetry: Culture and Science*, *1*(2), 171–183.

Bowman, J. L., Eshed, Y., & Baum, S. F. (2002). Establishment of polarity in angiosperm lateral organs. *Trends in Genetics*, *18*, 134–141. doi:10.1016/S0168-9525(01)02601-4

Bunak, V. V. (1957). Changes in the relative length of human extremity skeleton segments during the growth period. [in Russian]. *Izvestia Akademii Nauk RSFSR*, *84*, 33–45.

Clark, S. E. (2001). Meristems: Start your signaling. *Current Opinion in Plant Biology*, *4*, 28–32. doi:10.1016/S1369-5266(00)00131-X

Craigen, R. (1996). Hadamard matrices and designs. In C. J. Colbourn & J. H. Dinitz (Eds.), *CRC Handbook of Combinatorial Designs* (pp. 370-377). Boca Raton, FL: CRC Press.

Dinitz, J. H., & Stinson, D. R. (1992). A brief introduction to design theory. In J. H. Dinitz & D. R. Stinson (Eds.), *Contemporary design theory: A collection of surveys* (pp. 1-12). New York: Wiley.

Douady, S., & Couder, Y. (1992). Phyllotaxis as a physical self-organized growth process . *Physical Review Letters*, *68*, 2098–2101. doi:10.1103/PhysRevLett.68.2098

Jean, R. V. (1995). *Phyllotaxis: A systemic study in plant morphogenesis*. Cambridge: Cambridge Univ. Press.

Kappraff, J. (1990). *Connections, the geometric bridge between art and science*. New York: McGraw Hill.

Kappraff, J. (1992). The relationship between mathematics and mysticism of the golden mean through history. In I. Hargittai (Ed.), *Fivefold symmetry* (pp. 33-66). Singapore: World Scientific.

Koch, A. J., & Meinhardt, H. (1994). Biological pattern-formation–from basic mechanisms to complex structures. *Reviews of Modern Physics*, *66*, 1481–1507. doi:10.1103/RevModPhys.66.1481

Lee, H. W., & Levitov, L. S. (1998). *Universality in phyllotaxis: A mechanical theory. Symmetry in plants*. Singapore: World Scientific.

Levitov, L. S. (1991a). Energetic approach to phyllotaxis. *Europhysics Letters*, *14*, 533–539. doi:10.1209/0295-5075/14/6/006

Levitov, L. S. (1991b). Phyllotaxis of flux lattices in layered superconductors. *Physical Review Letters*, *66*, 224–227. doi:10.1103/PhysRevLett.66.224

Lindner, C. C., & Rodger, C. A. (1997). *Design theory*. Boca Raton, FL: CRC Press.

Luneburg, R. (1950). The metric of binocular visual space. *Journal of the Optical Society of America*, *40*(10), 627–642. doi:10.1364/JOSA.40.000627

Mandelbrot, B. B. (1983). *The fractal geometry of nature*. New York: Freeman.

Penrose, R. (1989). *The emperor's new mind.* Oxford: Oxford University Press.

Petoukhov, S. V. (1981). *Biomechanics, bionics, and symmetry.* Moscow: Nauka (in Russian).

Petoukhov, S. V. (1989). Non-Euclidean geometries and algorithms of living bodies. In I. Hargittai (Ed.), *Computers & Mathematics with Applications, 17*(4-6), 505-534. Oxford: Pergamon Press.

Petoukhov, S. V. (2003-2004). Attributive conception of genetic code, its bi-periodic tables, and problem of unification bases of biological languages. *Symmetry . Cultura e Scuola, 14-15*(part 1), 281–307.

Petoukhov, S. V. (2008a). *Matrix genetics, algebras of the genetic code, noise-immunity.* Moscow: RCD (in Russian).

Petten, B. M. (1959). *The human embryology.* Moscow: Medgiz (in Russian)

Rokhlin, D. G. (1936). *X-ray osteology and x-ray anthropology.* Moscow: Biomedgiz (in Russian)

Sachkov, V. N. (2004). *Introduction to combinatory methods of discrete mathematics.* Moscow: Binom.

Seberry, J., & Yamada, M. (1992). Hadamard matrices, sequences, and block designs. In J. H. Dinitz & D. R. Stinson (Eds.), *Contemporary design theory: A collection of surveys* (pp. 431-560). New York: Wiley.

Solovieva, F. I. (2006). *Introduction into theory of coding.* Novosibirsk: NGU (in Russian).

Stieger, P. A., Reinhardt, D., & Kuhlemeier, C. (2002). The auxin influx carrier is essential for correct leaf positioning. *The Plant Journal, 32*, 509–517. doi:10.1046/j.1365-313X.2002.01448.x

Thompson, d'Arcy W. (1942). *On growth and form.* Cambridge: Cambridge University Press.

Tolmachev, Y. A. (1976). *New optic spectrometers.* Leningrad: Leningrad University.

Waites, R., & Hudson, A. (1995). Phantastica: A gene required for dorsoventrality in leaves in Antirrhinum majus. *Development, 121*, 2143–2154.

Yaglom, I. M. (1988). *Felix Klein and Sophus Lie: Evolution of the idea of symmetry in the 19th century.* Boston: Birkhäuser.

Chapter 11
Physiological Cycles and Their Algebraic Models in Matrix Genetics

ABSTRACT

This chapter presents data about cyclic properties of the genetic code in its matrix forms of presentation. These cyclic properties concern cyclic changes of genetic Yin-Yang-matrices and their Yin-Yang-algebras (bipolar algebras) at many kinds of circular permutations of genetic elements in genetic matrices. These circular permutations lead to such reorganizations of the matrix form of presentation of the initial genetic Yin-Yang-algebra that arisen matrices serve as matrix forms of presentations of new Yin-Yang-algebras, as well. They are connected algorithmically with Hadamard matrices. New patterns and relations of symmetry are described. The discovered existence of a hierarchy of the cyclic changes of genetic Yin-Yang-algebras allows one to develop new algebraic models of cyclic processes in bioinformatics and in other related fields. These cycles of changes of the genetic 8-dimensional algebras and of their 8-dimensional numeric systems have many analogies with famous facts and doctrines of modern and ancient physiology, medicine, and so forth. This viewpoint proposes that the famous idea by Pythagoras (about organization of natural systems in accordance with harmony of numerical systems) should be combined with the idea of cyclic changes of Yin-Yang-numeric systems in considered cases. This second idea reminds of the ancient idea of cyclic changes in nature. From such algebraic-genetic viewpoint, the notion of biological time can be considered as a factor of coordinating these hierarchical ensembles of cyclic changes of the genetic multi-dimensional algebras.

INTRODUCTION AND BACKGROUND

This chapter continues an analysis of the genetic 8-dimensional Yin-Yang-algebra (bipolar algebra), which was described in Chapter 7. This analysis allows one to revelation of unknown properties of

DOI: 10.4018/978-1-60566-124-7.ch011

Figure 1. Some examples of inherited cyclic configurations in living matter: a leaf of fern, a cone, a shell of mollusk

this genetic algebra and its possible applications for deeper understanding of genetic and physiological systems including inherited physiological cycles.

One of the directions, where the results of analysis of genetic Yin-Yang-algebras can be useful, is related to a creation of algebraic models of inherited physiological cycles and rhythms in organisms. The statement that biological organisms exist in accordance with cyclic processes of environment and with their own cyclic physiological processes is one of the most classical statements of biology and medicine from ancient times (see for example (Dubrov, 1989; Wright, 2002)). Many branches of medicine take into account the time of day specially, when diagnostic, pharmacological and therapeutic actions should be made for individuals. The set of this medical and biological knowledge is usually united under names of chrono-medicine and chrono-biology. Many diseases are connected with disturbances of natural biological rhythms in organisms. The problem of internal clocks of organisms, which participate in coordination of all interrelated processes of any organism, is one of the main physiological problems. But cyclic principles are essential for spatial organization of living bodies as well. Biological morphogenesis gives many examples of a cyclic symmetric repetition of separate spatial blocks in constructions of organism bodies (Figure 1). Such biological "cyclomerism" has been studied from viewpoints of Euclidean and non-Euclidean geometries for a long time (Petoukhov, 1989).

Molecular biology deals with this problem of physiological rhythms and of cyclic re-combinations of molecular ensembles on the molecular level as well. Really, it is the well-known fact that in biological organisms proteins are disintegrated into amino acids and then they are re-built (are re-created) from amino acids again in a cyclic manner systematically. A half-life period (a duration of renovation of half of a set of molecules) for proteins of human organisms is approximately equal to 80 days in most cases; for proteins of the liver and blood plasma – 10 days; for the mucilaginous cover of bowels – 3-4 days; for insulin – 6-9 minutes (Aksenova, 1998, v. 2, p. 19). Such permanent rebuilding of proteins provides a permanent cyclic renovation of human organisms. Such cyclic processes at the molecular-genetic level are one of the parts of a hierarchical system of a huge number of interelated cycles in organisms. The phenomenon of repeated recombinations of molecular ensembles, which are carried out inside separate cycles, is one of the main problems of biological self-organization. This phenomenon draws additional attention to structural properties of recombinations and permutations of molecular elements of genetic code systems. We are studying these structural properties using a matrix language. One can mention here that in addition to cyclic renovation of proteins, heritable corporal forms of activity exist: cardio cycles, breath cycles, walking, run, crawling, swimming and so forth.

Do some structural connections of the genetic code systems with inherited physiological rhythms and with such cyclic processes exist? Matrix genetics proposes new mathematical data of structural analysis for a positive answer on the first question and for a creation of algebraic models of such hierarchical system of cyclic changes. These data were obtained on the basis of an analysis of the mentioned genetic 8-dimensional Yin-Yang-algebra. This algebra was revealed initially as a result of analysis of the genetic matrix $[C\ A; U\ G]^{(3)}$, where the symbol in parentheses means the third Kronecker power and the symbols C, A, U, G mark nitrogenous bases of the genetic code (cytosine, adenine, uracil, guanine).

This genetic 8-dimensional Yin-Yang-algebra (bipolar algebra) was described in Chapter 7. This matrix algebra defines the system of 8-dimensional numbers *YY* with 8 real coordinates $x_0, x_1, ..., x_7$:

$$YY = x_0*\mathbf{f}_0+x_1*\mathbf{m}_1+x_2*\mathbf{f}_2+x_3*\mathbf{m}_3+x_4*\mathbf{f}_4+x_5*\mathbf{m}_5+x_6*\mathbf{f}_6+x_7*\mathbf{m}_7 \quad (1)$$

The matrix form of presentation of these numbers *YY* is shown on Figure 3 and Figure 4 in Chapter 7. The multiplication table of the basic elements $\mathbf{f}_0$, $\mathbf{m}_1$, $\mathbf{f}_2$, $\mathbf{m}_3$, $\mathbf{f}_4$, $\mathbf{m}_5$, $\mathbf{f}_6$, $\mathbf{m}_7$ is demonstrated on Figure 6 in chapter 7. Multiplication of any two members of such octet numbers *YY* generates a new octet number of the same system. This situation is similar to the situation of real numbers (or of complex numbers, or of hypercomplex numbers) when multiplication of any two members of a numeric system generates a new member of the same numeric system. From the abstract mathematical viewpoint such numeric system can be used for modeling not only static systems but variable systems and processes as well. For such a case of variable processes, one should consider coordinates of *YY*-numbers in the expression (1) as variable functions of time: $x_0(t)$, $x_1(t)$, $x_2(t)$, $x_3(t)$, $x_4(t)$, $x_5(t)$, $x_6(t)$, $x_7(t)$. For example, these functions can be trigonometric functions like $\sin(w*t)$ and $\cos(w*t)$ or they can be Walsh-Hadamard functions, etc. Multiplication of any two such 8-dimensional superpositions of functions gives a new 8-dimensional superposition of functions, which corresponds to the expression (1) again. In special cases these variable functions $x_0(t), x_1(t), ..., x_7(t)$ can describe some permutations of elements in systems with variable compositions, etc.

We will continue using some special notions and terms, which were introduced in Chapter 7: female basic elements and coordinates, male basic elements and coordinates, quasi-real unit, etc. But the Yin-Yang-algebra, which was presented on Figure 3, Figure 4 and Figure 6 in Chapter 7 and which was marked by the symbol YY_8, will be marked by the symbol $YY_+^{[CAUG]}$ in this chapter. The reasons for this change are the following. Below we will meet with many kinds of Yin-Yang-algebras, which are produced by means of cyclic permutations of genetic elements in the symbolic genetic matrix $[C\ A; U\ G]^{(3)}$ In view of this, the upper index in the symbol $YY_+^{[CAUG]}$ or $YY_+^{[UCGA]}$ of a relevant algebra shows the kind of symbolic genomatrices $[C\ A; U\ G]^{(3)}$ or $[U\ C; G\ A]^{(3)}$, which is transformed into this kind of numeric Yin-Yang-matrices by means of the same algorithm of the Yin-Yang-digitization of 64 triplets (this algorithm was described in Chapter 7). The meaning of the lower index "+" will be explained in the second paragraph of this chapter.

The main aim of this chapter is a description of unexpected cyclic and other mathematical properties of genetic Yin-Yang-matrices in connection with biological phenomena of cyclic processes and self-developing.

REVEALING NEW GENETIC ALGEBRAS AS A RESULT OF CYCLIC PERMUTATIONS OF GENETIC ELEMENTS

Matrix genetics studies the genetic code in its matrix forms of presentation by means of matrix methods of the theory of discrete signals processing. This theory pays great attention to permutations of discrete elements in information processing (Ahmed, Rao, 1975; Trahtman & Trahtman, 1975). In view of this, the following question arises: what genetic matrices are produced by various cyclic permutations of genetic elements in the initial genetic matrices [C A; U G]$^{(3)}$? Do these new matrices possess interesting mathematical properties? One of the unexpected results of studying this question is the discovery that many kinds of such cyclic permutations generate new genetic matrices, which are related algorithmically with matrices of new kinds of 8-dimensional Yin-Yang-algebras as well (Petoukhov, 2008a-2008f). The transformation of these new genetic matrices into matrices of new Yin-Yang-algebras is carried out by means of the same alphabetic algorithm of the Yin-Yang-digitization of 64 triplets (see Chapter 7). This algorithm connects solidly each genetic triplet with one of the eight *YY*-coordinates x_1, x_2, x_3, x_4, x_5, x_6, x_7 (with its certain sign "+" or "-" in accordance with Figure 3 in chapter 7) and this connection is irrespective of a disposition of triplets in considered genetic (8x8)-matrices of 64 triplets.

We begin with the case of a circular permutation of the genetic letters C→A→G→U→C. This circular permutation means that in all triplets of the genomatrix [C A; U G]$^{(3)}$ the letter A is replaced by the letter C, the letter G is replaced by the letter A, etc. As a result a new genomatrix [U C; G A]$^{(3)}$ arises, which is shown on Figure 2. This symbolic genomatrix defines the appropriate numeric Yin-Yang-matrix $YY_+^{[UCGA]}$ by means of the algorithm of the Yin-Yang-digitization of 64 triplets. Numeric components of this matrix $YY_+^{[UCGA]}$ are disposed in the matrix cells on Figure 2. One can see that this new genomatrix [U C; G A]$^{(3)}$ can be obtained by another way at all. Really, the same matrix arises when the initial matrix [C A; U G]$^{(3)}$ (Figure 4 in chapter 7) is turned on 90^0 clockwise.

Figure 2. The genetic matrix [U C; G A]$^{(3)}$, which arises from the genetic matrix [C A; U G]$^{(3)}$ as a result of the circular permutation of the genetic letters C→A→G→U→C. All designations are the same as on Figure 4 in Chapter 7

UUU Phe, $-x_6$	UUC Phe, $-x_6$	UCU Ser, x_4	UCC Ser, x_4	CUU Leu, x_2	CUC Leu, x_2	CCU Pro, x_0	CCC Pro, x_0
UUG Leu, $-x_7$	UUA Leu, $-x_7$	UCG Ser, x_5	UCA Ser, x_5	CUG Leu, x_3	CUA Leu, x_3	CCG Pro, x_1	CCA Pro, x_1
UGU Cys, $-x_4$	UGC Cys, $-x_4$	UAU Tyr, $-x_6$	UAC Tyr, $-x_6$	CGU Arg, x_0	CGC Arg, x_0	CAU His, $-x_2$	CAC His, $-x_2$
UGG Trp, $-x_5$	UGA Trp, $-x_5$	UAG Stop, $-x_7$	UAA Stop, $-x_7$	CGG Arg, x_1	CGA Arg, x_1	CAG Gln, $-x_3$	CAA Gln, $-x_3$
GUU Val, x_2	GUC Val, x_2	GCU Ala, x_0	GCC Ala, x_0	AUU Ile, $-x_6$	AUC Ile, $-x_6$	ACU Thr, x_4	ACC Thr, x_4
GUG Val, x_3	GUA Val, x_3	GCG Ala , x_1	GCA Ala, x_1	AUG Met, $-x_7$	AUA Met, $-x_7$	ACG Thr, x_5	ACA Thr, x_5
GGU Gly, x_0	GGC Gly, x_0	GAU Asp, $-x_2$	GAC Asp, $-x_2$	AGU Ser, $-x_4$	AGC Ser, $-x_4$	AAU Asn, $-x_6$	AAC Asn, $-x_6$
GGG Gly, x_1	GGA Gly, x_1	GAG Glu, $-x_3$	GAA Glu, $-x_3$	AGG Stop, $-x_5$	AGA Stop, $-x_5$	AAG Lys, $-x_7$	AAA Lys, $-x_7$

If the same circular permutation C→A→G→U→C is repeated for the new genomatrix [U C; G A]$^{(3)}$, the genomatrix [G U; A C]$^{(3)}$ arises (Figure 3). The same matrix [G U; A C]$^{(3)}$ can be obtained by means of a turn of the matrix [U C; G A]$^{(3)}$ on 90^0 clockwise as well.

If the same circular permutation C→A→G→U→C is repeated for the new genomatrix [G U; A C]$^{(3)}$, the genomatrix [A G; C U]$^{(3)}$ arises (Figure 4). The same matrix [A G; C U]$^{(3)}$ can be obtained by means of a turn of the matrix [G U; A C]$^{(3)}$ on 90^0 clockwise as well.

Figure 3. The genetic matrix [G U; A C]$^{(3)}$, which arises from the initial genetic matrix [C A; U G]$^{(3)}$ as a result of the twice applications of the circular permutation of the genetic letters C→A→G→U→C. All designations are the same as on Figure 4 in Chapter 7

GGG Gly, x_1	GGU Gly, x_0	GUG Val, x_3	GUU Val, x_2	UGG Trp, $-x_5$	UGU Cys, $-x_4$	UUG Leu, $-x_7$	UUU Phe, $-x_6$
GGA Gly, x_1	GGC Gly, x_0	GUA Val, x_3	GUC Val, x_2	UGA Trp, $-x_5$	UGC Cys, $-x_4$	UUA Leu, $-x_7$	UUC Phe, $-x_6$
GAG Glu, $-x_3$	GAU Asp, $-x_2$	GCG Ala , x_1	GCU Ala, x_0	UAG Stop, $-x_7$	UAU Tyr, $-x_6$	UCG Ser, x_5	UCU Ser, x_4
GAA Glu, $-x_3$	GAC Asp, $-x_2$	GCA Ala, x_1	GCC Ala, x_0	UAA Stop, $-x_7$	UAC Tyr, $-x_6$	UCA Ser, x_5	UCC Ser, x_4
AGG Stop, $-x_5$	AGU Ser, $-x_4$	AUG Met, $-x_7$	AUU Ile, $-x_6$	CGG Arg, x_1	CGU Arg, x_0	CUG Leu, x_3	CUU Leu, x_2
AGA Stop, $-x_5$	AGC Ser, $-x_4$	AUA Met, $-x_7$	AUC Ile, $-x_6$	CGA Arg, x_1	CGC Arg, x_0	CUA Leu, x_3	CUC Leu, x_2
AAG Lys, $-x_7$	AAU Asn, $-x_6$	ACG Thr, x_5	ACU Thr, x_4	CAG Gln, $-x_3$	CAU His, $-x_2$	CCG Pro, x_1	CCU Pro, x_0
AAA Lys, $-x_7$	AAC Asn, $-x_6$	ACA Thr, x_5	ACC Thr, x_4	CAA Gln, $-x_3$	CAC His, $-x_2$	CCA Pro, x_1	CCC Pro, x_0

Figure 4. The genetic matrix [A G; C U]$^{(3)}$, which arises from the initial genetic matrix [C A; U G]$^{(3)}$ (Figure 4 in chapter 7) as a result of the triple applications of the circular permutation of the genetic letters C→A→G→U→C. All designations are the same as on Figure 4 in Chapter 7

AAA Lys, $-x_7$	AAG Lys, $-x_7$	AGA Stop, $-x_5$	AGG Stop, $-x_5$	GAA Glu, $-x_3$	GAG Glu, $-x_3$	GGA Gly, x_1	GGG Gly, x_1
AAC Asn, $-x_6$	AAU Asn, $-x_6$	AGC Ser, $-x_4$	AGU Ser, $-x_4$	GAC Asp, $-x_2$	GAU Asp, $-x_2$	GGC Gly, x_0	GGU Gly, x_0
ACA Thr, x_5	ACG Thr, x_5	AUA Met, $-x_7$	AUG Met, $-x_7$	GCA Ala, x_1	GCG Ala , x_1	GUA Val, x_3	GUG Val, x_3
ACC Thr, x_4	ACU Thr, x_4	AUC Ile, $-x_6$	AUU Ile, $-x_6$	GCC Ala, x_0	GCU Ala, x_0	GUC Val, x_2	GUU Val, x_2
CAA Gln, $-x_3$	CAG Gln, $-x_3$	CGA Arg, x_1	CGG Arg, x_1	UAA Stop, $-x_7$	UAG Stop, $-x_7$	UGA Trp, $-x_5$	UGG Trp, $-x_5$
CAC His, $-x_2$	CAU His, $-x_2$	CGC Arg, x_0	CGU Arg, x_0	UAC Tyr, $-x_6$	UAU Tyr, $-x_6$	UGC Cys, $-x_4$	UGU Cys, $-x_4$
CCA Pro, x_1	CCG Pro, x_1	CUA Leu, x_3	CUG Leu, x_3	UCA Ser, x_5	UCG Ser, x_5	UUA Leu, $-x_7$	UUG Leu, $-x_7$
CCC Pro, x_0	CCU Pro, x_0	CUC Leu, x_2	CUU Leu, x_2	UCC Ser, x_4	UUC Phe, $-x_6$	UUC Phe, $-x_6$	UUU Phe, $-x_6$

Dispositions of the 8 coordinates $x_1, x_2, \ldots, x_7$ are different in these four symbolic genomatrices [C A; U G]$^{(3)}$, [U C; G A]$^{(3)}$, [G U; A C]$^{(3)}$ and [A G; C U]$^{(3)}$. They define the four different numeric matrices $YY_+^{[CAUG]}$, $YY_+^{[UCGA]}$, $YY_+^{[GUAC]}$, $YY_+^{[AGCU]}$ by means of the same algorithm correspondingly. The beautiful fact is that the new three numeric matrices $YY_+^{[CAUG]}$, $YY_+^{[UCGA]}$, $YY_+^{[GUAC]}$, $YY_+^{[AGCU]}$ present appropriate 8-dimensional Yin-Yang-algebras as well by analogy with the matrix $YY_+^{[CAUG]}$ on Figure 4 in chapter 7. For example, let us consider the numeric matrix $YY_+^{[UCGA]}$, which is reproduced on Figure 5.

This matrix $YY_+^{[UCGA]}$ can be written in the linear form in accordance with the expression (1): $YY = x_0 * \mathbf{f}_0 + x_1 * \mathbf{m}_1 + x_2 * \mathbf{f}_2 + x_3 * \mathbf{m}_3 + x_4 * \mathbf{f}_4 + x_5 * \mathbf{m}_5 + x_6 * \mathbf{f}_6 + x_7 * \mathbf{m}$. In this case the basic matrices $\mathbf{f}_0$, $\mathbf{m}_1$, $\mathbf{f}_2$, $\mathbf{m}_3$, $\mathbf{f}_4$, $\mathbf{m}_5$, $\mathbf{f}_6$, $\mathbf{m}_7$ are shown on Figure 6.

This set of basic matrices is a closed set relative to multiplication: the result of multiplication of any two basic matrices is a matrix from the same set in accordance with the multiplication table on Figure 7.

This multiplication table defines an 8-dimensional algebra. The diagonal cells of this table contain no real units at all but they are occupied by elements "$\pm\mathbf{f}_6$" and "$\pm\mathbf{m}_7$". Thereby the set of the 8 basic matrices $\mathbf{f}_0$, $\mathbf{m}_1$, $\mathbf{f}_2$, $\mathbf{m}_3$, $\mathbf{f}_4$, $\mathbf{m}_5$, $\mathbf{f}_6$, $\mathbf{m}_7$ is divided into two equal subsets by the criterion of their squares. The first subset consists of elements with the even indexes: $\mathbf{f}_0$, $\mathbf{f}_2$, $\mathbf{f}_4$, $\mathbf{f}_6$. The squares of members of this $\mathbf{f}_6$-subset are equal to $\pm\mathbf{f}_6$ always. The second subset consists of elements with the odd indexes: $\mathbf{m}_1$, $\mathbf{m}_3$, $\mathbf{m}_5$, $\mathbf{m}_7$. The squares of members of this $\mathbf{m}_7$-subset are equal to $\pm\mathbf{m}_7$ always.

The basic element $\mathbf{f}_6$ possesses all properties of real negative unit "-1" in relation to the members of the $\mathbf{f}_6$-subset: $\mathbf{f}_6^2 = -\mathbf{f}_6$, $\mathbf{f}_6 * \mathbf{f}_0 = \mathbf{f}_0 * \mathbf{f}_6 = -\mathbf{f}_0$, $\mathbf{f}_6 * \mathbf{f}_2 = \mathbf{f}_2 * \mathbf{f}_6 = -\mathbf{f}_2$, $\mathbf{f}_6 * \mathbf{f}_4 = \mathbf{f}_4 * \mathbf{f}_6 = -\mathbf{f}_4$. But the element $\mathbf{f}_6$ does not possess the commutative property of real negative unit in relation to the members of the $\boldsymbol{m}_7$-subset: $\mathbf{f}_6 * \mathbf{m}_p \neq \mathbf{m}_p * \mathbf{f}_6$, where $p = 1, 3, 5, 7$. For this reason $\mathbf{f}_6$ is named "quasi-real negative unit of the $\mathbf{f}_6$-subset".

The basic element $\mathbf{m}_7$ possesses all properties of real negative unit "-1" in relation to the members of the $\mathbf{m}_7$-subset: $\mathbf{m}_7^2 = -\mathbf{m}_7$, $\mathbf{m}_7 * \mathbf{m}_1 = \mathbf{m}_1 * \mathbf{m}_7 = -\mathbf{m}_1$, $\mathbf{m}_7 * \mathbf{m}_3 = \mathbf{m}_3 * \mathbf{m}_7 = -\mathbf{m}_3$, $\mathbf{m}_7 * \mathbf{m}_5 = \mathbf{m}_5 * \mathbf{m}_7 = -\mathbf{m}_5$. But the element $\boldsymbol{m}_7$ does not possess the commutative property of real negative unit in relation to the members of the $\mathbf{f}_6$-subset: $\mathbf{m}_7 * \mathbf{f}_k \neq \mathbf{f}_k * \mathbf{m}_7$, where $k = 0, 2, 4, 6$. For this reason $\boldsymbol{m}_7$ is named "quasi-real negative unit of the $\mathbf{m}_7$-subset".

By definition, a Yin-Yang-algebra is a 2^n-dimensional algebra, a complete set of basic elements of which has no real unit at all but this set consists of two sub-sets of basic elements with 2^{n-1} elements in each and with the following feature: one of the basic elements of each sub-set possesses all the properties of the real positive unit "+1" or of the real negative unit "-1" relative to all basic elements of its sub-set

Figure 5. The numeric matrix $YY_+^{[UCGA]}$, which is numeric presentation of the genomatrix [U C; G A]$^{(3)}$ from Figure 2

$-x_6$	$-x_6$	x_4	x_4	x_2	x_2	x_0	x_0
$-x_7$	$-x_7$	x_5	x_5	x_3	x_3	x_1	x_1
$-x_4$	$-x_4$	$-x_6$	$-x_6$	x_0	x_0	$-x_2$	$-x_2$
$-x_5$	$-x_5$	$-x_7$	$-x_7$	x_1	x_1	$-x_3$	$-x_3$
x_2	x_2	x_0	x_0	$-x_6$	$-x_6$	x_4	x_4
x_3	x_3	x_1	x_1	$-x_7$	$-x_7$	x_5	x_5
x_0	x_0	$-x_2$	$-x_2$	$-x_4$	$-x_4$	$-x_6$	$-x_6$
x_1	x_1	$-x_3$	$-x_3$	$-x_5$	$-x_5$	$-x_7$	$-x_7$

Figure 6. The basic matrices for the numeric matrix $YY_+^{[UCGA]}$ from Figure 5

$$f_0 = \begin{vmatrix} 0&0&0&0&0&0&1&1 \\ 0&0&0&0&0&0&0&0 \\ 0&0&0&0&1&1&0&0 \\ 0&0&0&0&0&0&0&0 \\ 0&0&1&1&0&0&0&0 \\ 0&0&0&0&0&0&0&0 \\ 1&1&0&0&0&0&0&0 \\ 0&0&0&0&0&0&0&0 \end{vmatrix} ; \quad m_1 = \begin{vmatrix} 0&0&0&0&0&0&0&0 \\ 0&0&0&0&0&0&1&1 \\ 0&0&0&0&0&0&0&0 \\ 0&0&0&0&1&1&0&0 \\ 0&0&0&0&0&0&0&0 \\ 0&0&1&1&0&0&0&0 \\ 0&0&0&0&0&0&0&0 \\ 1&1&0&0&0&0&0&0 \end{vmatrix}$$

$$f_2 = \begin{vmatrix} 0&0&0&0&1&1&0&0 \\ 0&0&0&0&0&0&0&0 \\ 0&0&0&0&0&0&-1&-1 \\ 0&0&0&0&0&0&0&0 \\ 1&1&0&0&0&0&0&0 \\ 0&0&0&0&0&0&0&0 \\ 0&0&-1&-1&0&0&0&0 \\ 0&0&0&0&0&0&0&0 \end{vmatrix} ; \quad m_3 = \begin{vmatrix} 0&0&0&0&0&0&0&0 \\ 0&0&0&0&1&1&0&0 \\ 0&0&0&0&0&0&0&0 \\ 0&0&0&0&0&0&-1&-1 \\ 0&0&0&0&0&0&0&0 \\ 1&1&0&0&0&0&0&0 \\ 0&0&0&0&0&0&0&0 \\ 0&0&-1&-1&0&0&0&0 \end{vmatrix}$$

$$f_4 = \begin{vmatrix} 0&0&1&1&0&0&0&0 \\ 0&0&0&0&0&0&0&0 \\ -1&-1&0&0&0&0&0&0 \\ 0&0&0&0&0&0&0&0 \\ 0&0&0&0&0&0&1&1 \\ 0&0&0&0&0&0&0&0 \\ 0&0&0&0&-1&-1&0&0 \\ 0&0&0&0&0&0&0&0 \end{vmatrix} ; \quad m_5 = \begin{vmatrix} 0&0&0&0&0&0&0&0 \\ 0&0&1&1&0&0&0&0 \\ 0&0&0&0&0&0&0&0 \\ -1&-1&0&0&0&0&0&0 \\ 0&0&0&0&0&0&0&0 \\ 0&0&0&0&0&0&1&1 \\ 0&0&0&0&0&0&0&0 \\ 0&0&0&0&-1&-1&0&0 \end{vmatrix}$$

$$f_6 = \begin{vmatrix} -1&-1&0&0&0&0&0&0 \\ 0&0&0&0&0&0&0&0 \\ 0&0&-1&-1&0&0&0&0 \\ 0&0&0&0&0&0&0&0 \\ 0&0&0&0&-1&-1&0&0 \\ 0&0&0&0&0&0&0&0 \\ 0&0&0&0&0&0&-1&-1 \\ 0&0&0&0&0&0&0&0 \end{vmatrix} ; \quad m_7 = \begin{vmatrix} 0&0&0&0&0&0&0&0 \\ -1&-1&0&0&0&0&0&0 \\ 0&0&0&0&0&0&0&0 \\ 0&0&-1&-1&0&0&0&0 \\ 0&0&0&0&0&0&0&0 \\ 0&0&0&0&-1&-1&0&0 \\ 0&0&0&0&0&0&0&0 \\ 0&0&0&0&0&0&-1&-1 \end{vmatrix}$$

Figure 7. The multiplication table of the basic matrices of the numeric matrix $YY_+^{[UCGA]}$ from Figure 5

	f_0	m_1	f_2	m_3	f_4	m_5	f_6	m_7
f_0	$-f_6$	$-f_6$	$-f_4$	$-f_4$	$-f_2$	$-f_2$	$-f_0$	$-f_0$
m_1	$-m_7$	$-m_7$	$-m_5$	$-m_5$	$-m_3$	$-m_3$	$-m_1$	$-m_1$
f_2	f_4	f_4	$-f_6$	$-f_6$	f_0	f_0	$-f_2$	$-f_2$
m_3	m_5	m_5	$-m_7$	$-m_7$	m_1	m_1	$-m_3$	$-m_3$
f_4	f_2	f_2	$-f_0$	$-f_0$	f_6	f_6	$-f_4$	$-f_4$
m_5	m_3	m_3	$-m_1$	$-m_1$	m_7	m_7	$-m_5$	$-m_5$
f_6	$-f_0$	$-f_0$	$-f_2$	$-f_2$	$-f_4$	$-f_4$	$-f_6$	$-f_6$
m_7	$-m_1$	$-m_1$	$-m_3$	$-m_3$	$-m_5$	$-m_5$	$-m_7$	$-m_7$

but not relative to basic elements of another sub-set. One can see that the multiplication table on Figure 7 defines the 8-dimensional Yin-Yang-algebra $YY_+^{[UCGA]}$ and a relevant 8-dimensional numeric system. Concerning to multiplication of numbers of this new 8-dimensional system in their matrix forms of presentation, it means that both factors have the identical matrix disposition (Figure 5 in chapter 7) of their 8 parameters $x_0, x_1, \ldots, x_7$ (in the first factor) and $y_0, y_1, \ldots, y_7$ (in the second factor) and the final matrix has the same matrix disposition of its 8 relevant parameters $z_0, z_1, \ldots, z_7$. This genetic algebra for the case of the genomatrix $[U\ C;\ G\ A]^{(3)}$ is quite different from the genetic Yin-Yang-algebra $YY_+^{[CAUG]}$ with the multiplication table on Figure 6 in chapter 7.

This difference can be illustrated additionally by a numeric example. Let us consider an arbitrary 8-dimensional number:

$$YY = 3*f_0+4*m_1+7*f_2+2*m_3+2*f_4+1*m_5+9*f_6+8*m_7. \qquad (2)$$

If this number is considered from the viewpoint of the genetic Yin-Yang-algebra $YY_+^{[CAUG]}$ on Figure 3 and Figure 6 in chapter 7, its square is equal to $YY^2 = 117*\mathbf{f}_0 + 149*\mathbf{m}_1 + 69*\mathbf{f}_2 + 57*\mathbf{m}_3 - 15*\mathbf{f}_4 + 57*\mathbf{m}_5 + 117*\mathbf{f}_6 + 121*\mathbf{m}_7$. But if this number YY in the equation (2) with the same magnitudes of the coordinates $x_0, x_1, \ldots, x_7$ is considered from the viewpoint of the genetic Yin-Yang-algebra $YY_+^{[UCGA]}$ on Figure 7, its square is equal to quite another number: $YY^2 = -111*\mathbf{f}_0 - 127x_1*\mathbf{m}_1 - 195*\mathbf{f}_2 - 111*\mathbf{m}_3 - 39*\mathbf{f}_4 - 63*\mathbf{m}_5 - 231*\mathbf{f}_6 - 179*\mathbf{m}_7$.

One can note a certain similarity (or a symmetric relation) between the multiplication tables of both cases on Figure 6 in Chapter 7, and Figure 7 in the current chapter. Really, the internal contents of the first table can be transformed into the internal contents of the second table by means of a turn on 90^0 anticlockwise with a simultaneous inversion of all signs "+" and "-". Symmetric relations exist also among multiplication tables of many genetic Yin-Yan-matrices, which are mentioned below.

For the cases of the other two matrices $YY_+^{[GUAC]}$ and $YY_+^{[AGCU]}$, which correspond to the genomatrices $[G\ U;\ A\ C]^{(3)}$ (Figure 3) and $[A\ G;\ C\ U]^{(3)}$ (Figure 4), their appropriate sets of basic matrices and multiplication tables are constructed by analogy. Figure 8 and Figure 9 demonstrate final results in the form of their multiplication tables. One can see that these two multiplication tables define two 8-dimensional Yin-Yang-algebras as well.

The four set of basic matrices $\mathbf{f}_0, \mathbf{m}_1, \mathbf{f}_2, \mathbf{m}_3, \mathbf{f}_4, \mathbf{m}_5, \mathbf{f}_6, \mathbf{m}_7$ for the four matrix $YY_+^{[CAUG]}$, $YY_+^{[UCGA]}$, $YY_+^{[GUAC]}$ and $YY_+^{[AGCU]}$ are quite different, but the multiplication tables for the matrices $YY_+^{[CAUG]}$ and $YY_+^{[GUAC]}$ (Figure 6 in chapter 7 and Figure 8 in the current chapter) are identical to each other. These

Figure 8. The multiplication table of the Yin-Yang-algebra $YY_+^{[GUAC]}$

	f_0	m_1	f_2	m_3	f_4	m_5	f_6	m_7
f_0	f_0	m_1	f_2	m_3	f_4	m_5	f_6	m_7
m_1	f_0	m_1	f_2	m_3	f_4	m_5	f_6	m_7
f_2	f_2	m_3	$-f_0$	$-m_1$	$-f_6$	$-m_7$	f_4	m_5
m_3	f_2	m_3	$-f_0$	$-m_1$	$-f_6$	$-m_7$	f_4	m_5
f_4	f_4	m_5	f_6	m_7	f_0	m_1	f_2	m_3
m_5	f_4	m_5	f_6	m_7	f_0	m_1	f_2	m_3
f_6	f_6	m_7	$-f_4$	$-m_5$	$-f_2$	$-m_3$	f_0	m_1
m_7	f_6	m_7	$-f_4$	$-m_5$	$-f_2$	$-m_3$	f_0	m_1

matrices are transformed each into another by means of turn on 180^0 or by means of simultaneous permutations of complementary nitrogenous bases C↔ G and A↔ U in all triplets inside the appropriate genomatrices $[C\ A;\ U\ G]^{(3)}$ and $[G\ U;\ A\ C]^{(3)}$ (that is each codon is replaced by its anti-codon in this case). It means, that for the case of these "complementary" genomatrices $[C\ A;\ U\ G]^{(3)}$ and $[G\ U;\ A\ C]^{(3)}$, the same Yin-Yang-algebra possesses two different matrix forms of its presentation. The similar situation holds true for the second pair of the "complementary" genomatrices $[U\ C;\ G\ A]^{(3)}$ and $[A\ G;\ C\ U]^{(3)}$: their Yin-Yang-algebras $YY_+^{[UCGA]}$ and $YY_+^{[AGCU]}$ are identical to each another as well because of the identity of their multiplication tables (Figure 7 and Figure 9).

One should add that each of these four genomatrices $[C\ A;\ U\ G]^{(3)}$, $[G\ U;\ A\ C]^{(3)}$, $[U\ C;\ G\ A]^{(3)}$ and $[A\ G;\ C\ U]^{(3)}$ corresponds to its own Hadamard matrix by means of the general alphabetic algorithm. The claim is that each of these (8x8)-matrices possesses a black-and-white mosaic, which is transformed easily into a mosaic of an appropriate Hadamard (8x8)-matrix by means of the U-algorithm. This U-algorithm is described in Chapter 6 and it is based on objective molecular properties of the genetic alphabet. These four "genetic" Hadamard matrices, which are marked by $H_+^{[CAUG]}$, $H_+^{[GUAC]}$, $H_+^{[UCGA]}$, $H_+^{[AGCU]}$, are shown on Figure 10.

Figure 9. Multiplication table of the Yin-Yang-algebra $YY_+^{[AGCU]}$

	f_0	m_1	f_2	m_3	f_4	m_5	f_6	m_7
f_0	$-f_6$	$-f_6$	$-f_4$	$-f_4$	$-f_2$	$-f_2$	$-f_0$	$-f_0$
m_1	$-m_7$	$-m_7$	$-m_5$	$-m_5$	$-m_3$	$-m_3$	$-m_1$	$-m_1$
f_2	f_4	f_4	$-f_6$	$-f_6$	f_0	f_0	$-f_2$	$-f_2$
m_3	m_5	m_5	$-m_7$	$-m_7$	m_1	m_1	$-m_3$	$-m_3$
f_4	f_2	f_2	$-f_0$	$-f_0$	f_6	f_6	$-f_4$	$-f_4$
m_5	m_3	m_3	$-m_1$	$-m_1$	m_7	m_7	$-m_5$	$-m_5$
f_6	$-f_0$	$-f_0$	$-f_2$	$-f_2$	$-f_4$	$-f_4$	$-f_6$	$-f_6$
m_7	$-m_1$	$-m_1$	$-m_3$	$-m_3$	$-m_5$	$-m_5$	$-m_7$	$-m_7$

Figure 10. The four Hadamard matrices $H_+^{[CAUG]}$, $H_+^{[GUAC]}$, $H_+^{[UCGA]}$, $H_+^{[AGCU]}$, which are connected algorithmically with the genomatrices $[C\ A;\ U\ G]^{(3)}$, $[U\ C;\ G\ A]^{(3)}$, $[G\ U;\ A\ C]^{(3)}$, $[A\ G;\ C\ U]^{(3)}$ and their Yin-Yang-algebras. Each black (white) cell contains the element "+1" ("-1")

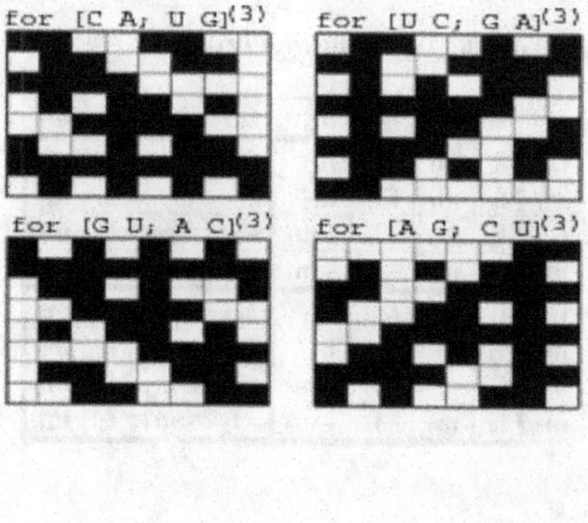

Each of the four genomatrices [C A; U G]$^{(3)}$, [U C; G A]$^{(3)}$, [G U; A C]$^{(3)}$ and [A G; C U]$^{(3)}$ can be transformed in a circular manner into the next genomatrix of this sequence by means of a turn on 90^0 clockwise. Correspondingly each of the four Yin-Yang-matrices $YY_+^{[CAUG]}$, $YY_+^{[UCGA]}$, $YY_+^{[GUAC]}$ and $YY_+^{[AGCU]}$ and each of the four Hadamard matrices $H_+^{[CAUG]}$, $H_+^{[GUAC]}$, $H_+^{[UCGA]}$, $H_+^{[AGCU]}$ can be transformed in a circular manner into the next matrix of their sequences by means of a turn on 90^0 clockwise. It is an important result that the cyclic permutations of genetic elements C→A→G→U→C lead to the appropriate cyclic changes of the Yin-Yang-algebras $YY_+^{[CAUG]} \rightarrow YY_+^{[UCGA]} \rightarrow YY_+^{[GUAC]} \rightarrow YY_+^{[AGCU]} \rightarrow YY_+^{[CAUG]}$ in the matrix forms of presentation of the genetic code. These cyclic permutations lead simultaneously to the appropriate cyclic changes of the genomatrices [C A; U G]$^{(3)}$→[U C; G A]$^{(3)}$→[G U; A C]$^{(3)}$ →[A G; C U]$^{(3)}$→[C A; U G]$^{(3)}$ and of the Hadamard matrices $H_+^{[CAUG]} \rightarrow H_+^{[GUAC]} \rightarrow H_+^{[UCGA]} \rightarrow H_+^{[AGCU]} \rightarrow H_+^{[CAUG]}$. Figure 11 illustrates this general situation for the sequence of the Yin-Yang-matrices only. Symbols of a clock and of the four parts of the world are disposed in the center of Figure 11 to cause heuristic associations and to reflect the thought that the spatial turns of these matrices (together with changes of their Yin-Yang-algebras and of their Hadamard matrices) can be carried out rhythmically in appropriate algebraic models of rhythmic physiological processes.

But these circular sequences of the genetic Yin-Yang-algebras and of the conjunct matrices are only a small part of a hierarchy of circular changes of different kinds of the genetic Yin-Yang-algebras together with conjunct matrices. This hierarchy, which is based on different kinds of circular permutations, should be studied step by step. Let us continue this study.

We begin with the genomatrix [C A; U G]$^{(3)}$ and the cyclic shifts 1-2-3→2-3-1→3-1-2→1-2-3 of three positions in all triplets there (for example, in the case of the triplet CAG such shifts produce the sequence CAG→AGC→GCA→CAG). Analogical cyclic shifts of the three positions 3-2-1→2-1-3→1-3-2→3-2-1 in the triplets at the reverse order of their reading are possible for analysis. Such cyclic changes produce the sequences of appropriate genomatrices $[C\ A;\ U\ G]_{123}^{(3)}$→ $[C\ A;\ U\ G]_{231}^{(3)}$→$[C\ A;\ U\ G]_{312}^{(3)}$ →$[C\ A;\ U\ G]_{123}^{(3)}$ and $[C\ A;\ U\ G]_{321}^{(3)}$→$[C\ A;\ U\ G]_{213}^{(3)}$→ $[C\ A;\ U\ G]_{132}^{(3)}$→$[C\ A;\ U\ G]_{321}^{(3)}$. Each of these six genomatrices is connected with its own Yin-Yang-algebras by the same algorithm (see Figures 6, 8 and 9 in chapter 7). Figure 12 shows one of the possible variants of schematic illustration of these two cyclic sequences of the mosaic Yin-Yang-matrices $YY_{+,123}^{[CAUG]}$, $YY_{+,231}^{[CAUG]}$, $YY_{+,312}^{[CAUG]}$, $YY_{+,321}^{[CAUG]}$, $YY_{+,213}^{[CAUG]}$, $YY_{+,132}^{[CAUG]}$.

A transposition of these Yin-Yang-matrices produces new relevant kinds of Yin-Yang-matrices (Petoukhov, 2008a). Each of these 12 Yin-Yang-matrices corresponds to its own kind of genetic Hadamard (8x8)-matrices.

By analogy the similar consideration of permutations and transpositions in the cases of other three initial genomatrices [U C; G A]$^{(3)}$, [G U; A C]$^{(3)}$ and [A G; C U]$^{(3)}$ leads to an appropriate increase of the total quantity of known Yin-Yang-matrices and their Hadamard genomatrices. We receive a further increase in this total quantity as well, if we consider other initial groups of such genomatrices, which are transformed cyclically each into another at the same cyclic permutation of genetic elements C→A→G→U→C. Examples of such groups are [C G; U A]$^{(3)}$→[U C; A G]$^{(3)}$→[A U; G C]$^{(3)}$→[G A; C U]$^{(3)}$→[C G; U A]$^{(3)}$, or [C A; G U]$^{(3)}$→[G C; U A]$^{(3)}$→[U G; A C]$^{(3)}$→[A U; C G]$^{(3)}$→[C A; G U]$^{(3)}$, or [G A; U C]$^{(3)}$→[U G; C A]$^{(3)}$→[C U; A G]$^{(3)}$→[A C; G U]$^{(3)}$→[G A; U C]$^{(3)}$, or

[G C; A U]$^{(3)}$→[A G; U C]$^{(3)}$→[U A; C G]$^{(3)}$→[C U; G A]$^{(3)}$→[G C; A U]$^{(3)}$. Such groups of matrices can be transformed each into the other by means of permutations of separate genetic elements like C→G→C or A→G→A, etc.

Figure 11. The circular sequence of the four Yin-Yang-matrices $YY_+^{[CAUG]} \rightarrow YY_+^{[UCGA]} \rightarrow YY_+^{[GUAC]} \rightarrow YY_+^{[AGCU]} \rightarrow YY_+^{[CAUG]}$*, which is based on the circular permutations of the genetic molecular elements* $C \rightarrow A \rightarrow G \rightarrow U \rightarrow C$

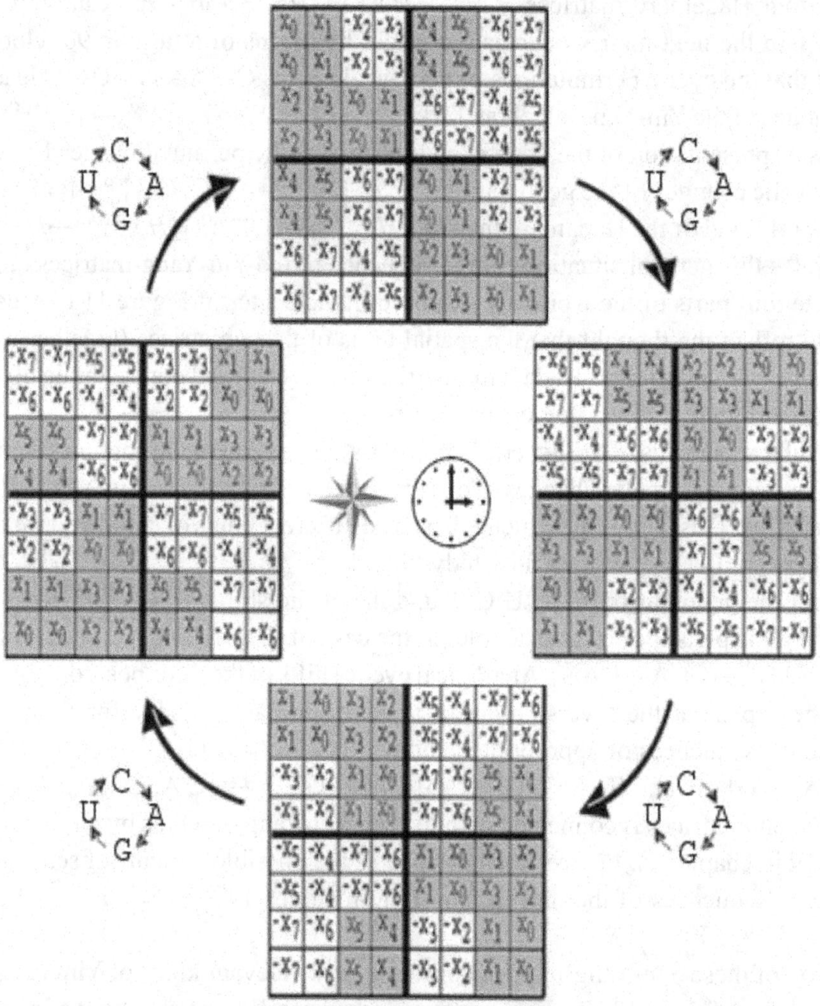

We have no possibility to demonstrate in this chapter all hierarchy of cyclic sequences of genetic Yin-Yang-algebras and of conjunct Hadamard matrices, which arise as a result of all possible kinds of cyclic permutations of genetic elements. Additional data about this hierarchy of "round dances" of Yin-Yang-algebras should be published separately. An important general result is that a considerable quantity of genetic Yin-Yang-matrices and of their Hadamard matrices exists and that these matrices are connected with different kinds of cyclic permutations of genetic elements. Each of these Yin-Yang-matrices can be transformed into the initial matrix $YY_{+,123}^{[CAUG]}$ by means of a relevant cyclic permutation of genetic elements.

In addition, a set of such hierarchies of cyclic metamorphoses of the genetic Yin-Yang-algebras allows one to model phenomena of metamorphoses of animals. For example, butterflies and moths have four stages of cyclic metamorphoses in their life: egg, larva (the caterpillar stage), pupa (the chrysalis

Figure 12. A schematic illustration of the two cyclic sequences of the mosaic Yin-Yang-matrices, which arise as a result of permutations of positions in triplets: $YY_{+,123}{}^{[CAUG]} \rightarrow YY_{+,231}{}^{[CAUG]} \rightarrow YY_{+,312}{}^{[CAUG]} \rightarrow YY_{+,123}{}^{[CAUG]}$ *and* $YY_{+,321}{}^{[CAUG]} \rightarrow YY_{+,213}{}^{[CAUG]} \rightarrow YY_{+,132}{}^{[CAUG]} \rightarrow YY_{+,321}{}^{[CAUG]}$*. Number over each matrix shows a relevant kind of permutations of positions in all triplets*

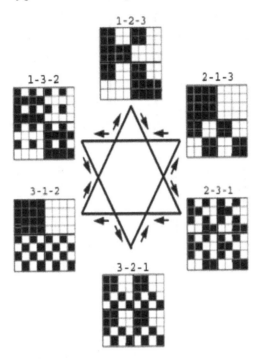

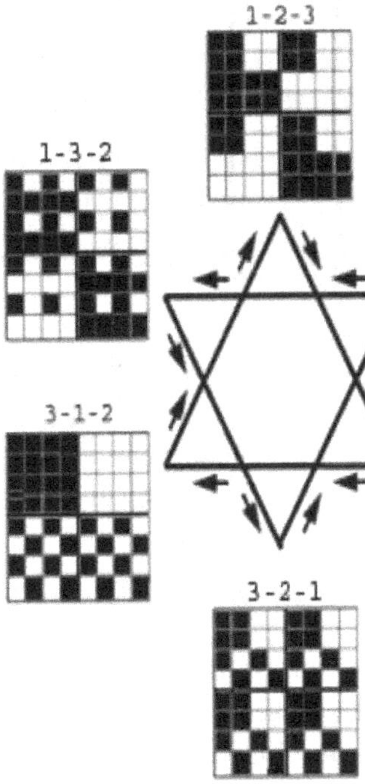

phase), and adult (Figure 21). One should note that in the chrysalis phase the biological organism does not eat at all; consequently atomic contents of the organism do not change practically, but its set of molecular compositions is reformed at this stage cardinally by means of complex permutations of groups of chemical elements. It reminds one strongly of the described change of a type of genetic Yin-Yang-algebra as a result of a simple cyclic permutation of genetic elements. In the proposed modeling approach, each of the named stages of metamorphosis is connected with forming its own kind of hierarchy (or of a colony) of the genetic Yin-Yan-algebras. Correspondingly a transition from one stage of biological metamorphosis to another stage is interpreted as a transition from one kind of hierarchy of the genetic Yin-Yang-algebras to another kind of their hierarchy (or as a transition from one colony of the genetic Yin-Yang-algebras to another colony of such genetic algebras).

THE OPPOSITIONAL CATEGORY OF GENETIC YIN-YANG-MATRICES

Let us return to the genetic matrix algebra $YY_+{}^{[CAUG]}$ (Figure 3 and Figure 6 in chapter 7). If all its Yang-coordinates are equal to zero ($x_1 = x_3 = x_5 = x_7 = 0$), the $YY_+{}^{[CAUG]}$-matrix becomes the matrix of the genoquaternion of the Yin-type (Figure 13, on the left side). We mark this "female" genoquaternion by the symbol G_f. If all Yin-coordinates are equal to zero ($x_0 = x_2 = x_4 = x_6 = 0$), the $YY_+{}^{[CAUG]}$-matrix becomes

Figure 13. On the left side: the matrix G_f of Yin-genoquaternion. On the right side: the matrix G_m of Yang-genoquaternion. Columns with null components are shown as well. Matrix cells with positive (negative) components are marked by black (white) colors as in all previous cases

x_0	0	$-x_2$	0	x_4	0	$-x_6$	0
x_0	0	$-x_2$	0	x_4	0	$-x_6$	0
x_2	0	x_0	0	$-x_6$	0	$-x_4$	0
x_2	0	x_0	0	$-x_6$	0	$-x_4$	0
x_4	0	$-x_6$	0	x_0	0	$-x_2$	0
x_4	0	$-x_6$	0	x_0	0	$-x_2$	0
$-x_6$	0	$-x_4$	0	x_2	0	x_0	0
$-x_6$	0	$-x_4$	0	x_2	0	x_0	0

0	x_1	0	$-x_3$	0	x_5	0	$-x_7$
0	x_1	0	$-x_3$	0	x_5	0	$-x_7$
0	x_3	0	x_1	0	$-x_7$	0	$-x_5$
0	x_3	0	x_1	0	$-x_7$	0	$-x_5$
0	x_5	0	$-x_7$	0	x_1	0	$-x_3$
0	x_5	0	$-x_7$	0	x_1	0	$-x_3$
0	$-x_7$	0	$-x_5$	0	x_3	0	x_1
0	$-x_7$	0	$-x_5$	0	x_3	0	x_1

the matrix of the genoquaternion of the Yang-type (Figure 13, on the right side). We mark this "male" genoquaternion by the symbol G_m.

Multiplication tables of the algebras of these genoquaternions were shown in Chapter 7 on Figure 7. The sum of these two matrices G_f and G_m gives the $YY_+^{[CAUG]}$-matrix (Figure 3, chapter 7):

$$G_f + G_m = YY_+^{[CAUG]} \qquad (3)$$

The typical feature of this $YY_+^{[CAUG]}$-matrix and of all other Yin-Yang-matrices, which have been considered in the previous paragraph and which possess the lower index "+", is that two halves of each of these matrices are mirror-antisymmetric in their black-and-white mosaics. For example, in the case of $YY_+^{[CAUG]}$-matrix (Figure 3 and Figure 4 in chapter 7) its left half and its right half are mirror-antisymmetric: each pair of cells, which are disposed mirror-symmetrically in these halves, have opposite colors. All these Yin-Yang-matrices can be produced by means of summation of relevant matrices of a Yin-qenoquaternion and of a Yang-qenoquaternion.

But what kind of a matrix arises in the case of subtraction of one genoquaternion from another genoquaternion? We mark this matrix by a symbol $YY_-^{[CAUG]}$ with the lower index "-":

$$G_f - G_m = YY_-^{[CAUG]} \qquad (4)$$

Figure 14 shows this new matrix $YY_-^{[CAUG]}$. One can see, that the black-and-white mosaic of this matrix possesses unexpectedly a relation of mirror-symmetry between the left half and the right half in contrast to the case of $YY_+^{[CAUG]}$-matrix. Figure 15 shows tessellations of a plane by these mosaics for an additional comparison

The second unexpected fact is that this $YY_-^{[CAUG]}$-matrix defines its own Yin-Yang-algebra as well. The multiplication table of this $YY_-^{[CAUG]}$-algebra is shown on Figure 16.

By analogy with the previous paragraph, one can analyze transformations of the genetic $YY_-^{[CAUG]}$-matrix, which are produced as a result of the same cyclic permutation of its genetic elements C→A→G→U→C. The same series of these cyclic permutations leads to a cyclic sequence of the following genetic matrices, which are shown on Figure 17: $YY_-^{[CAUG]} \rightarrow YY_-^{[UCGA]} \rightarrow YY_-^{[GUAC]} \rightarrow YY_-^{[AGCU]} \rightarrow YY_-^{[CAUG]}$. This Figure 17 is the analogue of Figure 11 for the genetic Yin-Yang-matrices, which were described in the previous paragraph.

Figure 14. The $YY_-^{[CAUG]}$-matrix

x_0	$-x_1$	$-x_2$	x_3	x_4	$-x_5$	$-x_6$	x_7
x_0	$-x_1$	$-x_2$	x_3	x_4	$-x_5$	$-x_6$	x_7
x_2	$-x_3$	x_0	$-x_1$	$-x_6$	x_7	$-x_4$	x_5
x_2	$-x_3$	x_0	$-x_1$	$-x_6$	x_7	$-x_4$	x_5
x_4	$-x_5$	$-x_6$	x_7	x_0	$-x_1$	$-x_2$	x_3
x_4	$-x_5$	$-x_6$	x_7	x_0	$-x_1$	$-x_2$	x_3
$-x_6$	x_7	$-x_4$	x_5	x_2	$-x_3$	x_0	$-x_1$
$-x_6$	x_7	$-x_4$	x_5	x_2	$-x_3$	x_0	$-x_1$

Figure 15. The tessellations of a plane by the mosaic of the $YY_+^{[CAUG]}$-matrix (on the left side) and by the mosaic of the $YY_-^{[CAUG]}$-matrix (on the right side)

Each of these new three matrices $YY_-^{[UCGA]}$, $YY_-^{[GUAC]}$ and $YY_-^{[AGCU]}$ is the Yin-Yang-matrix, which defines its relevant Yin-Yang-algebra. The multiplication tables of basic elements of these new Yin-Yang-matrices are shown on Figure 18.

All four genetic matrices $YY_-^{[CAUG]}$, $YY_-^{[UCGA]}$, $YY_-^{[GUAC]}$ and $YY_-^{[AGCU]}$ are connected with their own Hadamard matrices by means of the same algorithmic way (Figure 19).

By analogy with the previous paragraph, a total quantity of genetic Yin-Yang-matrices of this oppositional category (and a quantity of their relevant Hadamard matrices) increases significantly if one takes into consideration the same described permutations of the genetic elements including the permutations of positions in triplets, etc. In accordance with some preliminary estimation, a total quantity of Yin-Yang-matrices of both categories exceeds 1000 matrices considerably.

Figure 16. The multiplication table of the genetic $YY_-^{[CAUG]}$-algebra

	f_0	m_1	f_2	m_3	f_4	m_5	f_6	m_7
f_0	f_0	m_1	f_2	m_3	f_4	m_5	f_6	m_7
m_1	$-f_0$	$-m_1$	$-f_2$	$-m_3$	$-f_4$	$-m_5$	$-f_6$	$-m_7$
f_2	f_2	m_3	$-f_0$	$-m_1$	$-f_6$	$-m_7$	f_4	m_5
m_3	$-f_2$	$-m_3$	f_0	m_1	f_6	m_7	$-f_4$	$-m_5$
f_4	f_4	m_5	f_6	m_7	f_0	m_1	f_2	m_3
m_5	$-f_4$	$-m_5$	$-f_6$	$-m_7$	$-f_0$	$-m_1$	$-f_2$	$-m_3$
f_6	f_6	m_7	$-f_4$	$-m_5$	$-f_2$	$-m_3$	f_0	m_1
m_7	$-f_6$	$-m_7$	f_4	m_5	f_2	m_3	$-f_0$	$-m_1$

Figure 17. The cyclic sequence of the Yin-Yang-matrices $YY_-^{[CAUG]} \rightarrow YY_-^{[UCGA]} \rightarrow YY_-^{[GUAC]} \rightarrow YY_-^{[AGCU]} \rightarrow YY_-^{[CAUG]}$, which arises on the bases of the cyclic permutation of the genetic elements C→A→G→U→C in the matrices

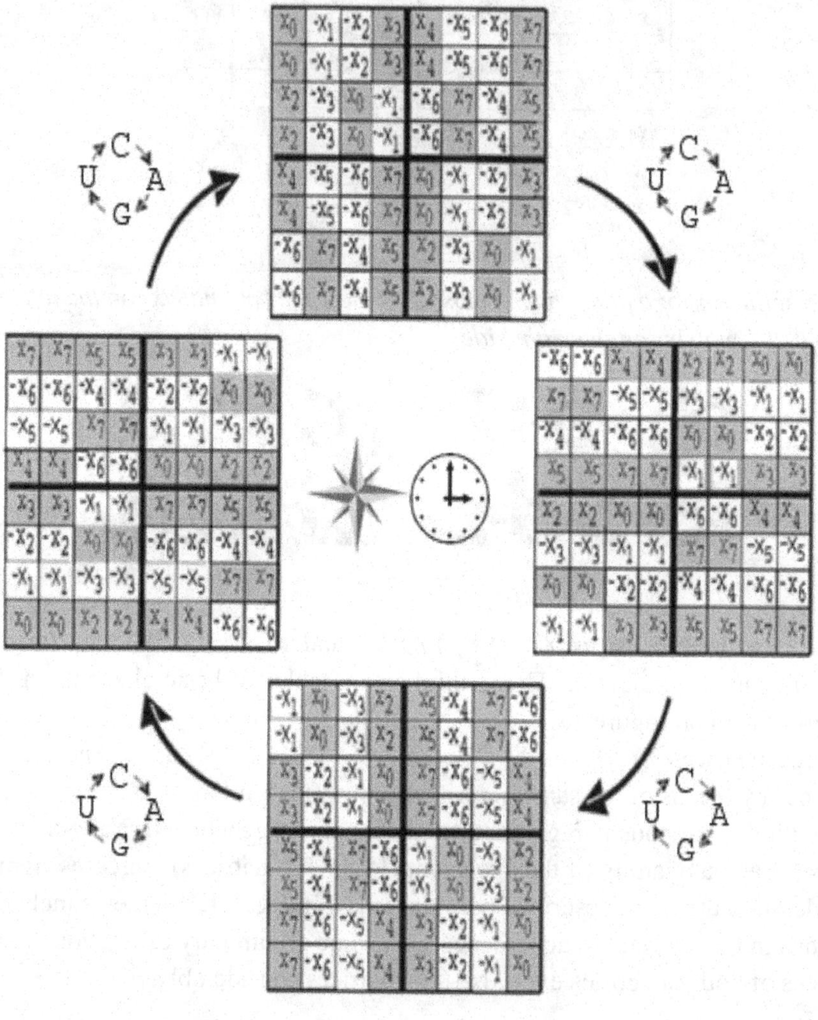

GENERATIVE AND SUPPRESSIVE PROPERTIES OF YIN-YANG-MATRICES OF THE TWO CATEGORIES

The connection of described Yin-Yang-matrices with their relevant Hadamard matrices seems to be interesting and prospective for study. This paragraph will consider a case when magnitudes of all coordinates of the Yin-genoquaternion G_f and of the Yang-genoquaternion G_m (Figure 13) are equal to 1 ($x_0 = x_1 = x_2 = x_3 = x_4 = x_5 = x_6 = x_7 = 1$). We will name conditionally such genoquaternions and their mentioned combinations ($G_f + G_m$) and ($G_f - G_m$) "elementary": the elementary Yin-genoquaternion, the elementary $YY_+^{[CAUG]}$-matrix, etc. This case is interesting specially because all components of Yin-Yang-matrices of both oppositional categories ($YY_+^{[CAUG]}$, $YY_-^{[CAUG]}$, etc.) are equal in this case to "+1" or "-1" like in Hadamard matrices. Simple changes of the signs "+" or "-" of some components of such elementary

Figure 18. The multiplication tables of the $YY_{-}^{[UCGA]}$-algebra (the upper table), the $YY_{-}^{[GUAC]}$-algebra (the middle table) and the $YY_{-}^{[AGCU]}$-algebra

	f_0	m_1	f_2	m_3	f_4	m_5	f_6	m_7
f_0	$-f_6$	f_6	$-f_4$	f_4	$-f_2$	f_2	$-f_0$	f_0
m_1	$-m_7$	m_7	$-m_5$	m_5	$-m_3$	m_3	$-m_1$	m_1
f_2	f_4	$-f_4$	$-f_6$	f_6	f_0	$-f_0$	$-f_2$	f_2
m_3	m_5	$-m_5$	$-m_7$	m_7	m_1	$-m_1$	$-m_3$	m_3
f_2	f_2	$-f_2$	$-f_0$	f_0	f_6	$-f_6$	$-f_4$	f_4
m_5	m_3	$-m_3$	$-m_1$	m_1	m_7	$-m_7$	$-m_5$	m_5
f_6	$-f_0$	f_0	$-f_2$	f_2	$-f_4$	f_4	$-f_6$	f_6
m_7	$-m_1$	m_1	$-m_3$	m_3	$-m_5$	m_5	$-m_7$	m_7

	f_0	m_1	f_2	m_3	f_4	m_5	f_6	m_7
f_0	f_0	m_1	f_2	m_3	f_4	m_5	f_6	m_7
m_1	$-f_0$	$-m_1$	$-f_2$	$-m_3$	$-f_4$	$-m_5$	$-f_6$	$-m_7$
f_2	f_2	m_3	$-f_0$	$-m_1$	$-f_6$	$-m_7$	f_4	m_5
m_3	$-f_2$	$-m_3$	f_0	m_1	f_6	m_7	$-f_4$	$-m_5$
f_4	f_4	m_5	f_6	m_7	f_0	m_1	f_2	m_3
m_5	$-f_4$	$-m_5$	$-f_6$	$-m_7$	$-f_0$	$-m_1$	$-f_2$	$-m_3$
f_6	f_6	m_7	$-f_4$	$-m_5$	$-f_2$	$-m_3$	f_0	m_1
m_7	$-f_6$	$-m_7$	f_4	m_5	f_2	m_3	$-f_0$	$-m_1$

	f_0	m_1	f_2	m_3	f_4	m_5	f_6	m_7
f_0	$-f_6$	f_6	$-f_4$	f_4	$-f_2$	f_2	$-f_0$	f_0
m_1	$-m_7$	m_7	$-m_5$	m_5	$-m_3$	m_3	$-m_1$	m_1
f_2	f_4	$-f_4$	$-f_6$	f_6	f_0	$-f_0$	$-f_2$	f_2
m_3	m_5	$-m_5$	$-m_7$	m_7	m_1	$-m_1$	$-m_3$	m_3
f_2	f_2	$-f_2$	$-f_0$	f_0	f_6	$-f_6$	$-f_4$	f_4
m5	m_3	$-m_3$	$-m_1$	m_1	m_7	$-m_7$	$-m_5$	m_5
f_6	$-f_0$	f_0	$-f_2$	f_2	$-f_4$	f_4	$-f_6$	f_6
m_7	$-m_1$	m_1	$-m_3$	m_3	$-m_5$	m_5	$-m_7$	m_7

Yin-Yang-matrices in accordance with the U-algorithm are enough to transform these elementary Yin-Yang-matrices into relevant Hadamard matrices.

The elementary genetic Yin-Yang-matrices of the two oppositional categories possess some beautiful properties relative to multiplication. If any elementary Yin-Yang-matrix $(G_f + G_m)$ is raised to the second power, the result is a tetra-reproduction of this Yin-Yang-matrix. For example, let us consider such exponentiation of each of the six Yin-Yang-matrices on Figure 12:

$$(YY_{+,123}^{[CAUG]})^2 = 4* YY_{+,123}^{[CAUG]};\ (YY_{+,231}^{[CAUG]})^2 = 4* YY_{+,231}^{[CAUG]};$$

$$(YY_{+,312}^{[CAUG]})^2 = 4* YY_{+,312}^{[CAUG]};\ (YY_{+,321}^{[CAUG]})^2 = 4* YY_{+,321}^{[CAUG]};$$

$$(YY_{+,213}^{[CAUG]})^2 = 4* YY_{+,213}^{[CAUG]};\ (YY_{+,132}^{[CAUG]})^2 = 4* YY_{+,132}^{[CAUG]} \quad (5)$$

This property can be illustrated graphically as giving rise to four identical matrices instead of one initial matrix (Figure 20). It generates some associations with the tetra-reproduction of gametal cells

Figure 19. The Hadamard matrices, which are connected algorithmically with the Yin-Yang-matrices $YY_{-}^{[CAUG]}$, $YY_{-}^{[UCGA]}$, $YY_{-}^{[GUAC]}$ *and* $YY_{-}^{[AGCU]}$. *Each black (white) cell contains the element "+1" ("-1")*

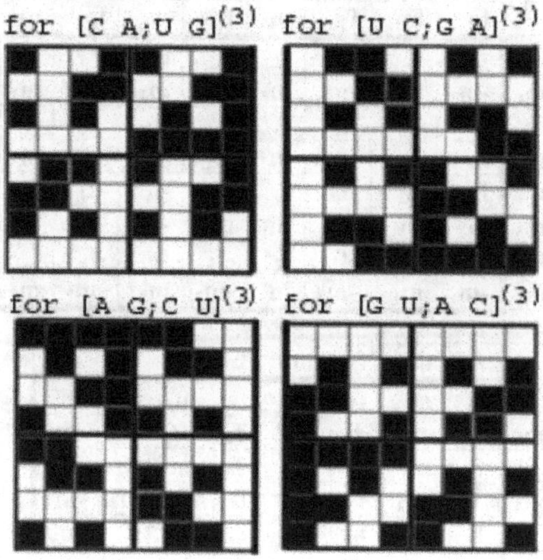

Figure 20. Schematic illustrations of the tetra-reproduction of the elementary Yin-Yang-matrices $YY_{+,123}^{[CAUG]}$; $YY_{+,231}^{[CAUG]}$; $YY_{+,312}^{[CAUG]}$; $YY_{+,321}^{[CAUG]}$; $YY_{+,213}^{[CAUG]}$; $YY_{+,132}^{[CAUG]}$ *by means of their rising into the second power*

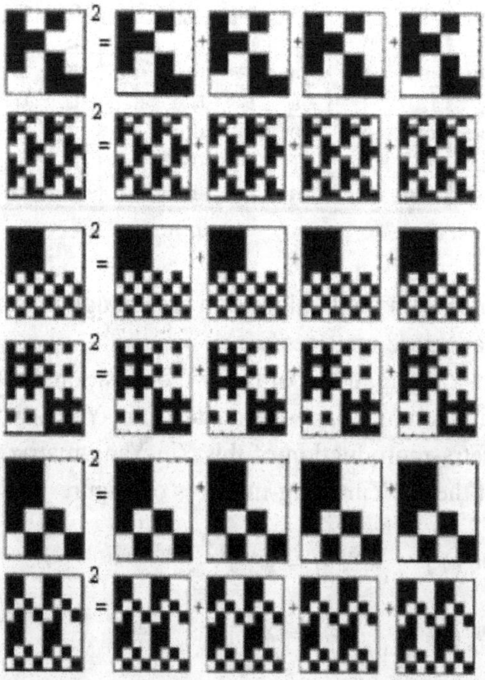

in a process of meiosis. In view of this, we name elementary Yin-Yang-matrices $(G_f + G_m)$ generative Yin-Yang-matrices or start-matrices. By the way, if the elementary Yin-genoquaternion G_f or if the elementary Yang-genoquaternion G_m is raised into the second power, the result is a double-reproduction of this elementary genoquaternion: $G_f^2 = 2*G_f$ and $G_m^2 = 2*G_m$. It reminds one of a double-reproduction (a dichotomy) of somatic cells in a process of mitosis.

On the contrary, elementary Yin-Yang-matrices $(G_f - G_m)$ possess a suppressive property: if any elementary Yin-Yang-matrix $(G_f - G_m)$ is raised to the second power, the result is the null matrix. For comparison with the expression (5), one can show the following results:

$$(YY_{-,123}^{[CAUG]})^2 = 0;\ (YY_{-,231}^{[CAUG]})^2 = 0\ ;$$

$$(YY_{-,312}^{[CAUG]})^2 = 0;\ (YY_{-,321}^{[CAUG]})^2 = 0;$$

$$(YY_{-,213}^{[CAUG]})^2 = 0;\ (YY_{-,132}^{[CAUG]})^2 = 0 \qquad (6)$$

In view of this, we name elementary Yin-Yang-matrices $(G_f - G_m)$ suppressive Yin-Yang-matrices or stop-matrices (or apoptosis-matrices). A generative Yin-Yang-matrix can be transformed into a suppressive Yin-Yang-matrix and vice versa by means of the simple inversion of the signs "+" and "-" in Yang-components (or in Yin-components) of this Yin-Yang-matrix. Such transformations can be defined as functions of time by means of a definition of Yin-Yang-coordinates as relevant functions of time.

Many interesting relations exist among generative and suppressive Yin-Yang-matrices and genoquaternions G_f and G_m. For example, the product of a generative Yin-Yang-matrix with a suppressive Yin-Yang-matrix is equal to the tetra-reproduction of the suppressive Yin-Yang-matrix: $(YY_{+,123}^{[CAUG]}) * (YY_{-,123}^{[CAUG]}) = 4*(YY_{-,123}^{[CAUG]})$, etc. On the contrary, the product of a suppressive matrix with a Yin-Yang-matrix $(G_f + G_m)$ is equal to the null matrix: $(YY_{-,123}^{[CAUG]}) * (YY_{+,123}^{[CAUG]}) = 0$. A set of such relations and properties gives new possibilities to create mathematical models of self-developing biological systems.

Do molecular-genetic facts exist about an important role of cyclic or circular principles in genetic systems? Yes, interesting phenomenological results of studying some circular principles of organizations of molecular-genetic systems are presented, for example, in the articles (Arques, Michel, 1996, 1997; Frey, Michel, 2003, 2006; Stambuk, 1999). One can hope that our algebraic-genetic results and these published phenomenological results will supplement each other for deeper understanding the genetic systems.

The described results about hierarchy of cyclic sequences of cyclic changes of genetic Yin-Yang-algebras show the following: a set of cyclic transformations and cyclic structures in living matter is connected with these cycles of algebraic changes or can be modeled by means of such changes of the genetic algebras. Nature has created the genetic code in such a manner that a wide set of cyclic permutations of molecular-genetic elements in genetic matrices (or in these matrix forms of presentation of the genetic code) leads to cyclic changes of their genetic Yin-Yang-algebras.

Can algorithmic principles of organization of many cyclic movements (walking, run, breath, cardio cycles, etc.) of a separate individual be presented as well in forms of mathematical models and algorithms, which are based on such changes of the Yin-Yang-algebras? In our opinion, it will be possible in the future. In such models, a transition of one kind of cyclic movement to another kind can be expressed as a transition from one cyclic sequence of changes of Yin-Yang-algebras to another cyclic sequence. One can also establish about a possible relation of our algebraic-genetic approach with the famous conception by Eigen about hypercycles in biological organizations (Eigen, 1979).

JOINING OF THE IDEA BY PYTHAGORAS AND THE IDEA OF CYCLIC CHANGES: ABOUT CELLULAR AUTOMATA, NEUROCOMPUTERS AND A NOTION OF BIOLOGICAL TIME

Pythagoras has formulated the famous idea: "All things are numbers". Such known slogans of Pythagoreans as "numbers operate the world", "the world is number" reflect representations of Pythagoreans. For Pythagoreans the systems of numbers expressed the "essence" of everything. In view of this idea, natural phenomena should be explained by means of systems of numbers; the systems of numbers play a role on uniting all things and expressing the harmony of nature (Kline,1980, p. 21, 24). Many prominent scientists and thinkers were supporters of this viewpoint. Not without reason B. Russell (1945) noted that he did not know any other person who would exert such an influence on the thinking of people as Pythagoras.

The history of science knows many thinkers who believed that all physics can be described in a language of some multi-dimensional numeric system or algebra. For example, W. Hamilton believed that all physics can be described in the language of his quaternions. These kinds of thoughts are in a line with Pythagorean idea. The data of matrix genetics about cyclic changes of the genetic Yin-Yang-algebras (or genetic Yin-Yang-numeric systems) provide materials for other kinds of thoughts or for broadening the idea by Pythagoras. The main point here is that a new idea about organization of living matter arises. In accordance with this idea, organization of living matter is based not on a single algebra but on cyclic changes of many algebras of a certain set (a set of genetic Yin-Yang-algebras). It means one that the idea by Pythagoras about numeric harmony of nature should be supplemented by another idea of cyclic changes of Yin-Yang-numeric systems. This new additional idea about cyclic changes of algebras in living matter lead to the idea about cyclic changes from the Ancient Chinese "The Book of changes" ("I Ching") which was written about Yin-Yang-systems a few thousand years ago (see Chapter 12). But instead of the quite wide notion of "a cyclic change" we use the strict mathematical notion of "a change of one genetic Yin-Yan-algebra into another genetic Yin-Yang-algebra". Can this idea about cyclic changes of Yin-Yang-algebras be applied to inanimate matter to some extent as well? The future will tell. The authors do not know any other theory in the field of mathematical natural sciences which is based on cyclic changes of multi-dimensional algebras. It seems that the genetic code leads us to the new category of theories of mathematical natural sciences, which are based on a conception of cyclic changes inside a bunch of multi-dimensional algebras depending on time and spatial features.

From the viewpoint of the proposed algebraic-cyclic conception about organization of living matter, the notion of "biological time" can be defined as a factor of a general coordinating (or a general synchronization) of many cycles of changes of genetic Yin-Yang-algebras inside the hierarchy of these changes. If an organism is a hierarchy of cyclic changes of genetic Yin-Yang-algebras, such a biological time is dispersed on all choruses of such cyclic processes of an organism. A dispersing of biological time along the whole organism reminds one of a dispersing of the feeling for music along the whole organism (our brain does not have a special center of music, and music appeals to the whole organism (Weinberger, 2004)).

The proposed algebraic-genetic approach to the problem of biological time provides the famous viewpoint (Whitrow, 1961) that biological time is internal time inside a spatial region of living matter (this spatial region is isolated to some extent from other regions of Universe).

Here we note that a wide class of cyclic and circadian phenomena in molecular biology is related to gene expressions. A problem of gene expression is one of the most important and the most difficult in bioinformatics. Nucleotide sequences in DNA are the same in all cells of our organism, but this inherited information gives different proteins and other results in different cells and in different time because of changes of gene expressions (or because of specifics of biological reading this DNA information). Gene expression is the process by which inheritable information from a gene is made into a functional gene product, such as protein or RNA. It is known that several steps in the gene expression process may be modulated, including the transcription step and translation step and the post-translational modification of a protein. Gene regulation gives the cell control over structure and function, and is the basis for cellular differentiation, morphogenesis and the versatility and adaptability of any organism. Gene regulation may also serve as a substrate for evolutionary change, since control of the timing, location, and amount of gene expression can have a profound effect on the functions (actions) of the gene in the organism. The cyclic and circadian types of gene expressions are connected closely to a problem of biological clock and relations of organisms with cyclic changes of environment. They take an important place in bioinformatics (Ceriani et al, 2002; Izumo et al, 2003; Matsumoto et al, 2007; McDonald & Rosbash, 2001; Oster et al., 2003; Panda et al, 2002).

In our view, cyclic ensembles of algebraic structures of the genetic code have a close relation to phenomenology of a huge chorus of cyclic and circadian gene expressions. Presented cyclic sets of the genetic Hadamard matrices and of genetic bipolar algebras give many new variants of applications of algebra, spectral analysis and symmetry theory for modeling and for studying interrelated systems of cyclic and circadian gene expressions. It is interesting that Hadamard matrices are already used in some mathematical investigations of gene expression (de Hoon et al, 2003).

In addition, it is important to note that Hadamard matrices and their Kronecker product are used in a problem which is a dream of biologists since the work of Darwin: to reconstruct the tree of evolution of living things. That tree could be the only scientific basis for classification. Hadamard conjugation for evolutionary trees was introduced in (Steel et al., 1992, 1993; Szekely et al, 1993, 1994; Szekely, Steel & Erdxs, 1993) where genetic sequences were analyzed.

Results of matrix genetics have interesting applications in the fields of cellular automata and neurocomputing. Concerning to cellular automata, one may form a new branch of studies on the basis of genomatrices. This branch may be called "genomatrix automata". Cellular automaton is a state machine that consists of an array of cells, each of which can be in one of a finite number of possible states. The cells are updated synchronously in discrete time steps, according to some interaction rules (Levy, 1993; Matthews, 2005; Wolfram, 2000).

A history of cellular automata ("machines by von Neumann") began with works by J. von Neumann, who considered these works as the most important among all he made in his life. It is known that cellular automata can function as universal computers and they can have properties of self-reproducing organisms. The most famous cellular automata are presented in a computer game "Life" by J. Conway. Cellular automata are utilized for modeling Navier-Stock's equations and turbulence, etc. Some scientists believe that theory of cellular automata makes a scientific breakthrough in understanding not only vital proccsses, but Universe as well (Matthews, 2005; Wolfram, 2000).

The described properties of genetic matrices, including first of all generative and suppressive kinds of Yin-Yang-matrices, are useful for creating a new class of cellular automata – genomatrix automata – with new relevant rules of interaction of ensembles of genomatrices as special collectives of cells. In contrast to classical cellular automata, a theory of genomatrix automata deals with not separate cells but with their genomatrix ensembles.

Another important application of matrix genetics is a creation of new kinds of neurocomputers. They may be called "genomatrix" neurocomputers which are controlled by a special set of logic operators. This set of operators corresponds to matrix features of the genetic code. One of famous neurocomputers called "Embryo" (Tsygankov, 2005) may be referred as a genomatrix neurocomputer. This genomatrix neurocomputer will reproduce some properties of living organisms in new and effective forms. It will be useful as well for deeper understanding the genetic system, for new decisions in the field of artificial intelligence (Russel, & Norvig, 2003), etc.

SOLUTIONS AND RECOMMENDATIONS

The discovery of the described properties of the genetic Yin-Yang-algebras, which are connected with cyclic permutations of genetic elements, gives new possibilities of a creation of mathematical models of cyclic biological processes. Such knowledge leads to new heuristic associations and directions of thought as well. A close connection of structural ensembles of the genetic code with special permutations of genetic elements receives new algebraic evidences because many permutations lead unexpectedly to new genetic Yin-Yang-algebras (from the viewpoint of matrix presentation of the genetic code systems). It can be essential for information processing in biological organisms. A result of mutual multiplication of any two 8-dimensional numbers from the expression (1) depends on a type of Yin-Yang-algebra, the multiplication table of which we use for the multiplication.

For example, if we use 10 kinds of Yin-Yang-algebras for such multiplication separately, 10 different results arise. The coordinates of these 8-dimensional numbers can reflect information parameters, a system processing of which can be done from the viewpoints of different Yin-Yang-algebras. It can be useful for the organizing of multi-channel processing of biological information and for using the same initial data for different biological tasks, each of which can be connected with its own genetic Yin-Yang-algebra (including a Yin-Yang-algebra of logic operators as well).

Various kinds of mosaic Yin-Yang-matrices give many types of a mosaic tessellation (for example, see Figure 15) of a plane or of flat surfaces like cylindrical or conical forms in biological objects. The full catalogue of such tessellations by means of Yin-Yang-matrices should be made. Can such mosaic tessellations be found in a history of ancient cultures or in real substances, for example, at a nano-sized level (by analogy with mosaics by Penrose, which were found in quasi-crystals)? This question should be researched additionally.

The study of cyclic processes in biological organisms has many aspects, mathematical modeling of which is needed. One example of these aspects concerns the famous concept of Ancient Oriental medicine about the cyclic nature of biological processes. According this concept, each organ has more or less a definite time interval for its culmination (its own time interval), when its activity is maximal, and each organ has a maximum sensitivity to pathogenic and medicinal influences just in this special time interval (Vogralik & Vogralik, 1978, p. 11). This phenomenological knowledge about the chronocyclic essence of biological organisms was used and tested during several thousand years by generations of oriental doctors, which were specially selected from many candidates in accordance with the criteria of their talents and of their brains. Many effective methods were constructed on the basis of this knowledge. (for example see (Cheng Xinnong, 1989; Needham, 1956)). One of them is the pulse diagnostics of Tibetan medicine. This pulse diagnostics was a universal method of diagnostics for an experienced doctor, who could determine not only many kinds of diseases, but report sometimes about physiological past and future of his patient (Tsydypov, 1988). This method shows additionally, that chronocyclic processes (pulse processes, etc.) in biological organisms carry astonishingly complete information about organism on the whole. One can hope that new knowledge about genetic Yin-Yang-algebras will be useful for deeper understanding these phenomena as well.

Here we point out that the main problem in biology, which was formulated by the prominent Russian physiologist A. Gurvich (1977, p. 27) in following: "*the main problem of biology is a supporting of a form at a permanent renovation of materials*". Results of matrix genetics presented in this book give new tools for mathematical modeling such properties and allow searching new invariant relations among elements of the genetic code systems. In particular, matrix genetics reveals that a set of cyclic shifts (permutations) of genetic elements in the genetic Yin-Yang matrices have invariant relations to the Yin-Yang types of algebras and to the genetic Hadamard matrices.

The fact of existence of the two types of oppositional Yin-Yang-matrices, elementary variants of which possess generative and suppressive properties, which were described above, can be utilized in mathematical models of self-developing biological systems. New useful bio-mathematical notions arise for such self-developing systems on the basis of cyclic changes of Yin-Yang-algebras, each of which is connected with its own "genovector calculation" and with its own anisotropic "genetic space" (see Chapter 7). One such new notion is "helical (or spiral) waves of growth" which can model phenomena of many helical configurations in biological bodies. The point is about extremely wide presence of helical configurations in biological morphology. Not without reason such helical structures were named "the curves of life" in the famous book (Cook, 1914). Producing helical constructions in many biological cases is realized in apical regions of growth of biological bodies. If one supposes that processes in a growth region are connected with appropriate Yin-Yang-algebras and that they depend on periodic cyclic changes of these algebras, a morphological construction of helical configurations can be generated. Such "helical waves" are interpreted as spatial movements of periodic cyclic changes of Yin-Yang-algebras.

The proposed algebraic-genetic version of biological time is useful for a deeper understanding of the hierarchy of interrelated processes in biological organisms. In the proposed case biological time is interpreted in relation to discrete cyclic changes in the hierarchy of genetic Yin-Yang-algebras. In view of this, such biological time is a discrete essence. The minimal time interval from one such change to the next change is a natural time unit of biological time in this conception. All other time intervals between adjacent changes of genetic Yin-Yang-algebras can be expressed as compositions of this time unit.

The history of science testifies that a development of new beautiful mathematical tools often leads to new theories and knowledge about natural systems and processes after some latent period. It is essential that we develop new mathematical approach, which is presented in this book for working with ensembles of genetic alphabets and multiplets, that can lead to a new language connected to matrix analysis and symmetries. New mathematical languages are important for development of scientific fields. A classical example is integral calculus and differential calculus. A main task of our book was a demonstration of new symmetrical patterns in the field of bioinformatics and also a demonstration of new phenomenological regularities and theoretical approaches. One can emphasize that our book shows how biological phenomena help in development of new mathematical tools and mathematical patterns including new types of multidimensional numbers and cyclic ensembles of algebras.

FUTURE TRENDS AND CONCLUSION

In this book we began with revealing the symmetrical mosaic structure of the single genomatrix [C A; U G]$^{(3)}$ (see Chapter 2, Figure 2). But now we have come to such great ensemble of genomatrices with symmetrical mosaics, members of which are interconnected by relation of permutation symmetries. In other words collective symmetries of genomatrices in great genomatrix ensemble are discovered in the field of matrix genetics. Systematic researches of these collective symmetries and of connections of these symmetrical genomatrices with Hadamard matrices are interesting for future progress in bioinformatics.

One can wait for an intensive development of algebraic models of biological phenomena of self-developing and of cyclic processes on the bases of generative and suppressive genetic Yin-Yang-matrices and on the bases of a hierarchy of cyclic changes of genetic Yin-Yang-algebras. These models will reveal deep connections of molecular-genetic systems with quantum mechanics and quantum computers where Hadamard matrices play important roles. The importance of permutations of molecular-genetic elements and the principle of molecular economy (see Chapter 6) for the genetic system will be clear more and more. On the basis of ideas about such permutations and about Yin-Yang-algebras, new models will arise including models of animal metamorphoses (Figure 21).

In our opinion, a variety of species of living matter is connected to a significant extent with the variety of types of genetic Yin-Yang-algebras and with the variety of cyclic permutations of the genetic elements. In particular, wide researches should be done about a natural division of all set of genes and proteins into special sub-sets by criteria of mutual cyclic transformations of members of each sub-set by means of cyclic permutations of their genetic elements (like the considered cyclic permutation C→A→G→U→C, which transforms any sequence of triplets into another sequence of triplets).

Special attention will be paid on the investigation of structural parallels between the theory of genetic Yin-Yang-algebras and the conceptions of the Ancient Chinese "The Book of Changes" ("I Ching"), where the basic role of cyclic changes in nature is considered in original manner (see Chapter 12).

Figure 21. Schematic presentation of four stages of cyclic metamorphoses of butterfly: egg, larva, pupa and adult

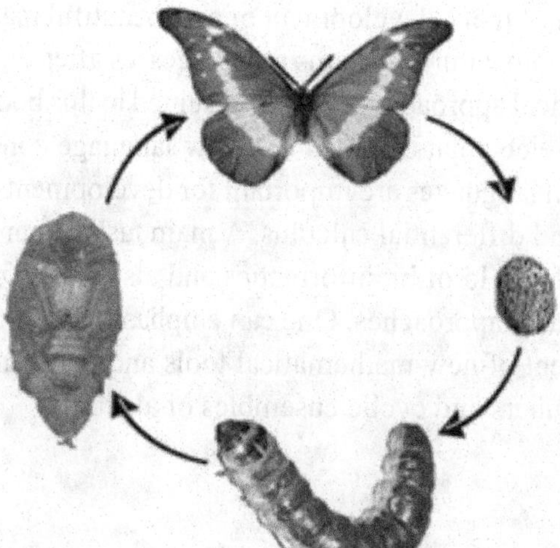

Cyclic changes of types of the genetic Yin-Yang-matrices are produced by means of cyclic permutations of genetic elements. Taking into account the existence of "male" genoquaternion and "female" qenoquaternion as two items (summands) of any 8-dimensional Yin-Yang-matrix, the opposite type of genetic Yin-Yang-matrices was discovered which is based on subtraction of these two genoquaternions. The mentioned hierarchy of Yin-Yang-algebras and of their changes can be utilized for a new algebraic definition of the notion of biological time and for creating a new category of mathematical models of self-developing and cyclic processes of biological objects. New patterns and symmetry relations are revealed by studying permutations of genetic elements in the matrix forms of presentation of the genetic code.

REFERENCES

Ahmed, N., & Rao, K. (1975). *Orthogonal transforms for digital signal processing*. New York: Springer-Verlag Inc.

Aksenova, M. (Ed.). (1998). *Encyclopedia of biology.* Moscow: Avanta+ (in Russian).

Arques, D., & Michel, C. (1996). A complementary circular code in the protein coding genes. *Journal of Theoretical Biology, 182*, 45–58. doi:10.1006/jtbi.1996.0142

Arques, D., & Michel, C. (1997). A circular code in the protein coding genes of mitochondria. *Journal of Theoretical Biology, 189*, 45–58. doi:10.1006/jtbi.1997.0513

Ceriani, M. F., Hogenesch, J. B., Yanovsky, M., Panda, S., Straume, M., & Kay, S. A. (2002). Genome-wide expression analysis in Drosophila reveals genes controlling circadian behavior. *The Journal of Neuroscience, 22*(21), 9305–9319.

Cheng, X. (Ed.). (1989). *Chinese acupuncture and moxibustion.* Beijing: Foreign Languages Press.

Cook, T. A. (1914). *The curves of life*. London: Constable and Co.

de Hoon, M. J. L., Imoto, S., Kobayashi, K., Ogasawara, N., & Miyano, S. (2003). Inferring gene regulatory networks from time-ordered gene expression data of Bacillus Subtilis using differential equations. In R. B. Altman, A. K. Dunker, L. Hunter & T. E. Klein (Eds.), *Proceedings of the 8th Pacific Symposium on Biocomputing* (pp. 17-28). Lihue, Hawaii: World Scientific.

Dubrov, A. P. (1989). *Symmetry of biorhythms and reactivity*. New York, London, Tokyo: Gordon & Breach Science Publishers.

Eigen, M. (1979). *The hypercycle-a principle of natural self-organization*. Berlin: Springer- Verlag.

Frey, G., & Michel, C. (2003). Circular codes in archaeal genomes. *Journal of Theoretical Biology, 223*, 413–431. doi:10.1016/S0022-5193(03)00119-X

Frey, G., & Michel, C. (2006). Identification of circular codes in bacterial genomes. *Computational Biology and Chemistry, 20*, 87–101. doi:10.1016/j.compbiolchem.2005.11.001

Gurvich, A. G. (1977). *Selected works*. Moscow, Russia: Meditsina (in Russian).

Izumo, M., Johnson, C. H., & Yamazaki, S. (2003). Circadian gene expression in mammalian fibroblasts revealed by real-time luminescence reporting: Temperature compensation and damping. *Proceedings of the National Academy of Sciences of the United States of America, 100*(26), 16089–16094. doi:10.1073/pnas.2536313100

Kline, M. (1980). *Mathematics. The loss of certainty.* New York: Oxford University Press.

MatsumotoA.Ukai-TadenumaM.YamadaR. G.HoulJ.UnoK.D .KasukawaT.DauwalderB.ItohT. Q.TakahashiK.UedaR.HardinP. E.TanimuraT.UedaH.

McDonald, M. J., & Rosbash, M. (2001). Microarray analysis and organization of circadian gene expression in Drosophila. *Cell, 107*(5), 567–578. doi:10.1016/S0092-8674(01)00545-1

Needham, J. (1956). *Science and civilization in China.* Cambridge: Cambridge University Press.

Oster, H., van der Horst, G. T., & Albrecht, U. (2003). Daily variation of clock output gene activation in behaviorally arrhythmic mPer/mCry triple mutant mice. *Chronobiology International, 20*(4), 683–695. doi:10.1081/CBI-120022408

Panda, S., Antoch, M. P., Miller, B. H., Su, A. I., Schook, A. B., & Straume, M. (2002). Coordinated transcription of key pathways in the mouse by the circadian clock. *Cell, 109*(3), 307–320. doi:10.1016/S0092-8674(02)00722-5

Petoukhov, S. V. (1989). Non-Euclidean geometries and algorithms of living bodies. *Computers & Mathematics with Applications (Oxford, England), 17*(4-6), 505–534. doi:10.1016/0898-1221(89)90248-4

Petoukhov, S. V. (2008a). *Matrix genetics, algebras of the genetic code, noise immunity.* M., RCD, 316 (in Russian).

Petoukhov, S. V. (2008b). The degeneracy of the genetic code and Hadamard matrices (pp. 1-8). Retrieved on February 22, 2008, from http://arXiv:0802.3366

Petoukhov, S. V. (2008c). Matrix genetics, part 1: Permutations of positions in triplets and symmetries of genetic matrices (pp. 1-12). Retrieved on March 6, 2008, from http://arXiv:0803.0888

Petoukhov, S. V. (2008d). Matrix genetics, part 2: The degeneracy of the genetic code and the octave algebra with two quasi-real units (the "yin-yang octave algebra") (pp. 1-23). Retrieved on March 23, 2008, from http://arXiv:0803.3330

Petoukhov, S. V. (2008e). Matrix genetics, part 3: The evolution of the genetic code from the viewpoint of the genetic octave Yin-Yang-algebra (pp. 1-22). Retrieved on May 30, 2008, from http://arXiv:0805.4692

Petoukhov, S. V. (2008f). Matrix genetics, part 4: Cyclic changes of the genetic 8-dimensional Yin-Yang-algebras and the algebraic models of physiological cycles (pp. 1-22). Retrieved on September 17, 2008, from http://arXiv:0809.2714

R. (2007). A functional genomics strategy reveals clockwork orange as a transcriptional regulator in the Drosophila circadian clock. *Genes Dev., 21*(13), 1687-1700.

Russell, B. (1945). *A history of western philosophy.* New York: Touchstone.

Stambuk, N. (1999). Circular coding properties of gene and protein sequences. *Croatica Chemica Acta, 72*(4), 999–1008.

Steel, M. A., Hendy, M. D., Szekely, L. A., & Erdxs, P. L. (1992). Spectral analysis and a closest tree method for genetic sequences. *Applied Mathematics Letters, 5*, 63–67. doi:10.1016/0893-9659(92)90016-3

Steel, M. A., Szekely, L. A., Erdxs, P. L., & Waddell, P. (1993). A complete family of phylogenetic invariants for any number of taxa. *New Zealand Journal of Botany, 31*, 289–296.

Szekely, L. A., Erdxs, P. L., & Steel, M. A. (1994). The combinatorics of reconstructing evolutionary trees. *J. Comb. Math. Comb. Computing, 15*, 241–254.

Szekely, L. A., Erdxs, P. L., Steel, M. A., & Penny, D. (1993). A Fourier inversion formula for evolutionary trees. *Applied Mathematics Letters, 6*, 13–17. doi:10.1016/0893-9659(93)90004-7

Szekely, L. A., Steel, M. A., & Erdxs, P. L. (1993). Fourier calculus on evolutionary trees. *Advances in Applied Mathematics, 14*, 200–216. doi:10.1006/aama.1993.1011

Trahtman, A. M., & Trahtman, V. A. (1975). *The foundations of the theory of discrete signals on finite intervals*. Moscow: Sovetskoie Radio (in Russian).

Tsydypov, C. T. (Ed.). (1988). *Pulse diagnostics of Tibetan medicine*. Novosibirsk: Nauka (in Russian).

Vogralik, V. G., & Vogralik, M. V. (1978) *Acupuncture*. Gor'kiy: Meditsina (in Russian).

Weinberger, N. M. (2004). Music and brain. *Scientific American, 291*(5), 88–95.

Whitrow, G. J. (1961). *The natural philosophy of time*. New York: Harper.

Wolfram, S. (2000). *A new kind of science*. London: Wolfram Publishing.

Wright, K. (2002). Times of our lives. *Scientific American, 287*(3), 58–65.

Chapter 12
Matrix Genetics and Culture

ABSTRACT

This chapter considers the topic of connections of the genetic code with various fields of culture and with inherited physiological properties which provide existence of these fields. Some examples of such physiological bases for branches of culture are described. These examples are related to linguistics, music, and physiology of color perception. Special attention is paid to connections between the genetic matrices and the system of the Ancient Chinese book "I Ching." The conception and its arguments are put forward that the famous table of 64 hexagrams of "I Ching" reflects notions of Ancient Chinese about music quint harmony as a universal archetype.

INTRODUCTION AND BACKGROUND

Results of researches of the genetic code are important, first of all, for genetics and biotechnology. However, a high attention to structural properties of the genetic code system is shown by the most different fields of science and culture as well. Many experts think that the genetic code is the bearer of keys to a solution of the phenomenon of life. Many researchers develop the theories and creative designs with taking into account properties and mechanisms of transfer of the hereditary information. For instance, knowledge about inherited properties of physiological systems of perception is important for engineers in the field of ergonomics and mechatronics systems which provide effectiveness, reliability and noise immunity of created machines and tools. This chapter presents some examples of possible applications of the research results, which were described in the previous chapters, in various fields of culture. A connection of the genetic code with musical harmony has been already considered in Chapter 4.

DOI: 10.4018/978-1-60566-124-7.ch012

Traditionally experts in various fields of science and culture are interested in studying living matter. Modern computerized technical systems need in reliability and noise-immunity in high extent relevant to "patents" of living nature. First of all, patterns and symbols of the genetic code of the double helix of DNA are utilized widely in many design solutions, art, etc. Creators of many cultural works would like to know theoretical bases, which can provide a physiological increasing of attention from the public to their productions and which can lead them to new effective works on a scientific platform. They believe that the person is a measure of all things (see this topic in (Teilhard de Chardin, 1959)). They need a deep understanding of connections of the genetic code with inherited physiological systems and phenomena.

Inherited physiological bases of linguistics, color perception, music perception, and others are studied in many interesting works (Andrews, 1990; Caglioti, Ramme, & Tscouvileva, 2006; Chomsky, 1980; Darvas, 2007; Hahn, 1989, 1998; Hargittai & Hargittai, 1994; He & Petoukhov, 2007; Jacob, 1974, 1977; Kappraff, 1990, 2000, 2002; Leyton, 1992; Loeb, 1971, 1993; Marcus, 1990, 2007; Nonnenmacher, Losa & Weibel,.1994; Petoukhov, 2001, 2008a; Shubnikov, & Koptsik, 1974; Smith, 1980; Teilhard de Chardin, 1959; Wehr, 1969). It has been known for a long time that various physiological systems have many structural analogies among them and that various branches of culture have many general structural features as well. For example, architecture was interpreted historically as non-movement music, and music was interpreted as dynamic architecture. Matrix genetics give new evidences that many inherited analogies among different physiological systems are based on the genetic code structures.

The possible relations of musical harmony with the genetic code and with the system of "I Ching" were suspected for decades already. Hungarian musician E. Tusa (1994) has paid attention to some structural analogies between the ancient Greek numerical table Lambdoma, which was used by ancient Greek theorists of music, and symbolical table of 64 hexagrams of "I Ching". In addition, Tusa has assumed that the musical table Lambdoma has connections simultaneously with the genetic code. This assumption was based on hypotheses, which existed since 1969 after a publication of the book (Stent, 1969), about the possible interrelation between the table of 64 hexagrams and the set of 64 genetic triplets. The article ended with the following statement: "*This summary is far from being complete. With my essay, I would like to stimulate everyone to search for further analogies and newer connections!*" (Tusa, 1994, p. 310). Some results of matrix genetics, which are described in our book, discovery such newer connections.

THE GENETIC CODE AND LINGUISTICS

Impressive discoveries in the field of the genetic code have been described by its researchers using the terminology borrowed from linguistics and the theory of communications. As experts in molecular genetics remark, "*the more we understand laws of coding of the genetic information, the more strongly we are surprised by their similarity to principles of linguistics of human and computer languages*" (Ratner, 2002, p. 203). Linguistics is one of the significant examples of existence and importance of ensembles of binary oppositions in information physiology.

Leading experts on structural linguistics have believed for a long time that languages of human dialogue were formed not from an empty place, but they are continuation of genetic language or, somehow, are closely connected with it, confirming the idea of information commonality of organisms. Analogies between systems of genetic and linguistic information are contents of a wide and important scientific sphere, which can be illustrated here in short only. We reproduce below some thematic thoughts by R. Jakobson (1985, 1987, 1999), who is one of the most famous experts and the author of a deep theory of binary oppositions in linguistics. Jakobson and others are holding the same views that we possess a language which is as old as life and which is alive among all languages. Among all systems of information transfer, the genetic code and linguistic codes only are based on the use of discrete components, which in itself makes no sense, but serve for the construction of the minimum units which make sense. In both cases of the genetic language and of a linguistic language, we deal with separate units which, taken in itself, have no sense, but they get a sense after their special grouping. (By the way, one can note here that matrix genetics deals with matrix forms of groupings of elements of genetic language successfully). A similarity between both information systems is not exhausted by this fact at all. According to Jakobson, all relations among linguistic phonemes are decomposed into a series of binary oppositions of elementary differential attributes (or traits). By analogy the set of the four letters of the genetic alphabet contains the three binary sub-alphabets, which were described in Chapter 1 and which allows to number columns and rows of the genetic matrices $[C\ A;\ U\ G]^{(n)}$, etc. As Jakobson stated, the genetic code system is the basic simulator, which underlines all verbal codes of human languages. "*The heredity in itself is the fundamental form of communications ... Perhaps, the bases of language structures, which are imposed on molecular communications, have been constructed by its structural principles directly*" (Jakobson, 1985, p. 396). These questions have arisen to Jakobson as a consequence of its long-term researches of connections between linguistics, biology and physics. Such connections were considered at a united seminar of physicists and linguists, which was organized by Niels Bohr and Roman Jakobson jointly at the Massachusetts Institute of Technology.

"*Jakobson reveals distinctly a binary opposition of sound attributes as underlying each system of phonemes... The subject of phonology has changed by him: the phonology considered phonemes (as the main subject) earlier, but now Yakobson has offered that distinctive attributes should be considered as "quantums" (or elementary units of language)... . Jakobson was interested especially in the general analogies of language structures with the genetic code, and he considered these analogies as indubitable*" (Ivanov, 1985). One can remind also of the title of the monograph "On the Yin and Yang nature of language" (Baily, 1982), which is characteristic for the theme of binary oppositions in linguistics.

Similar questions about a connection of linguistics with the genetic code excite many researchers. In addition a linguistic language is perceived by many researchers as a living organism. The book "Linguistic genetics" (Makovskiy, 1992) states: "*The opinion about language as about a living organism, which is submitted to the laws of a nature, ascends to a deep antiquity ... Research of a nature, of disposition and of reasons of isomorphism between genetic and linguistic regularities is one of the most important fundamental problems for linguistics of our time*".

One of the interesting questions is the existence of fractal images in linguistic and genetic texts. A number of publications are devoted to fractal features of linguistic and genetic texts (Gariaev, 1994; Jeffry, 1990; Yam, 1995, etc). Researches in this direction proceed all over the world.

We believe that achievements of matrix genetics and its mathematical notions and tools will be useful for revealing deep connections between genetic and linguistic languages. This matrix-genetic approach is capable of enriching its own arsenal of structural linguistics as a roughly developing science and to clear a problem of the unified bases of biological languages. It can be applied to researches on evolutionary linguistics, the analysis and synthesis of poetic forms, etc.

Figure 1. Some analogy between the binary sub-alphabets of the genetic code and the circle of binary-oppositional colors from physiology of inherited color perception

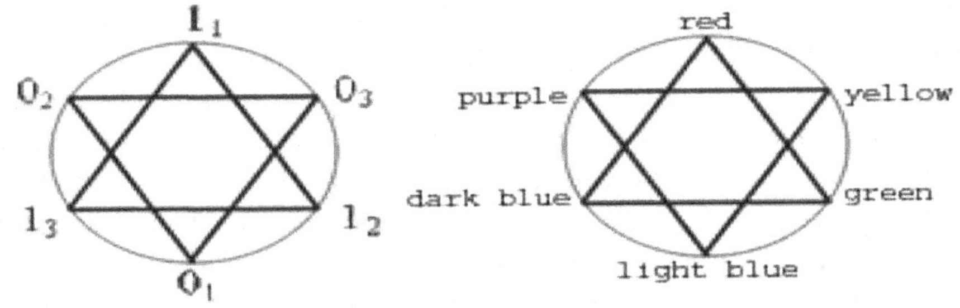

THE GENETIC CODE, A COLOR PERCEPTION AND A COLOR COMMUNICATION

This section describes an interesting analogy between the structures of the genetic code and a famous structure in physiology of color perception. Chapter 1 presented the three binary sub-alphabets of the genetic alphabet (Figure 2 in Chapter 1), which are based on the three types of molecular binary-oppositional attributes (or traits). From the viewpoint of each of these binary sub-alphabets, the set of the four letters of the genetic alphabet is divided into two pairs of equivalent letters, each of which corresponds to one of the binary symbols "0" and "1". By that, the three binary sub-alphabets contain six different binary symbols: 0_1, 1_1, 0_2, 1_2, 0_3, 1_3 (see Figure 2 in chapter 1). These symbols can be disposed into vertices of a six-vertex star (Star of David) in such way that binary oppositions 1_N and 0_N (where $N = 1, 2, 3$) take places on its opposite vertices (Figure 1, on the left side).

But the analogical six-vertex star is known for a long time in the physiology of inherited color perception where it presents the so called "color circle" (Figure 1, on the right side). Three pairs of binary-oppositional colors exist: "red-light blue", "green-purple" and "dark blue-yellow". These three pairs of colors can be interpreted as the ensemble of binary sub-alphabets of color perception. The colors of each pair are called traditionally complementary colors. By the way, the same construction of a six-vertex star was described in Chapter 11, Figure 12. One recalls the famous dictum by L. Bolzano here: "Cognition is a search of analogies".

This analogy of ternary ensembles of binary oppositions in the genetic code and in inherited fundamentals of color perception allows, in particular, proposing color portraits of genetic letters and the genetic triplets. For example, the oppositional symbols 1_1 and 0_1 of the first genetic sub-alphabet (Figure 2 in chapter 1) can be interpreted as the first oppositional pair of colors: "red-light blue" correspondingly. The oppositional symbols 1_2 and 0_2 of the second sub-alphabet – as the second oppositional pair: "green-purple". At last, the symbols 1_3 and 0_3 can be interpreted as the third oppositional pair of colors: "dark blue-yellow". From the viewpoint of the binary sub-alphabets of the genetic alphabet (Figure 2 in chapter 1), the genetic letter "C" presents a superposition of the symbols "$0_1 + 0_2 + 0_3$" and this letter "C" can be expressed by the sum of the three relevant colors: light blue (0_1) + purple (0_2) + yellow (0_3). By the same way the genetic letter "A" presents a superposition of the symbols "$1_1 + 0_2 + 1_3$" and it can be expressed by the sum of colors: "red + purple + dark blue". The genetic letter "G" is expressed in this scheme by the sum of colors: "red (1_1) + green (1_2) + yellow (0_3)". The genetic letter "U/T" is expressed in this scheme by the sum of colors: "light blue (0_1) + green (1_2)+ dark blue (1_3)".

In this approach each genetic multiplet can possess its own color expression on the bases of a superposition of colors of genetic letters, which form this multiplet. For example, the duplet GA is expressed by the superposition of colors of its letters G and A: "red + green + yellow + red + purple + dark blue".

In such a color expression of multiplets, all genomatrices [C A; U G]$^{(n)}$ of the Kronecker family take a form of the color mosaics, which reflect some "genetic" harmony. These color harmonic patterns of the Kronecker families of genetic matrices can have physiological activity, for example, to take special attention of people. In this case such color-genetic patterns will be useful in many fields, where color design is utilized: architecture, design of printed materials, toys, etc. The described analogy of ternary ensembles of binary oppositions in genetic coding and in color perception draws attention additionally to the deep value of chromaticity in information communication between different systems of an organism including a communication between proteins by means of their radiating and accepting "aerials" with a sharp selectivity to certain colors.

PARALLELS BETWEEN PATTERNS OF THE GENETIC CODE AND PATTERNS OF "I CHING"

G. Stent (1969, p. 64) published a hypothesis about a possible connection between genetic code structures and a symbolic system of the Ancient Chinese "The Book of Changes" (or "I Ching"). He is the famous expert in molecular genetics, and his thematic textbooks for students were translated into many countries including Russia (Stent, 1971). A few authors have supported him and his hypothesis later. For example, the Nobel Prize winner in molecular genetics F. Jacob (1974, p. 205) wrote as well: *"C'est peut-être I Ching qu'il faudrait étudier pour saisir les relations entre hérédité et langage"* (it means in English: perhaps, for revealing of relations between genetics and language it would be necessary to study them through the Ancient Chinese "I Ching"). In whole, a position about the necessity of the profound analysis of named parallels and their possible expansion exists in molecular genetics for 40 years. Our researches on matrix genetics give additional materials to this area and to the hypothesis by Stent. "I Ching", which is devoted to the binary-oppositional system of "Yin and Yang", declares a universality of a cyclic principle of organization in nature. Traditional Oriental medicine is based on positions of this book. Let us remind one of the known information about "I Ching" briefly.

A great number of literature sources are devoted to "I Ching". Our references include only a negligible part of a total quantity of these sources (Capra, 2000; Eremeev, 2005; Hesse, 1962; Kobzev, 1994; Needham, 1962; Petoukhov, 1999, 2001, 2005a, 2008a; Schonberger, 1976; Shchutskii, 1979, 1997; Stent, 1969; Tusa, 1994; Vinogrodskiy, 2002; Wilhelm, & Wilhelm, 1995; Yan, 1991).

Many of these sources label "I Ching" as one of the greatest and most mysterious human creations. From the viewpoint of the Chinese culture, which has generated it (this culture is the most ancient among all cultures, which continue their existence on the Earth), "I Ching" represents something even more considerable: the creation made by the Superperson, who has embodied a secret of the universe in special symbols and signs.

This book has had fundamental paradigmatic influence on the whole culture of traditional China and the adjacent countries. The ideas expressed in it have created original world-view and methodology. They have influenced to a huge extent the development of philosophy, religion, natural sciences, literatures and arts in Ancient China. In view of this, the Chinese culture became absolutely unique in the history

Figure 2. The national flag of South Korea, which contains the trigrams of "I Ching"

of world culture. Symbols and principles "I Ching" penetrated into all spheres of life of traditional China – from theoretical conceptions and high art to household subjects and decorations.

Many western scientists studied and used "I Ching". For example, the creator of analytical psychology C. Jung has developed his doctrine about a collective unconscious in connection with this book. According to Jung, the trigrams and the hexagrams of "I Ching" *"fix a universal set of archetypes (innate psychic structures)"* (Shchutskii, 1997, p. 12). Niels Bohr has chosen the symbol Yin-Yang as his personal emblem. Many modern physicists, who feel unity of the world, connect their theories with ideas of traditional Oriental culture, which unite all nature. For example, it has been reflected in the title "the eightfold way" of the famous book (Gell-Mann, Ne'eman, 1964). Intensive development of modern sciences about self-organizing and nonlinear dynamics of complex systems (synergetrics) promotes strengthening attention of western scientists to traditional eastern world-view (for example see (Capra, 2000)). Special groups study "Book of Changes" in many eastern and western universities. Annual scientific conferences on "I Ching" are conducted in Moscow (Russia) systematically during last few years. The great numbers of web-sites in Internet are devoted to the similar studies. The influence of "I Ching" is widely presented in the modern life of the countries of the East. For example, the national flag of South Korea bears symbols of trigrams (Figure 2). A great number of specialized schools work where instructors teach pupils methodological aspects of practical application of relevant ancient knowledge in medicine and in other fields. According to some versions, "I Ching" is a collection of knowledge, which was obtained by Ancient well-trained individuals by means of their practices of meditations and inspirational conditions.

One should note that "I Ching" was written some thousand years before the occurrence of modern Academies of sciences. It represents a set of dogmas of unclear origin. From the point of view of a modern science, the book confirms a universal conformity of these dogmas to a structure of nature without an appropriate substantiation of the dogmatic statements. Historically western academic science and its scientific laws (for example, Newton's laws, etc.) were developed without any connection with "I Ching" by means of another methodology (though inspirational conditions, in which the person suddenly receives the complete picture of the answer to his questions, are well-known in western science as well; it is enough to recollect the famous history of the Periodic table of chemical elements which was showed in a complete way to Mendeleyev in his sleep). For these reasons, many individuals consider "I Ching" as a bright fact of Ancient Oriental culture but not as a valuable source of knowledge.

The main table in "I Ching" is the famous table of 64 hexagrams in Fu-Xi's order, which is considered a universal natural archetype in Chinese tradition. Each hexagram is a pile of six broken and unbroken

Figure 3. The table of 64 hexagrams in Fu-Xi's order

	111	110	101	100	011	010	001	000
111	111111	111110	111101	111100	111011	111010	111001	111000
110	110111	110110	110101	110100	110011	110010	110001	110000
101	101111	101110	101101	101100	101011	101010	101001	101000
100	100111	100110	100101	100100	100011	100010	100001	100000
011	011111	011110	011101	011100	011011	011010	011001	011000
010	010111	010110	010101	010100	010011	010010	010001	010000
001	001111	001110	001101	001100	001011	001010	001001	001000
000	000111	000110	000101	000100	000011	000010	000001	000000

(solid) lines. Each broken line symbolizes Yin and each unbroken line symbolizes Yang. According western tradition, these broken and unbroken lines are shown in the form of the binary symbols "0" and "1" and each hexagram is shown as a sequence of such six binary symbols. Figure 3 demonstrates hexagrams in this form of six-digit binary numbers. Each position in all hexagrams has its own individual number: in the western numeric presentation of a hexagram, positions of its binary symbols are numbered left-to-right by numbers from 1 to 6 (in the Chinese graphical presentation, a numbering of the lines of each hexagram is read in the sequence bottom-up). Digrams (two-digit binary numbers or piles of two lines) and trigrams (three-digit binary numbers or piles of three lines) are considered as well. "*Trigrams, hexagrams and their components in all possible combinatory combinations form a universal hierarchy of classification schemes. These schemes in visual patterns embrace any aspects of reality – spatial parts, time intervals, the elements, numbers, colors, body organs, social and family conditions, etc.*" (Shchutskii, 1997, p. 10).

Let us mention one interesting historical moment. The creator of the first computer G. Leibniz, who had ideas of a universal language, was amazed by this table of 64 hexagrams when he became acquainted with it because he considered himself as the originator of the binary numeration system, which was presented in this ancient table already. Really, he saw the following fact. If each hexagram is presented as the six-digit binary number (by replacement of each broken line with the binary symbol "0" and by replacement of each unbroken line with the binary symbol "1"), this ancient sequence of 64 hexagrams in Fu-Xi's order was identical to the ordinal series of numbers from 63 to 0 in decimal notation. By analogy, a sequence of 8 trigrams in Fu-Xi's order is identical to the ordinal series of numbers from 7 to 0 in decimal notation. "*Leibniz has seen in this similarity the evidence of the pre-established harmony and unity of the divine plan for all epochs and for all people*" (Shchutskii, 1997, p. 12).

It's surprising that this ancient table of 64 hexagrams (Figure 3) is connected closely with the genomatrix $[C\ A;\ U\ G]^{(3)}$ of 64 genetic triplets (Figure 3 in Chapter 1). Really, as it was described in Chapter 1, if each triplet in the genomatrix $[C\ A;\ U\ G]^{(3)}$ is replaced by its coordinate six-digit number, which is a integration of binary three-digit numbers of its row and column, new numerical matrix arises (Figure 3 in Chapter 1). From the viewpoint of a decimal notation, the sequence of these coordinate numbers of triplets coincides with the ordinal series of decimal numbers from 0 to 63. One can see that the genomatrix on Figure 3 in Chapter 1 is identical to the ancient table on Figure 3 in its inverse order of presentation. This inessential difference disappears if we invert the binary symbols of the genetic letters A, C, G, U/T in the binary sub-alphabets of the genetic alphabet on Figure 2 in chapter 1. For example, in the case of such an inversion, the letter A is symbolized by the binary symbols 0_1, 1_2, 0_3 (instead of its symbols 1_1, 0_2, 1_3 on Figure 2 in chapter 1), etc. The reason, why we have chosen the variant of binary symbols in the binary sub-alphabets on Figure 2 in chapter 1, is the following. This chosen variant leads to the usual ascending numeration of columns and of rows of matrices in the theory of signal processing, if one reads the series of numbers 0, 1, 2, … in matrices left-to-right and top-down. The Chinese variant of reading is opposite.

Ancient Chinese culture knew nothing about the genetic code, which was discovered by western science recently. And our Kronecker construction of the genetic matrix $[C\ A;\ U\ G]^{(3)}$ in the field of matrix genetics was based on pure academic data of molecular genetics and used nothing from "I Ching". But as a result we see the formal coincidence of the described matrix structures. Consequently the described binary presentation of the genomatrix of 64 triplets is known a few thousand years already. Let us consider other parallels of the genetic code system with the system of "I Ching".

In Chinese tradition, Yin is symbolized mainly by means of not only the broken line but the number 6 as well. And Yang is symbolized mainly by means of not only the solid line but the number 9 as well (Shchutskii, 1979, 1997). But Figure 6 in Chapter 4 demonstrated already the visual realization of the inversion-symmetrical patterns 6 and 9 in a disposition of numbers of one kind in the multiplicative genomatrix $[C\ A;\ U\ G]^{(3)}$ of hydrogen bonds. These patterns 6 and 9 are invariants relative to algebraic operations with matrices of such a type (see Chapter 4).

In Chinese tradition, each hexagram is considered constructed from two independent trigrams – a bottom trigram and a top trigram (in western numeric presentation of hexagrams on Figure 3, these two trigrams correspond to the left three-digit half and to the right three-digit half of a six-digit number. For this reason the first half of six-digit numbers is marked on Figure 3 especially). For example, the book (Shchutskii, 1997, p. 86) states: "*The theory of "Book of Changes" considers that a bottom trigram concerns an internal life… and a top trigram concerns to an external world… . Similar positions in a top trigram and in a bottom trigram have the nearest relation to each other. In view of this, the first position relates by analogy to the fourth position, the second position – to the fifth position, and the third position relates by analogy to the sixth position… . If these correlative positions (1-4, 2-5, 3-6) are occupied by various lines, it is considered that "conformity exists" between them, and in the case when these correlative positions are occupied by identical lines, it is considered that "conformity is absent" between them*". (By the way, a six-digit coordinate number of each cell of the genomatrix $[C\ A;\ U\ G]^{(3)}$ (Figure 3 in Chapter 1) is also a binary hexagram arranged by two independent trigrams, which symbolize its row and its column). If one takes into account these conformities of the correlative positions 1-4, 2-5, 3-6, then a study of parallels between the genetic code and the system of "I Ching" reveals the possibility of a compact presentation of the (8x8)-table of 64 trigrams in the form of the third Kronecker power of a (2x2)-matrix of Chinese digrams. Let us explain it.

The following four digrams have a basic meaning in the system of "I Ching" (Figures 4, 5, 6, and 7).

One can construct a kernel matrix S from these digrams (Figure 8) by analogy with the kernel matrix [C A; U G] of the genetic alphabet on Figure 3 in chapter 1.

Exponentiation of this kernel matrix S into the third Kronecker power produces the matrix of 64 hexagrams in Fu-Xi's order, if one takes into account the mentioned meaning of the correlative positions 1-4, 2-, 4-6. The issue is that one should distinguish strongly the bottom half and the top half in digrams, trigrams and hexagrams (we will name all of them by a general name "multigrams"). And at

Figure 4. Old Yang

Figure 5. Old Yin

Figure 6. Young Yang

Figure 7. Young Yin

Figure 8. The kernel matrix of the Chinese digrams

Kronecker multiplication of the matrices S, each line of the bottom half (or of the top half) of the initial multigrams should be disposed in the bottom half (of the top half correspondingly) of a final multigram on the corresponding position. In another respect it is necessary to carry out the classical procedure of Kronecker multiplication of matrices. This Kronecker algorithm of construction of the Chinese matrix of 64 hexagrams corresponds to the traditional Chinese viewpoint: "*Hexagrams are not trigrams, which are alloyed together, but they are two trigrams, which are located on a vertical one over another*" (Shchutskii, 1997, p. 101).

The next parallel concerns the numbers 2 and 3. These numbers served as the base of Chinese numeric systems in Ancient China (Kobzev, 1994, p. 15) and they are named there as the number of the Earth and the number of Heaven correspondingly. But these two kinds of numbers – 2 and 3 – are presented in the molecular structures of the genetic code as numbers of hydrogen bonds of nitrogenous bases. In view of this, all numerical features of the quint genomatrices of the hydrogen bonds, which were described in Chapter 4, are reproduced in the matrices of "I Ching" as well. In particular, these ancient matrices are connected with the golden section $\varphi = (1+5^{0.5})/2 = 1.618\ldots$. We will demonstrate this below.

The set of the four Chinese digrams of Old and Young Yin and Yang (Figure 8) is divided into two sub-sets. One sub-set contains two digrams (Young Yin and Young Yang), each of which consists of the various types of lines – broken and unbroken. The second sub-set contains two digrams (Old Yin and Old Yang), each of which consists of an identical type of line – broken or unbroken. Let us replace in the matrix S (Figure 8) the digrams of the first sub-set by number 2 and the digrams of the second sub-set by number 3. And let us do the same replacement of digrams on the correlative positions 1-4,

2-5, 3-6 of each hexagram in the ancient table of 64 hexagrams with subsequent multiplication of three introduced numbers for each hexagram (as usual by Kronecker multiplication of numeric matrices). As a result we get two quint matrices $P_{MULT}^{(1)}$ and $P_{MULT}^{(3)}$, which were shown on Figure 2 in Chapter 4 already. Chapter 4 has demonstrated the connection of these matrices with the golden matrices, all components of them are equal to the golden section in integer powers (Figure 3 in Chapter 4). In this direction of researches we come to the thought that the ancient table of 64 hexagrams "I Ching" should be considered not as a simple table but as a matrix, which is connected with some other matrices by means of the usual matrix operations.

The numbers 8 and 12, which characterize the canonical sub-sets of amino acids in Chapter 3 (Figure 1 in Chapter 3), were wide known in Ancient China as "*a standard measure of alternative partitioning of space-time on Chinese chrono-topograms… . The numbers 8 and 12 are interconnected also as key parameters of a cube and an octahedron. A cube possesses 8 tops and 12 edges and an octahedron possesses 8 sides and 12 edges. These both types of polyhedrons served for the Chinese thinkers as one of basic simulators in their understanding of the Universe since ancient times*" (Kobzev, 1994, p. 39, 40).

The four named digrams (Figure 8) were symbolized in Ancient China by the numbers 6, 7, 8, 9 as well (Shchutskii, 1997, p. 22, 522). It is interesting that these four numbers characterize the quantity of protons in the chemical elements, which form the molecules of the genetic letters (or of the nitrogenous bases): the carbon C has 6 protons (its ordinal number 6 in the Mendeleev's table), the nitrogen N has 7 protons, the oxygen O has 8 protons and amino group (amides) NH_2 has 9 protons. The same four numbers 6, 7, 8, 9 occupy all cells of the genetic matrix $[C\ A;\ U\ G]^{(3)}$ of 64 triplets, if each triplet is presented by the sum of numbers 2 and 3 of hydrogen bonds of its nitrogenous bases (see details and some other examples in (Petoukhov, 2001)).

Attempts of a biunique comparison of 64 genetic triplets and 64 hexagrams of "I Ching" are known in literature sources. A review of these attempts exists in the book (Petoukhov, 2001). Authors of these attempts issued from the Chinese table of 64 hexagrams and then they utilized additional assumptions to guess a biunique correspondence between each hexagram and each genetic triplet. But the quantity of possible variants of such a correspondence is huge (Chapter 1 has considered the number $64! \approx 10^{89}$ of possible variants of a disposition of 64 triplets inside an octet matrix). In view this, it is impossible practically to find an adequate correspondence by such a way. Matrix genetics suggests another way to find such a correspondence by means of researching the matrix presentations of the genetic code without any initial connections with the "I Ching". But as a result of this research, the correspondence of the genomatrix of the 64 triplets to the ancient table of the 64 hexagrams appeared suddenly. This correspondence gives new possibility for a biunique correspondence between each hexagram and each genetic triplet. In addition, from the viewpoint of the binary numeration of its rows and columns, the ancient table of 64 hexagrams is connected with the matrix of diadic shifts, if one takes into account the modulo-2 addition of these binary numbers as it was described in Chapter 1 (Figure 5).

One can note that the table of 64 hexagrams can be transformed into a numeric Yin-Yang-matrix by means of an analogue of the algorithm, which was described in Chapter 7 to construct the genetic Yin-Yang-matrix (see Figure 4 in Chapter 7). Really, the digrams, which occupy the first correlative positions 1-4 and 2-5 in hexagrams, can be interpreted as numeric symbols α (if the digram possesses two lines of an identical type) or β (if the digram possesses two lines of different types). Digrams, which occupy the last correlative positions 3-6 in hexagrams, can be interpreted as numeric symbols γ (if the digram is Old Yang or Young Yin) or δ (if the digram is Young Yang or Old Yin). In addition one should use a rule of signs "+" and "-" for hexagrams, which is analogical to the rule of these signs, which was described in Chapter 7 to yield the genetic Yin-Yang-matrix (Figure 4 in Chapter 7). In such a way we obtain a Yin-Yang-algebraic presentation of the ancient table of 64 hexagrams.

MUSICAL HARMONY IN THE MATRICES OF THE GENETIC CODE AND OF "I CHING"

Ancient Chinese possessed the highest musical culture. Long before Pythagoras they used the musical system known in Europe under the name Pythagorean musical scale. Due to the cosmic meaning of music in ancient China, this quint scale was preferable for ancient Chinese in comparison with an equal temperament scale which was known for them many centuries ago as well. This Chinese system was borrowed by Pythagoras in many aspects (the analysis of these questions is presented in detail, for example, in the book (Needham, 1962, v.4). The Pythagorean doctrine (about a key role of numbers in the organization of the world) has arisen not at empty place. This doctrine was developed under an influence of the more ancient Oriental doctrines in some aspects. For example, according to Chinese notions, the even number 2 is the female number and the odd number 3 is the male number. But the same notion was typical for the Pythagorean School, in particular, for Pythagorean theory of musical harmony: "*the integer 2 was considered to be the female number which can give birth to no new tones without the participation of the male number 3*" (Kappraff, 2006, p. 303).

History has not saved data on how the founders of the system of "I Ching" have created the table of 64 hexagrams and why they declared its universal archetypical meaning for nature. It seemed that answers to these questions are inaccessible to modern people. But our investigations on matrix genetics about analogies among the genetic code, Pythagorean musical scale and the system of "I Ching" have led to an interesting conception about musical bases of the ancient table of 64 hexagrams. According to this new conception, this ancient table reflects relations of musical quint harmony (like the Pythagorean musical scale). This natural conception removes the mystery of the origin of the system of this table to a high extent. We shall present materials for this conception.

Music was the cornerstone in the Chinese civilization, which is the longest living culture in history. Even the national system of measures and weights has been constructed in connection with musical instruments. This fact has no precedents in history of civilizations (Needham, 1962; Eremeev, 2005, p. 76). Old Chinese music was used intensively in ancient therapy with success according to many literature sources. It is interesting because this music is based on such a music scale which is connected with the genetic code in its matrix form of presentation (as it was described in Chapter 4).

According to notions of Ancient Chinese, music is present at the origin of the world and plays a space role: music represents a microcosm reflecting a structure of the Universe. Fine music possesses strictly a certain structure which cannot be broken as it is impossible to violate the law. Musical rules are reproduced in different fields of culture. Briefly speaking, music in the Ancient China was considered as a general natural archetype.

Rules on painting or architecture products were created similar to a rhythm in music. In essence music in Ancient China was considered as general natural archetype.

One of the significant examples of the connections of the Pythagorean harmonic doctrine with the system of "I Ching" is the following: the Chinese table of 64 hexagrams coincides with the quint genomatrix $[3\ 2; 2\ 3]^{(3)}$ (Figure 2 in Chapter 4) and belongs to the Kronecker family of quint matrices $[3\ 2; 2\ 3]^{(n)}$ in the case of a replacement of the Chinese digrams by the mentioned numbers of Earth and

Heaven as discussed above. In this case this ancient table has an obvious connection with the numeric triangle by Nicomachus of Gerasa (see Chapter 4 and the expression 4.1), with the golden section and with all the Pythagorean doctrine about musical harmony and aesthetics of proportions.

Taking all similar facts into account, we think that the system of "I Ching" is a collection of ancient knowledge and notions about music. In addition, from the viewpoint of matrices of "I Ching", a direct indication of the connection of the bases of music with the numbers 2 and 3 exists in a classical work of literature "Spring and Autumn" by Lu Bu We from chapter on music: "*The origins of music lie far back in the past. Music arises from Measure and is rooted in the great Oneness. ... Music is founded on the harmony between heaven and earth, on the concord of obscurity and brightness*" (this citation is taken from the book (Hesse, 1962, p. 31), the book author was a Nobel Prize winner and an expert of "I Ching"). These words about Earth and Heaven relate directly to the numbers 2 and 3 which symbolized Earth and Heaven in Ancient China (including Chinese music) and which transform the table of 64 hexagrams into the quint matrix [3 2; 2 3][(3)]. The words about the concord of obscurity and brightness correspond to the known fact that the Chinese musical system contained 12 sounds, each of which had a magic sense: the odd sounds personified brightness, active forces of Heaven; the even sounds personified obscurity, passive forces of Earth (Eremeev, 2005).

On the basis of the knowledge about such a correspondence between the ancient table to the quint matrix, a musical presentation of the table of 64 hexagrams is possible for the case of the described replacement of the diagrams by numbers 2 and 3 (Figure 10 in Chapter 4).

From the viewpoint of our proposed conception, the table of 64 hexagrams in Fu-Xi's order is constructed as a code of knowledge about musical harmony and as a special presentation of the quint musical scale. The statement of ancient Chinese about a universal correspondence of the given table to nature and about its status of a general natural archetype is a repetition of their understanding of music as the universal organizing beginning, which concerns all aspects of life.

FUTURE TRENDS AND CONCLUSION

Many phenomena of culture have connections with inherited physiological peculiarities of our systems of perception and communications or, in other words, they have physiological bases. The study of connections of genetic code structures with features of relevant physiological phenomena is important. Matrix genetics proposes new mathematical methods and approaches using matrix presentation and analyses to study these connections with their patterns, symmetries, etc.

The genetic language is considered usually as a language and as a possible basis of linguistic languages and of communication physiology. One of the main problems for all languages is the following: how discrete components of a language, each of which in itself has no sense, acquire a sense in their grouping into ensembles? A grouping of discrete elements by means of the described matrix way is one possible variant of grouping, but this way has demonstrated its effectiveness in studying the genetic code systems and their evolution. In view of this, matrix genetics should be developed intensively.

In particular the study of connections of genetic code structures with various fields of culture and with physiological bases of cultural phenomena concerns the interesting theme of archetypes. The creator of analytic psychology C. Jung has proposed ideas about congenital archetypical notions. According to Jung, a universal set of congenital mental structures is inherent in individuals. And according to Jung, the system of trigrams and hexagrams of "I Ching" fixes the universal set of such archetypes.

According to Jung, the universal set of archetypes - congenital mental structures is inherent in people. Jung was the expert and the connoisseur of this book. He named it as the great and unique product. Visualization of the form of number 69 (Figure 6 in Chapter 4) in the matrix of genetic triplets, which is connected with hexagrams of "I Ching", forces one to pay additional attention to this circle of questions about archetypes. For example, why do individuals utilize the symbols of 6 and 9 in their written languages? Data of matrix genetics allow one to think that habitual forms of the symbols of 6 and 9 have not been invented by our ancestors in an arbitrary manner, but they are ones of universal archetypes, which were simply reproduced in written languages.

Another example of universal archetypes, which has been described by Jung himself, is the archetype of the quaternary set: a universal archetype, which is a logical prerequisite of every entire judgment. According to Jung, it frequently has a structure 3+1, one of the elements of which takes a special place or possesses a different nature. Just the fourth element, adding to the others, makes them a single whole, which symbolizes the universal set (see Wehr, 1969). This archetypes coincides with the structure of the genetic alphabet with its four letters, where the letter U is opposite to the other three genetic letters A, C, G by their molecular peculiarities. In matrix genetics this genetic quaternary set leads to the transformation of the genetic matrices [C A; U G] and $[C A; U G]^{(3)}$ into Hadamard matrices by means of the U-algorithm as described in Chapter 6. One can think that the same archetype is realized in the tetra-reproduction of gametal cells at meiosis.

It is obvious that a study of connections of the genetic Yin-Yang-algebras with cultural phenomena, where the binary opposition of male and female beginnings is one of the main themes from ancient time, gives many additional materials. This topic is under wide investigation now.

The described study in the field of matrix genetics of the connections of the genetic code structures with various fields of culture including the culture of Ancient China is only in an initial investigation stage. This study can give many new and unexpected results. In view of this, it will be continued in different directions including structural phenomena of inherited physiological systems and processes, which are related to bioinformatics. There is no doubt, new beautiful symmetrical patterns will be found on this way. Special attention will be paid to the mentioned study from the viewpoint of the genetic Yin-Yang-algebras and relevant Hadamard matrices, which allow modeling phenomena of inheritance using formalisms of quantum mechanics and quantum computers. Simultaneously new ideas and simulators in the fields of structural linguistics, physiology of perceptions, theory of musical harmony and others will arise due to materials of matrix genetics. In addition one can suppose that bioinformatics will discover many other useful things as well in ancient Oriental medicine and culture.

It is known that ancient Chinese connected the 8 trigrams of "I Ching" with 8 kinds of special energy. Perhaps one can hope that 8-parametric genetic matrices of the Yin-Yang algebras will be useful to understand bases of these ancient ideas about special forms of energy.

The Kronecker family of matrix forms of presentation of the genetic code systems is useful for the detection of many interesting connections of the genetic code structures with structures from various fields of culture including the culture of Ancient China. New patterns and relations of symmetry are found in studying analogies of the genetic code structures with patterns of linguistics, color perception, architecture, musical harmony, ancient patterns of "I Ching", etc. This study should be continued.

REFERENCES

Andrews, E. (1990). *Markedness theory: The union of asymmetry and semiotics in language, sound, and meaning*. Durham: Duke University Press.

Baily, C. J. L. (1982). *On the yin and yang nature of language*. London: Ann Arbor.

Caglioti, G., Ramme, G., & Tscouvileva, T. (2006). Futurism and chromatic architectures of music in space-time. *Symmetry: Culture and Science, 17*(1-2),

Capra, F. (2000). *The Tao of physics: An exploration of the parallels between modern physics and eastern mysticism*. New Jersey: Shambhala Publications, Inc.

Chomsky, N. (1980). On cognitive structures and their development: A reply to Piaget. In M. Piattelli-Palmarini (Ed.), *Language and learning: The debate between Jean Paigaet and Noam Chomsky*. Cambridge, MA: Harvard University Press.

Darvas, G. (2007) *Symmetry*. Basel: Birkhäuser Book.

Eremeev, V. E. (2005). *Symbols and numbers of "book of changes."* Moscow: ASN (in Russian).

Gariaev, P. P. (1994). *Wave genome*. Moscow: ASM (in Russian).

Gell-Mann, M., & Ne'eman, Y. (1964). *The eightfold way*. New York: W.A. Benjamin.

Hahn, W. (1989). *Symmetrie als entwicklungsprinzip in natur und kunst*. Königstein: Langewiesche.

Hahn, W. (1998). *Symmetry as a developmental principle in nature and art*. Singapore: World Scientific.

Hargittai, I., & Hargittai, M. (1994). *Symmetry:Aa unifying concept*. Bolinas, CA: Shelter Publications.

He, M., & Petoukhov, S. V. (2007). Harmony of living nature, symmetries of genetic systems, and matrix genetics. *International journal of integrative medicine, 1*(1), 41-43.

Hesse, H. (1962). *The glass bead game*. New Jersey: Pergamon Press.

Ivanov, V. V. (1985). *Selected works on semiotics and history of culture*. Moscow: Yazyki Russkoy Kultury (in Russian).

Jacob, F. (1974). Le modele linguistique en biologie. *Critique, Mars, 30*(322), 197-205.

Jacob, F. (1977). The linguistic model in biology. In D. Armstrong & C. H. van Schooneveld (Ed.). *Roman Jakobson: Echoes of his scholarship*. Lisse: Peter de Ridder (pp. 185-192).

Jakobson, R. (1987). *Selected works*. Moscow: Progress (in Russian).

Jakobson, R. (1987). *Language in literature*. Cambridge: MIT Press.

Jakobson, R. (1999). *Texts, documents, studies*. Moscow: RGGU (in Russian).

Jeffry, H. J. (1990). Chaos game representation of gene structure. *Nucleic Acids Research, 18*(8), 2163–2170. doi:10.1093/nar/18.8.2163

Kappraff, J. (1990). *Connections, the geometric bridge between art and science.* New York: McGraw Hill.

Kappraff, J. (2000). The arithmetic of Nichomachus of Gerasa and its applications to systems of proportions. *Nexus Network Journal, 2*(4). Retrieved on October 3, 2000, from http://www.nexusjournal.com/Kappraff.htmlKappraff, J. (2002). *Beyond measure: Essays in nature, myth, and number*. Singapore: World Scientific.

Kappraff, J. (2006). The lost harmonic law of the Bible. In J. Sharp (Ed.). *Proceedings of London-Bridges 2006.* London: Bridges (pp. 300-312).

Kobzev, A. I. (1997). *Studies on symbols and numbers in Chinese classical philosophy*. Moscow: Vostochnaia Literatura (in Russian).

Leyton, M. (1992). *Symmetry, causality, mind.* Cambridge, MA: MIT Press.

Loeb, A. L. (1971). *Color and symmetry*. New York: Wiley.

Loeb, A. L. (1993). *Concepts & images: Visual mathematics*. Boston: Birkhäuser.

Makovskiy, M. M. (1992). *Linquistic genetics*. Moscow: Nauka (in Russian).

Marcus, S. (1990). *Algebraic linguistics; analytical models*. Bucharest: Academic.

Marcus, S. (2007). *Words and languages everywhere*. Bucharest: Polimetrica.Needham, J. (1962). *Science and civilization in China.* Cambridge: Cambridge University Press.

Nonnenmacher, T. F., Losa, G. A., & Weibel, E. R. (Eds.). (1994). *Fractals in biology and medicine*. Basel: Birkhäuser.

Petoukhov, S. V. (1999). Genetic code and the ancient Chinese "book of changes." . *Symmetry: Culture and Science, 10*, 211–226.

Petoukhov, S. V. (2001). *The bi-periodic table of genetic code and the number of protons*. Moscow: MKC (in Russian).

Petoukhov, S. V. (2005a). The rules of degeneracy and segregations in genetic codes. The chronocyclic conception and parallels with Mendel's laws. In M. He, G. Narasimhan & S. Petoukhov (Eds.), *Advances in Bioinformatics and its Applications, Series in Mathematical Biology and Medicine, 8*, 512-532. Singapore: World Scientific.

Petoukhov, S. V. (2008a). *Matrix genetics, algebras of the genetic code, noise-immunity*. Moscow: RCD (in Russian).

Ratner, V. A. (2002). *Genetics, molecular cybernetics*. Novosibirsk: Nauka.

Schonberger, M. (1976). *The I Ching and the genetic code.* New York: ASI Publishers Inc.

Shchutskii, I. K. (1979). *Researches on the I Ching*. Princeton: Princeton University Press

Shchutskii, I. K. (1997). *The Chinese classical "the book of changes."* Moscow: Vostochnaya literatura (in Russian).

Shubnikov, A. V., & Koptsik, V. A. (1974). *Symmetry in science and art*. New York: Plenum Press.

Smith, C. S. (1980). *From art to science*. Cambridge: MIT Press.

Stent, G. S. (1969). *The coming of the golden age*. New York: The Natural History Press.

Stent, G. S. (1971). *Molecular genetics*. San Francisco: W.H.Freeman and Company.

Teilhard de Chardin, P. (1959). *The phenomenon of man*. New York: Harper and Row.

Tusa, E. (1994). Lambdoma-"I Ging"-genetic code. *Symmetry: Culture and Science, 5*(3), 305–310.

Vinogrodskiy, B. B. (2002). *Taoism's practices of immortality*. Moscow: Sophia (in Russian).

Wehr, G. (1969). *C. Jung in Selbstzeugnissen und Bilddokumenten*. Berlin: Rowohlt.

Wilhelm, H., & Wilhelm, R. (1995). *Understanding the I Ching: The Wilhelm lectures on the "book of changes."* Princeton: Princeton University Press

Yam, P. (1995). Talking trash (Linguistic patterns show up in junk DNA). *Scientific American, 272*(3), 12–15.

Yan, J. F. (1991). [*Ching: The Tao of life*. New York: North Atlantic Books.]. *DNA (Mary Ann Liebert, Inc.)*, I.

About the Authors

Sergey Petoukhov, Ph.D., is a chief scientist of the Department of Biomechanics, Mechanical Engineering Research Institute of the Russian Academy of Sciences, Moscow. He is Full Professor and Grand Ph.D. from the World Information Distributed University in 2004. Dr. Sergey Petoukhov has been awarded as an academician of European Academy of Informatization since 2004. He is a Laureate of the State Prize of the USSR (1986) for his achievements in biomechanics. Dr. Sergey Petoukhov graduated Moscow Physical-Technical Institute in 1970 and post graduated in 1973 with a specialty of "biophysics". He received a Golden Medal of the National Exhibition of Scientific Achievements (VDNH) in 1973 in Moscow for his physical model of human vestibular apparatus. He received his first scientific degree in the USSR in 1973: Candidate of Biological Sciences in specialty "biophysics". He received his second scientific degree in the USSR in 1988: Doctor of Physical-Mathematical Sciences in two specialties - "biomechanics" and "crystallography and crystallo-physics". He was an academic foreign stager of the Technical University of Nova Scotia, Halifax, Canada in 1988. He was elected as an academician of Academy of Quality Problems (Russia) in 2000. Dr. Sergey Petoukhov is a Director of Department of Biophysics and Chairman of Scientific-Technical Council in "Scientific-Technical Center of Information Technologies and Systems (INTES)", Moscow, Russia. He was Vice-President of the International Society "ISIS-Symmetry" ("International Society for the Interdisciplinary Study of Symmetry") in 1989-2000. He is Chairman of International Advisory Board of "International Symmetry Association (ISA)" since 2000 (its headquarters is at Budapest, Hungary, www.symmetry.hu). Dr. Sergey Petoukhov is an Honorary Chairman of Board Directors of "International Society of Symmetry in Bioinformatics (ISSB)" since 2000. He is Vice-president and academician of National Academy of Intellectual and Social Technologies (Russia) since 2003. Dr. Sergey Petoukhov is academician of the International Diplomatic Academy (Belgium, www.bridgeworld.org). He is Russian chairman (chief) of an official scientific cooperation between Russian and Hungarian Academies of Sciences in the theme "Non-linear models in biomechanics, bioinformatics and in theory of self-organizing systems". Dr. Sergey Petoukhov has published over 150 research papers (including 5 books) in biomechanics, bioinformatics, mathematical and theoretical biology, theory of symmetries and its applications, and mathematics. He is a member of editorial board of international journals: "Journal of Biological Systems" and "Symmetry: Culture and Science". He was a guest editor of special issues (in bioinformatics topic) of international journal "Journal of Biological Systems" (2004) of World Scientific Publishing. Dr. Sergey Petoukhov is the book editor of "Symmetries in Genetic Informatics" (2001), "Advances in Bioinformatics and its Applications" (2004), a Russian edition (2006) of the book by Canadian Professor R.V. Jean "Phyllotaxis: A Systemic Study in Plant Morphogenesis (Cambridge University Press, Cambridge, 1994). He is a co-organizer of international conferences in theory of symmetries and its applications (Budapest,

Hungary, 1989; Hirosima, Japan, 1992; Washington, USA, 1995; Haifa, Izrael, 1998; Budapest, Hungary, 2003 and 2006; Moscow, Russia, 2006). He was Chairman of the International Program Committee of "International Conference on Bioinformatics and its Applications–2004" (Fort Lauderdale, Florida, USA). He is a co-chairman of the organizing committees of the international conferences "Modern science and ancient Chinese "The Book of Changes" (I-Ching)", Moscow, Russia, 2003, 2004, 2005, 2006. He is a teacher of a course of architectural bionics in the Russian University of People Friendship (Moscow). He is actively involved in promoting science, education, and technology.

Matthew He, Ph.D., is Full Professor and Director of the Division of Math, Science, and Technology of Nova Southeastern University (NSU), Florida, USA. He is Full Professor and Grand Ph.D. from the World Information Distributed University in 2004. He has been awarded as an academician of European Academy of Informatization since 2004. Dr. Matthew He received his Ph. D. in Mathematics from University of South Florida in 1991. Dr. Matthew He was a research associate at the Department of Mathematics and Theoretical Physics, Cambridge University, Cambridge, England in 1986 and at the Department of Mathematics, Eldgenossische Technische Hochschule, Zurich, Switzerland in 1987. Dr. Matthew He was also a visiting professor at National Key Research Lab of Computational Mathematics of Chinese Science of Academy and University of Rome, Italy in 1998. Dr. Matthew He has published over 90 research papers in mathematics, computer science, information theory and bioinformatics. Dr. Matthew He is an invited series editor of Biomedical and Life Sciences of Henry Stewart Talk on "Using Bioinformatics in Exploration in Genetic Diversity". Dr. Matthew He is an editor of International Journal of Biological Systems, an editor of International Journal of Cognitive Informatics and Natural Intelligence, an editor of International Journal of Integrative Biology, and an editor of International Journal of Software Science and Computational Intelligence. Dr. Matthew He is the book editor of Advances in Bioinformatics and its Applications and a guest editor of a special issue of the Journal of Biological Systems of World Scientific Publishing in 2004. Dr. Matthew He is an Associate Editor of the Proceedings of the International Conference on Mathematics and Engineering Techniques in Medicine and Biological Sciences (2002, 2003, and 2004). Dr. Matthew He received the World Academy of Sciences Achievement Award in recognition of his research contributions in the field of computing in 2003. Dr. Matthew He is a Chairman of International Society of Symmetry in Bioinformatics and a member of International Advisory Board of "International Symmetry Association (ISA). Dr. Matthew He received Professor of the Year Award in Excellence of Teaching and Research in 2002 at NSU. Dr. Matthew He is a member of American Mathematical Society, Association of Computing Machinery and IEEE Computer Society. Dr. Matthew He serves as a member of board directors of several international associations and is actively involved in promoting science, education, and technology.

Index

D

E

F

G

H

I

J

K

L

M

N

O

P

Q

R

S

T

U

V

W

Y